AF334142

Engineering Wireless-Based Software Systems and Applications

For a listing of recent titles in the *Artech House Computing Library,* turn to the back of this book.

To my wife Tracy and my lovely son Kevin
To my parents Ming Gao and Ye-Fang Qin
—Jerry Zeyu Gao

To my Mom and Dad
—Simon Shim

To my wife Teresa, my son Wesley, and my daughter Emily
—Hsing Mei

To Tao and our precious one—Jolie
To my parents Mingzhao Su and Yunjun Xu
—Xiao Su

Contents

Preface   xvii

PART I
Introduction   1

CHAPTER 1
Engineering in the Wireless World   3
1.1   Moving to the Wireless World   3
      1.1.1   Wireless Markets and Needs   3
      1.1.2   Living in the Wireless World   6
1.2   Understanding Wireless-Based Application Systems   8
      1.2.1   Basic Definitions   8
      1.2.2   Basic Components   8
      1.2.3   Special Features and Properties   10
      1.2.4   Essential Requirements and Expectations   11
      1.2.5   Benefits, Issues, Challenges, and Limitations   12
      1.2.6   Classification of Wireless-Based Application Systems   13
1.3   Engineering Wireless-Based Application Systems   14
      1.3.1   Engineering Perspectives   16
      1.3.2   Engineering Issues, Challenges, and Needs   17
1.4   Summary   20
      References   21

PART II
Mobile Technologies and Operation Environments   23

CHAPTER 2
Mobile Devices and Operating Systems   25
2.1   An Overview of Mobile Devices   26
      2.1.1   Limitations   26
      2.1.2   Evolution and Integration   30
2.2   Mobile Operating Systems   35
      2.2.1   Symbian OS   39
      2.2.2   Windows CE/Mobile   41
      2.2.3   Embedded Linux (Mobile Linux)   44
2.3   Mobile Application Platforms   45

	2.3.1 Java Platform	46
	2.3.2 BREW	48
2.4	Summary	51
	References	51

CHAPTER 3

Wireless Application Protocols and Technology — 53

3.1	Introduction to WAP	53
	3.1.1 Brief History	53
	3.1.2 WAP Background	55
3.2	Overview of WAP Architecture and Protocols	56
	3.2.1 WAP Programming Model	56
	3.2.2 WAP Protocol Stack	57
	3.2.3 WAP Service Architecture	59
	3.2.4 From Web Proxy to WAP Proxy Gateway	62
	3.2.5 WAP Push Mechanism	66
3.3	WAE	68
	3.3.1 WML	68
	3.3.2 WMLScript	70
	3.3.3 UAProf	71
3.4	WAP and Messaging Services	72
	3.4.1 SMS	72
	3.4.2 MMS	74
	3.4.3 MMS Works Together with WAP and SMS	74
3.5	Future of WAP—OMA	75
	3.5.1 Converging to WAP 2	75
	3.5.2 OMA	76
3.6	Summary	78
	References	79

CHAPTER 4

Technologies for Mobile Client Software — 81

4.1	XML-Based Technologies	81
	4.1.1 XML Syntax	82
	4.1.2 XML Validation: Document Type Definition/Schema	83
	4.1.3 XML Styling: Cascading Style Sheets and Extensible Stylesheet Language	86
	4.1.4 XML-Based Markup Languages	89
4.2	Mobile Web Client	90
	4.2.1 XHTML Basic	90
	4.2.2 The XHTML Mobile Profile	92
4.3	Java Mobile Client	94
	4.3.1 CLDC Connection Framework	94
	4.3.2 MIDP Development	95
4.4	i-mode	98
	4.4.1 i-mode Infrastructure	98

  4.4.2 i-mode Client Technology 99
  4.4.3 Server-Side Technologies 100
 4.5 Summary 104
  References 104

CHAPTER 5
Wireless Multimedia Technologies **107**

 5.1 Multimedia Streaming 108
  5.1.1 Interactive Mobile Services 109
  5.1.2 Extensible Markup Language 109
  5.1.3 SMIL 110
  5.1.4 Mobile Streaming Architecture 110
  5.1.5 Backend Server 111
  5.1.6 Streaming Client 116
  5.1.7 Solution Providers 118
 5.2 Wireless Network Technologies 119
  5.2.1 Network Protocols 119
  5.2.2 QoS and Scalability 121
  5.2.3 Proxy Requirement 121
  5.2.4 Unicast Versus Multicast 122
  5.2.5 Wireless Multimedia Streaming Standards 122
 5.3 Codec 123
  5.3.1 MPEG 123
  5.3.2 MPEG-4 Architecture 123
  5.3.3 MPEG-4 Visual Profiles 124
  5.3.4 MPEG-4 Encoding and Decoding 124
  5.3.5 MPEG-4 File Format 124
  5.3.6 Audio Compression Standard 124
 5.4 Conclusion 125
  References 125
  Selected Bibliography 126

PART III
Wireless Networks **129**

CHAPTER 6
Wireless Local Area Networks **131**

 6.1 Wireless LAN Protocols 131
 6.2 Wireless LAN Standards 132
  6.2.1 The 802.11 Physical Layer 133
  6.2.2 The 802.11 MAC Sublayer 137
 6.3 Wireless LAN Infrastructures 142
  6.3.1 Basic Service Areas 142
  6.3.2 Extended Service Areas 143
  6.3.3 Mobility 144
 6.4 Future for Wireless LAN Networks 144
  6.4.1 Bluetooth 145

CHAPTER 7

Wireless Wide Area Networks   149

Wireless Personal Area Networks   171

PART IV

Engineering Wireless-Based Application Systems   197

6.4.2 Ultrawideband	145
6.4.3 IEEE 802.16 Broadband Wireless	146
6.4.4 3G Cellular Networks	147
6.4.5 Discussions	147
6.5 Summary	147
References	148
Selected Bibliography	148
7.1 Wireless WAN Overview	149
7.1.1 Common Wireless WAN Architecture	149
7.1.2 Mobility Management	151
7.1.3 Communication Background	153
7.2 Cellular Mobile Networks—Up to 2.5G	155
7.2.1 1G Networks: AMPS	156
7.2.2 2G Networks: D-AMPS, GSM, cdmaOne, and More	157
7.2.3 2.5G Networks: General Packet Radio System and Enhanced Data Rates for Global/GSM Evolution	160
7.3 Cellular Mobile Networks—3G and Beyond	162
7.3.1 3G Networks: CDMA2000 and WCDMA	163
7.3.2 4G Networks and Beyond	165
7.3.3 Standardization Organizations	167
7.4 Summary	168
References	169
8.1 Wireless PAN Overview	171
8.1.1 IEEE 802.15	172
8.1.2 Near-Field Communication	174
8.2 Bluetooth	175
8.2.1 Introduction	175
8.2.2 Protocol Layers	177
8.2.3 States and Operations	181
8.2.4 Application Profiles	183
8.3 Device Coordination (Wireless Access Protocol)	188
8.3.1 Jini	188
8.3.2 Universal Plug-and-Play	191
8.3.3 Home Service Gateway	193
8.4 Summary	194
References	195

CHAPTER 9

System Requirements Engineering, Analysis, and Modeling

9.1   Understanding the Wireless Application Domain   199
9.2   Engineering System Requirements   200
       9.2.1   Requirements Engineering Process   200
       9.2.2   Requirements Elicitation for Wireless-Based Software Systems 202
9.3   System Analysis and Modeling   207
       9.3.1   System Infrastructure and Connectivity Modeling   208
       9.3.2   System User Analysis and Modeling   209
       9.3.3   System Function Analysis and Modeling   211
       9.3.4   System Data and Object Analysis and Modeling   212
       9.3.5   System Dynamic Behavior Analysis and Modeling   220
9.4   Summary   225
       References   226

CHAPTER 10

System Architectures for Wireless-Based Application Systems

10.1   Basic Concepts About System and Software Architectures   227
        10.1.1   What Is System Architecture?   227
        10.1.2   What Is Software Architecture?   230
        10.1.3   Why Is Software Architecture Important?   230
        10.1.4   Software Architecture Presentation and Specification   231
        10.1.5   Architecture Design Process   231
        10.1.6   Quality Factors in Architectural Design   233
10.2   The Network-Based System Infrastructure Classification   234
        10.2.1   Application Systems Based on Wireless LAN Networks   234
        10.2.2   Wireless Personal Network–Based Application Systems   234
        10.2.3   Application Systems Based on Wide Area Networks   235
        10.2.4   Wireless Application Systems Based on Heterogeneous
                  Networks   237
        10.2.5   Mobile Commerce Application Systems Based on Sensor
                  Networks   238
10.3   Wireless Internet-Based Application System Architectures   239
        10.3.1   The i-mode Application System Architecture   239
        10.3.2   Mobile-Aware Application Architecture   242
        10.3.3   Mobile-Transparent Application Architecture   244
        10.3.4   Mobile-Interactive Application Architecture   245
        10.3.5   Advantages and Limitations of Different Architectures   246
        10.3.6   Comparison of Different Wireless Internet Application
                  Architectures   247
10.4   Smart Mobile Application Architectures   249
        10.4.1   Advantages and Limitations of Smart Mobile Application
                  Systems   250
10.5   Wireless Enterprise Application Architectures   251
        10.5.1   Wireless Enterprise Application Architecture Based on
                  Microsoft Mobile Technology   252
        10.5.2   Java-Based Wireless Enterprise Application Architectures 253

10.6   Summary   257
       References   258

CHAPTER 11

Wireless Security: Introduction   259

11.1   Secret Key Cryptography   260
       11.1.1   Basic Operations   260
       11.1.2   Block Cipher Standards   263
       11.1.3   Stream Cipher and RC4   267
       11.1.4   Confidentiality   268
       11.1.5   Authentication   268
       11.1.6   Data Integrity   269
11.2   Public Key Cryptography   270
       11.2.1   The RSA Algorithm   271
       11.2.2   Confidentiality   273
       11.2.3   Authentication   274
       11.2.4   Data Integrity   275
11.3   Wireless Security Attacks   275
       11.3.1   Attacks Against Confidentiality   276
       11.3.2   Attacks Against Authentication   277
       11.3.3   Attacks Against Data Integrity   279
       11.3.4   Attacks Against Availability   280
11.4   Summary   281
       References   282

CHAPTER 12

Wireless Security Solutions   285

12.1   Security Threats and Solutions for Wireless LANs   285
       12.1.1   Overview of IEEE 802.11 WEP Standard   285
       12.1.2   Security Threats for Wireless LANs   286
       12.1.3   Security Countermeasures for Wireless LANs   288
       12.1.4   IEEE 802.11i Security Standard   289
12.2   Security Threats and Solutions for Wireless PANs   293
       12.2.1   Overview of Bluetooth Security Mechanisms   293
       12.2.2   Security Threats for Bluetooth-Based Wireless PANs   296
       12.2.3   Security Countermeasures for Bluetooth-Based Wireless PANs 298
12.3   Security Threats and Solutions for Cellular Networks   299
       12.3.1   Overview of Security Mechanisms in GSM Cellular Networks 300
       12.3.2   Security Threats for GSM Cellular Networks   303
       12.3.3   Security Solutions in 3G Cellular Networks   305
12.4   Summary   307
       References   308

CHAPTER 13

Design of Mobile Client Software   311

13.1   Developing Mobile Client Software on Mobile Devices   311

      13.1.1 Understanding Mobile Accesses and Mobile Users'
               Expectations   312
      13.1.2 Enable Mobile Technologies   312
      13.1.3 Design Issues and Challenges   313
  13.2 Classification of Mobile Clients and Architecture Styles   314
      13.2.1 Classification of Mobile Clients   314
      13.2.2 Unimodal Versus Multimodal Mobile User Interfaces   318
  13.3 Design for Mobile Client Software   320
      13.3.1 Mobile Client Development Process and Methods   321
      13.3.2 Design Principles, Guidelines, and Tips   322
  13.4 Mobile Client Design Issues and Solutions   325
      13.4.1 Design for Reliable Mobile Client Software   325
      13.4.2 Design Unified Mobile Client Interfaces   328
      13.4.3 Adding Mobile Links to Online Systems   330
  13.5 Application Examples of Mobile Client Software Design   333
  13.6 Summary   337
      References   339

PART V
Mobile Commerce Systems   341

CHAPTER 14
Introduction to Mobile Commerce Systems   343
  14.1 M-Commerce Versus E-Commerce   344
  14.2 Wireless Device Constraints, Application Usability, and Interface
      Issues   345
      14.2.1 Device Constraints   346
      14.2.2 Mobile Application Usability   347
      14.2.3 User Interface Issues for M-Commerce   348
      References   350
      Selected Bibliography   350

CHAPTER 15
Multimedia Messaging Service   351
  15.1 MMS Overview   351
      15.1.1 MMS Protocol   352
      15.1.2 Message Types   353
      15.1.3 Message Format   353
      15.1.4 Addressing Model   354
  15.2 Multimedia Presentation in MMS   354
      15.2.1 WML   355
      15.2.2 SMIL   355
      15.2.3 XHTML   356
  15.3 MMS Client   356
  15.4 MMS Content Delivery   357
  15.5 Client State Diagram   358
  15.6 MMS Server   359

15.7   Server State Diagram   360
15.8   Case Study: Nokia   360
15.9   Case Study: Alcatel and Intel   361
15.10 Comparison   362
15.11 Conclusion   363
       References   363
       Selected Bibliography   363

CHAPTER 16

Wireless Advertising and Marketing Systems   365

16.1   Understanding Wireless Advertising and Marketing   366
    16.1.1   Wireless Ads   366
    16.1.2   Business Models   369
    16.1.3   Business Issues and Technical Challenges   370
16.2   Engineering Wireless Advertising Solutions   373
    16.2.1   Enterprise Processes and Workflows in Wireless Advertising   373
    16.2.2   Wireless Ad Targeting   377
    16.2.3   Wireless Ad Delivery   378
    16.2.4   Wireless Advertisement Tracking   378
    16.2.5   Wireless Advertising Payment   380
    16.2.6   Performance Measurement in Wireless Advertising   381
16.3   Major Players and Their Solutions   381
    16.3.1   SkyGo   381
    16.3.2   AvantGo   382
    16.3.3   Vindigo   383
    16.3.4   Avesair   384
16.4   Summary   384
       References   384

CHAPTER 17

Mobile Payment Systems   387

17.1   Wireless Payment   387
    17.1.1   Basic Requirements of Wireless-Based Payment Systems   388
    17.1.2   Payment Schemes   389
    17.1.3   Mobile Payment Process   390
    17.1.4   Security Solutions for M-Payment Systems   390
17.2   Different Types of Wireless-Based Payment Systems   391
    17.2.1   Account-Based Payment Systems   391
    17.2.2   Mobile Wallets   393
    17.2.3   SET-Based Mobile Wallet   393
17.3   Major Players in Wireless Payment Systems   394
    17.3.1   Paybox   394
    17.3.2   Ultra's M-Pay   394
    17.3.3   Encorus PaymentWorks   395
    17.3.4   SNAZ   395
17.4   Payment Models, Challenges, and Issues   396
    17.4.1   Business Challenges   396

17.4.2   Technical Challenges   396
17.4.3   Interoperability Challenges   397
17.5   Conclusion   398
References   398
Selected Bibliography   399

About the Authors   401
Index   403

Preface

The wide deployment of wireless networks and mobile technologies, along with the significant increase in the number of mobile device users, have created a very strong demand on various wireless-based, mobile-based software application systems and enabling technologies. This provides many new business opportunities and challenges to wireless and networking service providers, mobile technology vendors, content providers, and solution integrators. Living in a wireless world changes and enhances people's lives in many areas, such as mobile communications, wireless information sharing and learning, m-commerce, home environment, and entertainment. Today, business organizations and government agencies face new pressure for technology updates in network infrastructures and enterprise solutions to support wireless connectivity and mobility.

What Is This Book About?

This comprehensive resource gives you a thorough understanding of the software engineering processes and methods needed to tackle the complexities of building software for wireless systems. It focuses on engineering perspectives of wireless-based application systems to address design and construction issues in system analysis and modeling, system application architectures, mobile client design, design principles, and development techniques. The book gives clear guidelines that explain design trade-offs that are needed to meet user and system requirements. The book also includes an in-depth examination of security issues and solutions that ensure the design of reliable and safe software.

This book offers tutorials in mobile technologies, mobile supporting platforms, wireless networking, wireless security, and wireless multimedia to provide a solid knowledge base in engineering wireless-based application systems. Furthermore, the book offers a solid grasp of networking theories and concepts, and explains the technical intricacies of 2G to 4G cellular networks, wireless local area networks (LANs), and wireless personal area networks (PANs). To help in understanding and building diverse wireless application systems, this practical reference not only covers the theory of mobile commerce solutions and wireless application systems, but also shows how to build such wireless-based application systems, such as wireless advertising and marketing and mobile payment systems.

Why Did We Write This Book? Why Do You Need This Book?

To meet the increasing demand on various reliable wireless-based software application systems, engineers and students need to know how to engineer quality wireless-based application systems using a cost-effective approach. Since the first book on mobile computing was published in 1996, many books have been published to help readers understand wireless networks, and mobile technologies, programming, and commerce. In addition, there are a number of books focusing on specific topics, such as wireless instant messaging, information services, and wireless multimedia. However, there are no books on the current market focusing on engineering topics for developing wireless-based software systems. Engineers need a comprehensive snapshot of engineering issues, methods, and application solutions to help them understand how to develop new wireless-based application systems. They need to master the system development processes, system analysis and design, wireless application architecture design, and domain-specific solutions in wireless applications. In addition, before tackling engineering and design issues, engineers need to have a basic understanding and comparative views about wireless networks and mobile platforms and technologies.

This book is written in an attempt to fill this gap. It intends to focus on common engineering issues and solutions in developing wireless-based software and application systems. It prepares readers with basic fundamental concepts and background about wireless-based application systems, in the fields of wireless networking, mobile technology and platforms, wireless information services, and mobile commerce applications. Unlike other books, this book is written to help readers understand engineering issues and solutions in developing wireless-based application systems, including system analysis and design, infrastructures, and wireless security. It covers a comprehensive tutorial on engineering issues and solutions in building wireless e-commerce systems, wireless information systems, wireless advertising systems, and moblie payment systems.

Book Organization

The book consists of five parts. We briefly describe their contents here.

- *Part I: Introduction*—This includes the introductory chapter of the book, which presents the market needs, user demands, and future trends for wireless application systems and services. It covers the fundamental concepts, properties, and components of wireless-based software systems, and presents engineering issues and challenges.
- *Part II: Mobile Technologies and Operation Environments*—There are four chapters that cover different mobile technologies and operation platforms. Chapter 2 provides an overview of mobile devices, focuses on different mobile operating systems (including Symbian OS, Windows CE/Mobile, Embedded Linux, and Palm OS), and discusses the major mobile application platforms. Chapter 3 provides a tutorial about wireless application protocol (WAP), including the WAP application environment (WAE). In addition, it discusses

messaging services, such as short message service (SMS), multimedia messaging service (MMS), and the future of WAP. Chapter 4 offers a comprehensive coverage of extensible markup language (XML), and different types of mobile technologies, including Web-based and Java-based mobile technologies, as well as i-mode technology. Chapter 5 focuses on wireless multimedia technologies. It covers topics in mobile media streaming and architecture, supporting technologies, standards, and the major companies involved.

- *Part III: Wireless Networks*—This part includes three chapters that discuss different types of wireless networks. Chapter 6 presents the basic concepts of wireless LAN networks, discusses wireless LAN infrastructures, protocols, standards, current issues, and future developments. Although wireless WANs include many networks, Chapter 7 focuses on cellular-based wireless technologies, including cellular mobile networks from 1G to 3G and beyond. Wireless personal area networking topics are covered in Chapter 8. It provides an overview of wireless PANs, discusses Bluetooth network protocols, operations, and application profiles. It also covers other wireless PAN technologies (such as IrDA, Zigbee, and NFC) and device coordination.

- *Part IV: Engineering Wireless-Based Application Systems*—This part discusses common engineering issues in building wireless-based application systems. There are five chapters. Chapter 9 covers the topics in engineering requirements, system analysis, and design modeling for wireless-based application systems. Application examples are used to demonstrate how to perform system requirement engineering for wireless-based application systems, and how to carry out system analysis and modeling tasks using well-known software engineering methods. Chapter 10 focuses on wireless-based system infrastructure and mobile application architectures. It presents the basic concepts and classifications of wireless-based application system infrastructures. It discusses and compares different wireless Internet application architecture models, presents smart application systems on wireless networks, and covers wireless enterprise application architectures. Since security is always a major concern in wireless application systems, two chapters are included to address wireless security topics. Chapter 11 presents the fundamental concepts of wireless security, covers secret and public key cryptography, and discusses the topics in wireless security attacks. Chapter 12 provides a comprehensive snapshot on existing wireless security solutions. These include security threats and solutions for different types of wireless networks, including LANs, PANs, and cellular networks. Mobile client design topics are covered in Chapter 13. It provides a design process for building mobile client software, and classifies and compares mobile client types and two mobile interaction modes. It offers design principles, guidelines, and solutions to deal with the technical challenges in mobile client design.

- *Part V: Mobile Commerce Systems*—This part consists of four chapters. Chapter 14 provides an introduction to mobile commerce and systems. It covers the essential concepts and models, such as service-based and location-based commerce models. Chapter 15 discusses the topic of multimedia messaging services. Chapter 16 is dedicated to wireless advertising and marketing issues from the engineering perspective. It addresses the business models, presents the essential processes in wireless marketing and advertising, and discusses the necessary engineering solutions, major players, and standards in wireless advertising. Wireless payment systems and solutions are covered in Chapter 17, in which wireless payment processes, mobile payment schemes, protocols, and different payment systems and major players are summarized.

How to Use This Book

This book is written for professionals and students who need to know how to engineer wireless-based software and application systems over a wireless network, using appropriate mobile technologies and platforms. It can be used as a textbook for a university undergraduate-level or graduate-level course that focuses on the design of wireless-based software application systems or mobile computing. This book also can be used as a textbook for a course on wireless electronic commerce systems.

Introduction

The wide deployment of wireless networks and mobile technologies, along with the significant increase in the number of mobile device users, have created a very strong demand on various wireless-based, mobile-based software application systems and enabling technologies. This not only provides many new business opportunities and challenges to wireless and networking service providers, mobile technology vendors, and software industry and solution integrators, but also changes and enhances people's lives in many areas, including communications, information sharing and exchange, commerce, home environment, education, and entertainment. Business organizations and government agencies face new pressure for technology updates to upgrade their networking infrastructures with wireless connectivity to enhance enterprise-oriented systems and solutions.

CHAPTER 1

Engineering in the Wireless World

This chapter provides an introduction to the engineering of wireless-based software and application systems. The chapter first discusses the current and future market needs and user demands in wireless-based software and application systems. Then, it covers the basic concepts of wireless-based software and application systems to help readers understand their essential components, basic system requirements and user expectations, distinct properties, and major benefits and limitations. In addition, it classifies various wireless-based application systems. The rest of the chapter provides an introduction to engineering perspectives of wireless-based software and application systems, including tasks, development processes, and typical engineering issues and challenges for wireless-based software and application systems. Finally, a summary is given as a last section.

1.1 Moving to the Wireless World

The wide deployment of wireless networks and mobile technologies, along with the significant increase in the number of mobile device users, have created a very strong demand on new mobile-based technologies and diverse wireless-based application systems.

1.1.1 Wireless Markets and Needs

With the fast deployment of wireless networks and advances of mobile technologies, the number of mobile devices in the United States has been increasing at a very fast pace in the recent years. The comparison of the number of cell phones in use in the United States, given by Neuharth [1], shows the increasing magnitude of this phenomenon:

- 1985: 340,000;
- 1995: 33.8 million;
- 2004: 165.9 million.

This phenomenon also occurred around the world. According to Cyberatas and WindWire [2], the number of mobile users worldwide exceeded 468 million in 2000—a much higher number than the 365 million people using the Internet that year. Figure 1.1 shows the market projection by eTForecasts in 2005 of the number of cellular subscribers by region for the next 5 years [3]. Clearly, we will expect that Asia is the region with the most significant increase in the number of cell phone users in the next few years.

Today, with the deployment of third generation (3G) cellular wireless networks and additional new mobile technologies, more new mobile subscribers will join the wireless world. According to In-Stat/MDR's report (http://www.instat.com) on June 15, 2004, China had 260 million mobile subscribers in 2003, and more than 4 million new subscribers were added every month in 2003. They found that the number of mobile subscribers in China would grow from 268.69 million in 2003 to 497.86 million by 2008, growing at a compound annual rate of 11.7%, and reaching a penetration rate of 37.6%. Moreover, In-State/MDR forecasts that there will be 118 million 3G wireless subscribers in China by 2008.

The fast increase in the number of mobile users creates many new business opportunities, and brings a strong demand on diverse mobile services and applications in many areas.

Wireless Advertising and Marketing

Jupiter Media Metrix predicted that the wireless advertising market will reach $700 million annually in the U.S. market over the next 4 years, while the Kelsey Group projects revenues that will be five times higher—$3.9 billion [2]. Studies by wireless media research companies, such as WindWire (http://www.Windwires.com) and SkyGo (http://www.SkyGo.com), indicated that permission-based alerts delivered to wireless phones capture consumers' attention, drives response actions, and builds brand awareness. Microsoft, Yahoo, AOL, and other large companies have created subsidiaries to focus on the wireless advertising market.

Wireless Multimedia Services

With the promise of greater bandwidth on 3G networks, rapid advances in compression technologies, and new multimedia handset capabilities, wireless networks and mobile video technologies are ready to provide mobile users with various multimedia services, such as mobile video and content services, multimedia instant messaging, and chatting. By 2009, mobile video services are expected to generate

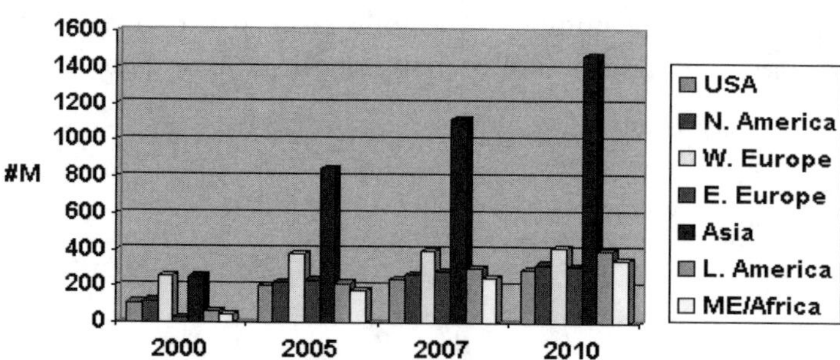

Figure 1.1 Cell phone subscribers by regions (in millions), projected by eTForecasts.

$5.4 billion in annual revenue. Based on a recent survey conducted by In-Stat/MDR, 13.2% of U.S. wireless subscribers are extremely or very interested in purchasing video services for their wireless phones. Moreover, interest in mobile video is higher than for all other mobile multimedia services covered in this survey, including games and music services. In addition, the In-Stat/MDR report [4] suggests that by 2009, 22.3 million Americans will be viewers of mobile video content, and 31.1 million will use video messaging services. Based on this report, mobile video services will account for approximately 14.9% of total wireless data revenues.

Wireless Messaging and Data Services

As indicated in [5], wireless message and data services will go through a migration path from text-based short messaging services (SMS) to multimedia messaging services (MMS). In-Stat/MDR's Consumer Mobility Study reported on April 19, 2004, that 54% of the respondents currently use one or more wireless data services. In addition, the survey result also indicated that mobile data usage is now a mainstream phenomenon across multiple demographic groups. In-Stat/MDR found that wireless data customers use 42% more voice minutes than nondata users. According to In-Stat/MDR, SMS appears to be the leader among mobile data services used by survey respondents; however, Internet access services, ringtones, and mobile games all had strong showings. The report given in [5] has made a prediction of market revenues for multimedia messaging services.

Location-Based Application Services

Today, more than 77 million mobile customers use location-based services. Due to the significant increase in mobile device users, a similar trend in location-based service revenue, which could reach more than $19 billion by 2006, is also expected. Figure 1.2 shows a detailed market projection on location-based service revenues in the U.S. market [6].

Today, there are three major driving forces in the current wireless market. The first driving force is the wireless networking and mobile technologies. It is driven by the fast emergence of wireless network standards, ubiquity and interconnection,

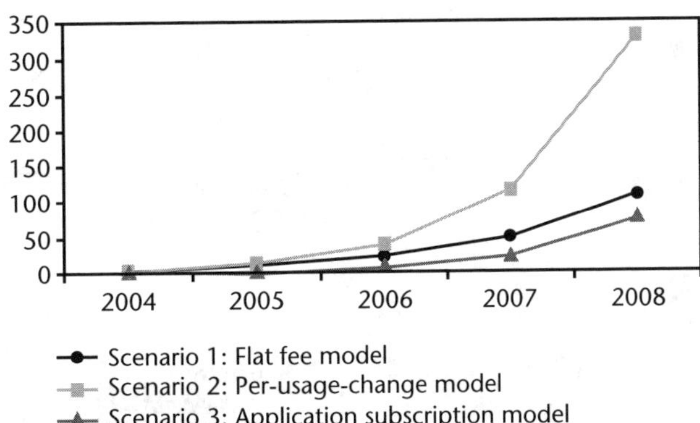

Figure 1.2   U.S. location-based service revenue projection. (*From:* [7]. © 2003 Ovum Research. Reprinted with permission.)

and mobile technology advances. The emergence of affordable and powerful terminal devices provides another strong push. Wireless-related business opportunities form the second driving force, due to the strong demand in the following aspects:

- Diverse and innovative wireless applications and services on mobile devices;
- New business models and opportunities on the wireless supplier value chain;
- Direct wireless channels reaching customers for business marketing, advertising, trading, and services;
- New business expansions and services to new customers through wireless networks.

The third driving force is coming from mobile users. With the fast increase in the number of mobile device users, more mobile access and mobility needs will be demanded from the public.

1.1.2 Living in the Wireless World

To meet the strong growing demand of wireless applications, we need more innovative application systems and solutions to provide mobile users with diverse application functions and services. As predicted by many experts, these wireless systems will change lifestyles by enriching their mobile experience. Let's take a look at what kind of mobile experience a businessman will have on his future business travel. Figure 1.3 shows Scott Chen's mobile experience on the trip. As shown in Figure 1.3, Scott started his trip to Beijing, China, at the San Francisco International Airport by checking in himself, using his mobile phone to access an airline's reservation and checking system through the wireless Internet. When he arrived in Beijing, he used his mobile phone to access a location-based mobile portal system to book a hotel room. Then, he

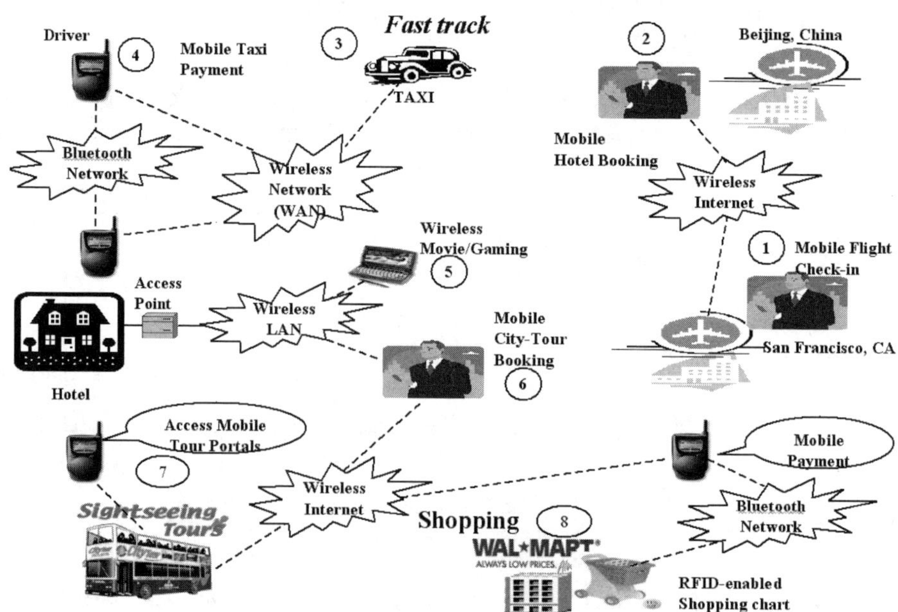

Figure 1.3 A businessman's mobile experience in a wireless world.

took a taxi on a highway through Fast Track to arrive at his booked hotel. To pay his taxi fare, he used his mobile phone to make a peer-to-peer mobile payment, by communicating with the taxidriver's mobile phone for a mobile payment transaction supported by a wireless payment system. In the hotel, he played a number of mobile games and enjoyed a live show of the Peking Opera on his smart mobile phone by accessing the hotel's wireless LAN. On the second day, after meeting with his customers, he decided to take a city tour. Using his mobile phone, he found a good city tour, using a wireless travel portal to make his reservation. On his city tour bus, he took many photos using his mobile phone. He shared his pictures and videos with his wife in San Francisco, using a wireless multimedia instant messaging system. On the return to his hotel, he stopped at a supermarket, where he used his mobile phone to interact with the RFID-enabled shopping chart to make his purchasing transaction.

Clearly, moving from today's wired world to tomorrow's wireless world not only needs wireless networks and mobile technologies, but also many wireless application systems. Here, we list the major players who can contribute to or provide these solutions.

- *Wireless service providers* can provide their mobile device users with diverse digital wireless applications and services with new wireless-based software application systems, for example, wireless-based instant messaging, digital chatting, and multimedia services.
- *Content providers, portal businesses, and publishers* can provide mobile device users with location-based mobile portals and mobile sites, using wireless-based portal systems and mobile contents.
- *Internet service providers* can connect the Internet to wireless networks for their Web users, by offering them various wireless Internet application systems and services, for example, digital e-mails and short messages between the Internet and wireless networks.
- *Government agencies and organizations* can build wireless channels between government offices and the public, offering mobile users with diverse administration functions and public services, for example, election and tax-reporting services.
- *Businesses* can link their enterprise-oriented systems and organizations to customers and suppliers, using the wireless connectivity to enhance their enterprise-oriented solutions, including both interorganizational and intraorganizational business workflows.
- *Product manufacturers and merchants* can set up wireless-based electronic commerce systems, supporting presales, trading, payment, and postsales services on mobile devices.
- *Schools and educational institutions* can establish wireless campuses, providing students with wireless e-service systems, mobile-based education-oriented portals, virtual classrooms, and distant learning services.
- *Transportation companies* can provide their passengers with wireless connectivity, so the passengers can access location-oriented mobile content and travel information while they travel in a mobile environment, such as in trains, airplanes, and subways.

- *Entertainment businesses* can provide dynamically downloadable digital music, movies, and games, allowing users to play them using mobile devices.
- *Banks, financial businesses, and payment solution providers* can offer users wireless-based payment solutions, supporting mobile payment, billing, and trading transactions.
- *Travel agencies and businesses* can offer users location-based travel portal systems on mobile devices, providing virtual travel guides, sightseeing spot illustrations, and picture- and video-sharing and integration.

1.2 Understanding Wireless-Based Application Systems

What are the primary engineering issues, challenges, solutions, and needs in engineering wireless-based software systems? What kinds of knowledge background on wireless networking and mobile technology must engineers have before constructing wireless-based software application systems? Before engineering wireless-based software and application systems, we must understand their essential features and properties, basic system requirements and elements, and benefits and limitations. This section covers these topics to provide readers with the basic understanding of wireless-based application systems.

1.2.1 Basic Definitions

These are the basic definitions for engineering wireless-based software systems.

- *Wireless-based application systems* refer to the application systems that are developed using mobile technologies and deployed on wireless networks to provide mobile application functions and services to mobile device users.
- *Wireless Internet application systems* refer to the application systems that are developed using mobile technologies and deployed over a global network, which connects wireless networks and the Internet together, to offer mobile application functions and services to WAP-enabled mobile device users.
- *Wireless information systems* refer to the information systems that are developed using mobile technologies and deployed on wireless networks to offer diverse mobile information and data access services to mobile device users.
- *Mobile commerce systems* refer to electronic commerce systems that are developed based on wireless networks and mobile technologies to support customers and merchants using mobile devices to conduct various mobile commerce transactions and activities, including presales, trading, and postsales activities.

1.2.2 Basic Components

The major objective of developing wireless-based software and application systems is to take advantage of wireless networking and mobile technology to provide diverse wireless application solutions and services to mobile device users.

The essential components of a wireless-based software application system can be classified into four layers.

- The *mobile client layer* consists of five classes of software stored on mobile devices, executed on the appropriate operating environments:
  - *Mobile client interface software* refers to a user interface program that supports the interactions between the system and users on mobile devices.
  - *Mobile client application programs* refer to mobile programs that provide application functions to mobile device users.
  - *Mobile browsers* refer to microbrowsers on a mobile device, which receives and processes the mobile users' requests, and sends the system's responses to its wireless server.
  - *Wireless network connectivity components and APIs* refer to different wireless network connectivity application interface programs and supporting components on mobile devices. Typical examples are Bluetooth, GSM, GPRS, and MMS connectivity APIs.
  - *Mobile databases and access components on mobile devices* support mobile data accesses, caching, migration, and synchronization. For each wireless-based application system, there may be a mobile database that stores mobile application data on mobile devices.
- The *networking layer* consists of wireless network–related software components and systems on wireless networks.
  - *Wireless data exchange servers* refer to data exchange servers on wireless networks that support mobile users to share messages and data.
  - *Wireless data synchronization servers* refer to data synchronization servers (or systems) on the wireless networks that support mobile users performing data synchronization using mobile devices.
  - *Wireless router and gateway software* refers to wireless networking software and components that support wireless routers and gateways on wireless networks.
- *Server layer*
  - A *wireless server* supports wireless communications with wireless mobile browsers on mobile devices, similar to the operation of a Web server. Its major functions include receiving and processing users' requests from mobile device users, and sending back the responses from the application servers.
  - *Middleware components* facilitate the client-server interactions between mobile client software and wireless application systems.
  - *Application servers* refer to application programs that provide domain-specific application functions and services for mobile device users.
- *Data store layer*
  - *Central service databases* store large-scale mobile applications and service data to support wireless-based application systems. Typical service

data includes mobile device user information, such as user membership data, security code, and mobility data.

- *Database management systems* manage central application databases, and control and support different database access transactions for application servers.
- *Application database access programs* refer to database access interfaces between application servers and database systems to supporting application data access transactions.

1.2.3 Special Features and Properties

Wireless-based software systems support diverse mobile computing applications, and share several common features [7]. These common properties are discussed next.

- *Wireless communications:* Since wireless networks and connectivity protocols are the fundamental infrastructures of wireless-based software systems, they carry the essential properties of wireless networks.
  - *Heterogeneous network:* A conventional software application system over a wired network includes computers that are connected to a single network. In a wireless-based software system, mobile devices (run as mobile computers) usually are connected to heterogeneous network connections in several ways. One popular approach is to provide wireless-based software systems that support different types of mobile device users based on different wireless networks, such as a wireless LAN or a WAN (e.g., a GPRS network). The other popular approach is to add wireless connectivity into an Internet-based software application system to form a wireless Internet-based system to support mobile users' global accesses.
  - *Unreliable network connections:* Compared to wired networks, current wireless networks and mobile technology provide poor reliability in network communications, due to users' mobility, high signal collision rate, and limited network capacity in network bandwidth and communication volumes in each service zone. This feature causes more challenges in developing reliable wireless-based information systems.
- *Mobile access:* Mobile device users are allowed to access the system at any time and anywhere in the online or off-line mode. In the online mode, a mobile device user is allowed to dynamically access the functional services and information content in application servers. In the off-line mode, a mobile device user not only can access preloaded functions and mobile data on wireless devices, but also can synchronize the programs and mobile data on them.
- *Mobility:* Mobility is a distinct feature of wireless networks for mobile users. It refers to the ability to allow mobile users to change their system access locations, so that location-aware functions, services, and data can be provided to mobile users while connected to the network. This mobility feature not only increases the volatility of some information in the system, but also increases

the complexity in supporting mobility-based functions. It introduces several design issues and requirements in system construction, including mobile user support, mobile information migration, location-based information accesses, and location-dependent functional services.

- *Accessibility:* Mobile devices are the major operating platforms for mobile users in wireless-based software systems. Compared with stationary computers, mobile devices have three characteristics in mobile accesses [8].

  - *Accessible:* Wireless devices are rapidly becoming personal devices because they are portable and available for use at all times.

  - *Personal:* The typical wireless device belongs to a single person and thus becomes uniquely identified with that individual.

  - *Location-aware:* If a wireless device is on and connected, it can be used to track a user's physical location.

1.2.4   Essential Requirements and Expectations

Before constructing wireless-based software systems, engineers must understand the essential system requirements and expectations from mobile device users. Here we list a number of essential requirements for a wireless-based software system.

- *Reliability:* The system must support reliable communications in heterogeneous networks, so that it delivers reliable information content, transactions, and functional services to its mobile device users.

- *Performance:* The system must provide mobile device users with fast system responses, efficient and highly available functional services, and quick transaction processing. The mobile client software of the system also must effectively utilize the resources of mobile devices.

- *Portability:* Since wireless handheld devices are portable, personal, and mobile for users, the wireless-based software system must be accessed in a fast and easy way in a mobile environment.

- *Interoperability:* The interoperability of wireless-based software systems has two features: (1) device-oriented interoperability, and (2) network-oriented interoperability. The device-oriented interoperability refers to the degree of the system capability of supporting users to access the same set of system functions and services using various mobile devices (such as mobile phones, PDAs, and Pocket PCs) from various manufacturers. For instance, mobile phone users expect that they can access the same wireless portal using cell phones from different manufacturers. The device-oriented interoperability raises strong interoperability requirements and challenges in building mobile client software for wireless information systems and wireless commerce applications. The network-oriented interoperability refers to the degree of the system capability of supporting the same set of system functions and services in mobile data and wireless (or wireless Internet) communications crossing different heterogeneous networks. This brings challenging interoperability requirements for supporting various mobile communications, transactions, and mobile information access to cope with diverse network bandwidths, data transfer speeds, standards, and protocols.

- *Security:* The system must be developed with appropriate wireless and wired security solutions to meet various security requirements. Typical security requirements and expectations are given here:
  - Privacy and confidentiality of mobile device users;
  - Secured end-to-end transactions between peers;
  - Secured communications between mobile clients and its application server;
  - Authentication and/or certification for all involved parties.
- *Scalability:* The system scalability can be demonstrated in the following aspects:
  - Support of the quickly increasing number of mobile device users;
  - Coping with dynamic changes in mobile data volumes and communication traffic caused by the mobility feature of the system;
  - Accommodation of the increasing number of application servers and services.
- *Privacy and confidential:* The system must be able to protect the privacy and confidential of all mobile users' information and their driven transactions.

1.2.5   Benefits, Issues, Challenges, and Limitations

Compared to conventional software application systems (such as online application systems), wireless-based software systems have three distinct advantages.

- *Convenient mobile accesses:* The first distinct advantage is that they allow mobile users to access information, applications, and services, anytime and anywhere.
- *Location-based applications and services:* The second advantage is that they can be equipped with various location-based applications and services, which are known as the "killer applications." For example, a wireless advertising system can deliver location-based product ads and promotion messages to mobile device users according to their current locations.
- *Personal-based mobile interactions and messaging services:* The third major advantage is that they can be used as a cost-effective mobile personal communication channel to support one-to-many (or one-to-one) mobile interactions among mobile device users, allowing them to conduct peer-to-peer mobile digital communications, information sharing, and exchanges. They can also be equipped with various content delivery and messaging services to deliver various kinds of multimedia content to mobile devices based on individual interests.

However, there are a number of limitations in building wireless-based software and application systems. They can be classified into the following two groups.

- *Limitations of mobile devices*
  - *Limited CPU power and operation time*: This is due to the limited lifetime of a mobile device's battery.
  - *Small display screen:* There are three types of mobile devices: mobile phones, PDAs, and pocket PCs. A mobile phone, such as Nokia 3650, has a small display screen (e.g., 176 × 208 pixels). The small screen usually has 5 to 6 lines, and usually is 15 to 20 characters wide. A PDA has a standard screen size ranging from 208 × 320 to 240 × 320 pixels.
  - *Limited storage space:* All mobile devices have limited storage spaces. For example, the Nokia 3650 phone has only 3.4 MB internal memory.
- *Limitations of wireless networks*
  - *Limited bandwidth and capacity*: Compared with wired networks, wireless networks have a lower data capacity and limited bandwidth. For example, the earliest schemes for sending data rate over the Global System for Mobile Communications (GSM) had a rate from 9.6 to 14.4 Kbps. The data capacity over the General Packet Radio Service (GPRS) is from 72 to 115.2 Kbps.
  - *Unreliable network connectivity:* Since wireless-based application systems are usually constructed using heterogeneous networks, including wireless and wired networks (such as the Internet), the heterogeneous networking feature causes the network connection to be unreliable and vulnerable, due to the higher rate of signal conflicts and communication failures, and the inconsistency of data communication speeds and bandwidths.

1.2.6   Classification of Wireless-Based Application Systems

Wireless networking and mobile technologies can be used to develop diversified wireless-based software and application systems. They can be classified based on wireless networking into the following classes: (1) Bluetooth-based, (2) wireless LAN-based, (3) wide-area network-based, (4) wireless Internet-based, and (5) hybrid-based.

In addition, we can classify wireless-based application systems based on application types. Here, we list a number of popular wireless-based application types.

- *Wireless-based e-commerce systems* refer to electronic commerce systems based on wireless networks and mobile technologies. On the one hand, they allow service providers, content providers, and merchants to perform automatic presales, trading, and postsales transactions based on mobile device users' requests. On the other hand, wireless-based e-commerce systems provide mobile devices users with goods, products, services, advertisements, and postsale information. Typical examples of these systems are wireless-based advertising systems, location-based trading and sales systems, and mobile payment systems.
- *Wireless-based information messaging systems* provide mobile users with different kinds of information messaging services, to allow them to access, share, and exchange mobile information with different media formats, using wire-

less connectivity protocols over wireless networks. Mobile users can perform one-to-one and/or one-to-many information delivery, chats, and exchanges using static or real-time dynamic approaches. Mobile information could be presented in a text format, such as instant short messages, or in a multimedia format, such as audio or video. Typical application examples are: (1) wireless-based (or wireless Internet-based) instant messaging and text chatting systems, (2) wireless multimedia information and content service systems, and (3) wireless streaming-based mobile talking systems.

- *Wireless-based entertainment systems* provide diverse entertainment services to users, allowing them to download, access, play, and view various types of digital entertainment resources using mobile devices. The typical entertainment resources are digital movies, games, and music.

- *Wireless-based personal groupware* refer to a set of mobile software groupware on mobile devices, allowing users to access their personal address books, calendars and meeting schedules, e-mail boxes, and other facilities.

- *Wireless-based enterprise application systems* add new wireless connectivity and solutions to existing enterprise-oriented information systems and business management solutions. They allow enterprise users and customers to exchange information efficiently, and access internal workflows and business operations effectively.

- *Wireless-based real-time application systems* refer to sensor-based real-time control and application systems based on wireless networks and mobile technologies. In general, they receive real-time signals and/or inputs from wireless-based sensors, and process them to generate real-time outputs and/or signals using domain-specific application processes.

- *Wireless-based portal systems*, similar to online portal systems, allow content providers to offer well-classified domain-specific mobile contents to mobile device users.

- *Mobile search engines and mobile Yellow Pages systems*, similar to online search engines, allow users to search specific information content using mobile devices from many mobile sites over the global wireless networks. Similar to online Yellow Pages systems, mobile Yellow Page systems provide various Yellow Pages directory services for mobile users through mobile devices.

- *Wireless location-based application systems* refer to the application systems that provide mobile location-based services to users by detecting the current locations (e.g., geographical coordinates) of mobile users. A typical example is a navigation system in a car. Another example is a location-based mobile portal that provides users with mobile information (e.g., news, shops, maps, and sports), based on their current locations.

1.3 Engineering Wireless-Based Application Systems

Engineering wireless-based software and application systems encompasses many different perspectives. As shown in Figure 1.4, engineers must understand and deal with issues in the following six areas.

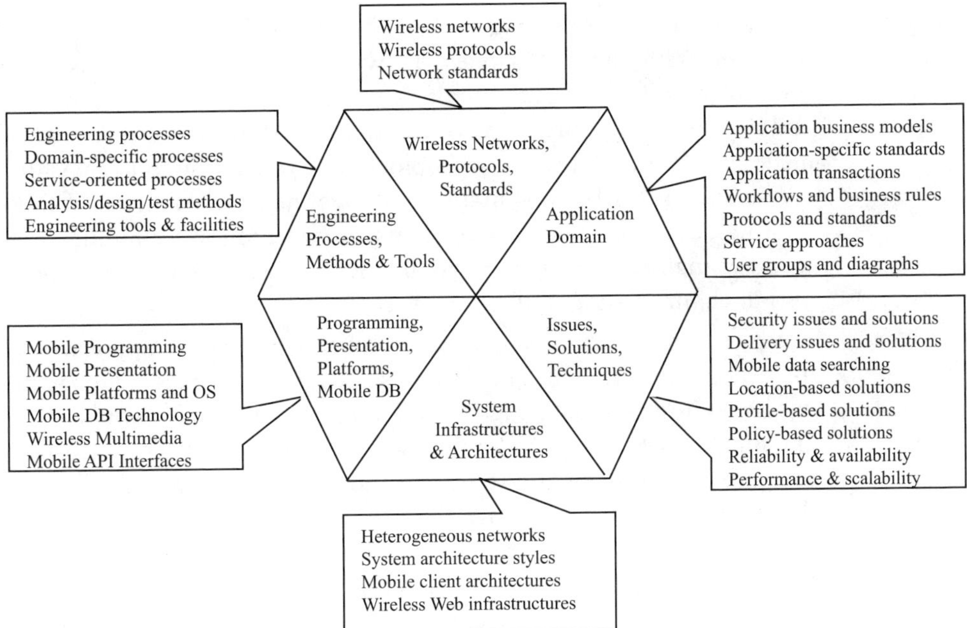

Figure 1.4   Engineering perspectives for building wireless-based application systems.

- *Wireless networks, protocols, and standards:* Engineers need to understand wireless network structures, protocols, and standards to know their features, strengths, limitations, and applicability. To help readers gain the basic concepts and knowledge on wireless networking (such as GSM, GPRS, 2G, and 4G), we include three chapters in Part III to cover three different types of wireless networks, including wireless wide area networks, wireless local area networks, and wireless personal area networks.

- *Wireless programming, mobile platforms, presentation technologies, and mobile databases:* To create effective mobile client software providing a rich mobile experience, engineers must understand the pros and cons of different mobile platforms (e.g., Symbian OS, Palm OS, Window CE, and J2ME); basic wireless programming languages; distinct features and limitations of various presentation technologies (e.g., WML, J2ME, XHMTL, and cHTML); mobile database support; and related multimedia technology. Five chapters of Part II of this book are written to assist readers in gaining these basics.

- *Engineering processes, methods, and tools:* To provide engineers with details about engineering processes, analysis and design methods, and basic engineering principles, Part IV of this book discusses system engineering processes, system analysis models, and design methods. In addition, a design procedure, design principles, and guidelines are provided to help engineers construct user-friendly mobile client software with good reliability and consistent formats.

- *Network infrastructures and system architectures:* System architecting is the first and most important task in the system design phase. This book includes two chapters to assist system architects in this job. One chapter is written to discuss six different types of system architectures, with their strengths and limitations for constructing wireless-based application systems based on the basic understanding of wireless network infrastructures. This chapter also provides an overview about major vendor-based wireless enterprise solutions and their infrastructures. The other chapter focuses on mobile client development and their architecture styles, including thick, thin, and smart mobile client styles, and their advantages and weakness. In addition, unimodal and multimodal mobile interface concepts and interaction techniques are covered, to allow engineers to support mobile user accesses in more than one input and output channel using multimedia formats.

- *Issues, solutions, and techniques:* Engineering wireless applications systems may encounter some unique design issues and challenges. The basic engineering issues and new challenges in building wireless-based application systems are highlighted in the following section. More detailed discussions about design issues and existing solutions and techniques are covered in Part IV of this book. For example, Part IV includes two chapters addressing wireless security topics. One chapter covers wireless security concepts in cryptographic techniques, authentication, confidentiality, and data integrity. The other chapter discusses current wireless security solutions in different wireless networks. The other chapters in Part IV cover design issues and solutions in system architectures, mobile client styles and structures, mobile data delivery, system reliability, and mobile interoperability.

- *Wireless application domain:* To successfully build wireless application systems, engineers must learn and understand domain-specific business models, processes and workflows, application standards (e.g., transactions and presentations standards), customers and service models, and policies. To provide a basic mobile application knowledge, Part V of this book consists of four chapters discussing basic mobile commerce concepts and different mobile commerce applications, including wireless marketing, advertising, and payments. For instance, the chapter on wireless payment systems covers mobile payment business models and processes, major players, and methods. It discusses and compares different types of wireless payment systems, including account-based mobile payment solutions, mobile wallets, and POS-based payment systems.

1.3.1 Engineering Perspectives

The major tasks for an engineering team can be classified into the following types.

- *System analysis tasks*
  - Understanding wireless networks with connectivity protocols;
  - Collecting, analyzing, and specifying networking connectivity requirements, infrastructures, and components;
  - Understanding and comparing different mobile technologies and operating environments;
  - Understanding wireless domain application knowledge;

- Analyzing and specifying various system application functions, nonfunction requirements, service features, and system-user interaction requirements.

- *System design tasks*

  - Conducting design for network infrastructure, network connectivity, and system architectures, including communication protocols and connectivity interfaces;

  - Generating database design for mobile databases and system repositories;

  - Selecting the suite mobile technologies and targeted mobile devices, as well as related operating environments;

  - Performing design for mobile client software on mobile devices, including mobile data and mobile user interfaces;

  - Designing for the required domain-specific application requirements, such as a particular wireless security solution.

- *Implementation tasks*

  - Implementing network connectivity components and communication application interfaces based on the selected network protocols;

  - Implementing mobile data management systems and/or components to support required mobile data accesses and transactions;

  - Implementing mobile client software using the selected mobile technologies under target operating environments;

  - Implementing the required solutions and components to meet the domain-specific application requirements, such as wireless security.

- *Testing and deployment tasks*

  - Performing system validation for the required functions and service features;

  - Conducting system testing and system performance evaluation, including system installation, stress testing, security testing, performance, and usability testing, with special attention to system performance, reliability, availability, mobility, and scalability;

  - Deploying the system into a targeted wireless networking infrastructure with the specified connectivity protocols and mobile devices.

1.3.2 Engineering Issues, Challenges, and Needs

Besides the common limitations challenges in constructing wireless-based software and application systems, there are many other engineering issues, challenges, and needs in the engineering production cycle. Here, we highlight common challenges for all wireless-based software and application systems, as well as engineering issues in terms of software development phases.

Common Engineering Challenges

There are several common engineering challenges in constructing a wireless-based software or application system. In the past years, there have been a number of published articles addressing these challenges [9–12]. We classify and summarize them next.

- *Reliability:* Due to the vulnerability of wireless network connectivity and communications, designing highly reliable system functions and services to mobile device users is the first engineering challenge. The system reliability depends on the reliability of networking communications, mobile data transactions, mobile client and server functions, and application services.
- *Mobility:* Since system mobility is one of the unique features of wireless-based software systems, designing system mobility is another engineering challenge. This encompasses cost-effective solutions to support mobile user accesses, mobile data migration, and synchronization.
- *Performance:* Since mobile device users usually access the system in a mobile environment, they expect the system to respond to their requests and provide diverse functional services in a fast and easy manner. Therefore, designing and achieving good system performance must be one of the engineering challenges.
- *Standardization:* Engineering wireless-based software systems involves many standards, including wireless network standards (e.g., infrastructures, network protocols, and connectivity interfaces); mobile technology standards (e.g., mobile APIs and mobile presentation standards); and domain-specific wireless application standards (e.g., wireless payment protocols and wireless advertising standards). Thus, system standardization must be another engineering challenge.
- *Scalability:* System scalability is another concern in the engineering of wireless-based software and application systems. The fast and continuous increase in the number of mobile device users in recent years creates strong demand on system scalability to support scale-up functionalities and services, growing demands on user mobility, and increasing volumes of mobile data accesses.
- *Interoperability:* System interoperability of a wireless-based application system demonstrates how well the system can be deployed and operated on different mobile platforms and network connectivity.

System Analysis Issues

The primary system analysis issues for wireless-based software systems are listed here:

- How to understand, analyze, and specify the required wireless networking connectivity and system infrastructures;
- How to understand, analyze, and specify mobile user requirements, as well as system-user interactions, including their profiles, mobility, accessibility, and system-user interactions;

- How to understand, analyze, and specify mobile data requirements, including their attributes and relationships, with special attention paid to mobile data delivery modes, synchronization, and migration requirements;
- How to understand, analyze, and specify wireless-based system security requirements for mobile users, network communications, application transactions, and application servers;
- How to understand, analyze, and specify domain-specific functional features and services for wireless-based application systems;
- How to analyze and specify nonfunctional requirements of wireless-based software systems, including mobility, interoperability, performance, scalability, and reliability.

System Design Issues and Challenges

There are a number of system design issues for wireless-based software systems:

- How to design and document wireless networking connectivity for a selected wireless network infrastructure based on the standardized protocols;
- How to select appropriate mobile technologies and platforms on mobile devices;
- How to design and/or use cost-effective wireless security solutions to deal with the limitations of wireless networks and mobile devices;
- How to design and specify user-friendly mobile interfaces for wireless-based applications;
- How to select and/or define appropriate mobile data transaction models for mobile databases;
- How to design and specify cost-effective mobile application databases and central service databases.

The major design challenges are listed here:

- How to achieve good mobile portability to support diverse mobile devices;
- How to provide mobile users with good personality and customization;
- How to cope with the mobile database design issues in data consistency and concurrency, data synchronization and conflict, and data security;
- How to design efficient mechanisms for data caching and migration for mobile databases;
- How to ensure that the generated security solution will deliver secured transactions and will guarantee the confidence and privacy of mobile device users.

System Testing Issues and Challenges

- How to validate functions and application services of wireless-based software systems to address their special features, such as wireless-based application communications, mobile transactions, system mobility, and security;

- How to define and/or select performance testing and evaluation models and metrics to check system performance, reliability, availability, and scalability for wireless-based software and application systems;
- How to develop (or deploy) test tools to support test automation for wireless-based software and application systems.

1.4 Summary

With the wide deployment of modern wireless network technology, today's business market has a very strong demand for diverse wireless-based application systems and services to meet the needs of mobile device users. In order to design these systems, engineers must have a basic understanding of wireless-based software and application systems, including basic concepts, essential components, distinct features and properties, major requirements, user expectations, primary advantages, and limitations. In addition, they must know the engineering perspectives in developing wireless-based software application systems, including development processes, tasks, engineering issues, and challenges. As an introductory chapter, this chapter covers these subjects.

To conduct successful engineering projects for developing wireless-based application systems, engineers and managers must do the following.

- Engineers and managers must have a basic understanding of wireless networks, including network infrastructures, communication protocols, and connectivity standards. The chapters in Part II of this book cover these subjects. In addition, Part III prepares readers to understand the advantages, limitations, and applications of different wireless networks.
- Engineering and managers must have a good understanding of various mobile technologies and operating environments, including mobile presentation technologies, mobile programming basics, and mobile operating systems. Part II of this book includes the chapters addressing these subjects, and WAP technology, as well as wireless multimedia technologies. Moreover, Part II helps readers compare and select mobile technologies for specific applications.
- Engineering and managers must apply well-defined software engineering processes, principles, and methods to conduct system analysis and design to deal with engineering issues and technical challenges in system infrastructure, mobile user interactions, mobile data accesses and services, and wireless security. Part IV of this book includes chapters covering these important subjects to prepare readers to conduct system analysis and design for wireless-based application systems, and define wireless-based system infrastructures and security solutions.
- Engineers and managers must comprehend a specific wireless application domain and related engineering issues, including business models and business workflows, and be able to define the solutions in the specific application domain. Part V of this book is devoted to popular applications in wireless commerce and wireless information services. They offer readers the necessary domain knowledge and related engineering perspectives for different types of

wireless commerce systems and wireless Internet systems, such as wireless payment, location-based portals, wireless advertising, and wireless information services.

References

[1]   Neuharth, A., "How Can We Control Cellphone Addiction?" *USA Today*, July 9, 2004.

[2]   Yunos, H. M., J. Z. Gao, and S. Shim, "Wireless Advertising's Challenges and Opportunities," *IEEE Computer*, Vol. 36, No. 5, May 2003, pp. 30–37.

[3]   eTForecast, "USA Leads Broadband Subscriber Top 15 Ranking," November 14, 2005, http://www.etforecasts.com/pr/pr1105.htm.

[4]   In-Stat/MDR, "Mobile Video Services in the United States, 2004–2009," #IN0401657MCD, May 24, 2004, http://www.instat.com/press.asp?ID=9768CSKU=IN0401657MCD.

[5]   Alcatel, *Multimedia Message Service Solution*, Technical Report, retrieved October 2005, at http://www.intel.com/business/bss/solutions/blueprints/pdf/alcatel_sb.pdf.

[6]   *Market Analysis: U. S. Cellular Location-Based Services 2004–2008 Forecast and Analysis: Mapping a Mobile Future*, IDC, 2003.

[7]   Betti, D., J. Delaney, and N. Murrel, "MMS and SMS: Multimedia Strategies for Mobile Messaging," *Ovum Research*, April 2003, http://www.ovum.com/go/content/007972.htm.

[8]   Forman, G. H., and J. Zahorjan, "The Challenges of Mobile Computing," *IEEE Computer*, Vol. 27, No. 4, April 1994, pp. 38–47.

[9]   Dornan, A., *The Essential Guide to Wireless Communications Applications*, Upper Saddle River, NJ: Prentice Hall, 2001.

[10] Bharghavan, V., "Challenges and Solutions to Adaptive Computing and Seamless Mobility over Heterogeneous Wireless Networks," *Int. J. on Wireless Personal Communications: Special Issues on Mobile Wireless Networking*, March 1997.

[11] Foreman, G. H., and J. Zahorjan, "The Challenges of Mobile Computing," *IEEE Computer*, Vol. 27, No. 4, April 1994, pp. 38–47.

[12] Imielinski, T., and B. R. Badrinath, "Mobile Wireless Computing: Challenges in Data Management," *Communications of the ACM*, Vol. 37, 1994, pp. 18–29.

PART II

Mobile Technologies and Operation Environments

Mobile devices and technologies have been evolving rapidly for the past two decades. The advancements of hardware technology have triggered the birth of the mobile operating systems and many other new applications, such as microbrowsers, personal information management, and multimedia support on mobile devices. Due to the efforts by the WAP Forum (and continued by the OMA), Internet technology has extended to mobile devices and wireless networks. The WAP technology uses the existing wireless infrastructure to optimize information delivery between mobile devices and the Internet. The standard format for describing and exchanging data in the mobile Internet is XML. WAP uses the XML-based XHTML for content presentation. Other XML-based technologies, such as SyncML for synchronization and SMIL for streaming multimedia, are widely used in the mobile software environment. In addition to WAP-based mobile Web technology, the Java Platform, Micro Edition (Java ME, also called J2ME), is another critical technology for mobile Internet software development. It is believed that mobile clients supporting WAP, Java, and messaging services will dominate the market in the near future.

There are four chapters that cover different mobile technologies and operation platforms. Chapter 2 provides an overview of mobile devices, and focuses on different mobile operating systems (including Symbian OS, Windows CE/Mobile, Palm OS, and Embedded Linux); discusses the major mobile application platforms (including Java Platform, BREW, and Mophun); and presents personal information management and SyncML. Chapter 3 provides a tutorial about the wireless application protocol (WAP), including WAP architecture, protocols, technology, and the WAP application environment (WAE). In addition, it discusses messaging services, such as SMS, MMS, and the future of WAP. Chapter 4 offers comprehensive coverage of XML and different types of mobile technologies, including Web-based, and Java-based mobile client software technologies, as well as i-mode technology. Chapter 5 focuses on wireless multimedia technologies. It covers the topics of mobile media streaming and architecture, supporting technologies, standards, and major players.

CHAPTER 2

Mobile Devices and Operating Systems

Of all the components in a wireless network system, the mobile device gets the most attention. End users can only interact with the system through mobile devices. Mobile devices feature a low-power transceiver that is typically designed to transmit voice and data. The mobile device connects to the network infrastructure wirelessly through an available access point or base station.

Mobile devices have been evolving rapidly for the past two decades. This evolving process originated from the development of both information technology (IT) and the telecommunication industries. The advancement of integrated circuit (IC) technology in the IT industry has reduced the size of computers, and created new types of portable computers, such as laptop notebooks and handheld devices. On the other hand, the advancement of cellular technologies in the telecommunication industry has produced mobile phones. Through the enhancements made in features and performance on both sides, the boundary between handheld computers and mobile phones began to blur. A new generation of mobile devices was made to function not just as computers or mobile phones, but also as mobile equipment with both computing and communication capabilities. This chapter focuses on these handheld mobile devices.

The operating system (OS) is an important element between mobile device hardware and mobile applications. Historically, mobile devices executed proprietary operating systems that were developed by device manufacturers. This restriction was very restrictive for the users, developers, wireless operators, and device manufacturers themselves. As the device capabilities increased, proprietary operating systems were no longer practical. Open operating systems were released to manage a wide range of mobile devices. Application development platforms that were specifically designed for mobile devices, and which were aimed at producing cross-system applications more efficiently, were also proposed. However, the evolution of mobile operating systems and application environments are still in the early stage.

This chapter provides an overview of mobile devices and mobile operating systems, with an emphasis on trends in evolution and applications. It is organized as follows. In Section 2.1, we describe the limitations, evolution, and trends of mobile devices. In Section 2.2, we present three mobile operating systems—Symbian OS, Windows CE/Mobile, and Embedded Linux (Mobile Linux). Section 2.3 describes two popular mobile application development platforms—Java and binary runtime environment for wireless (BREW). Section 2.4 summarizes the key points in this chapter.

2.1 An Overview of Mobile Devices

Handheld mobile devices evolved mainly from personal digital assistants (PDAs) and mobile phones, into a variety of forms to address a broad range of applications and user needs. Mobile devices differ in features, performance, and size. This section provides an overview on the history, limitations, and future trends of mobile devices.

In recent years, manufacturers have released new mobile devices more frequently than ever. The history of mobile devices is actually not that long—less than 25 years for the mobile phone, and less than 15 years for the PDA. Most of today's mobile devices are not simple mobile phones or PDAs. Their evolution consists of three stages: the mobile phone stage, the PDA stage, and the smart phone stage [1].

To support additional software applications, the smart phone has a powerful application processor. An open mobile OS is needed to manage all its resources and activities. More discussions on mobile operation systems and application development platforms are presented in Sections 2.2 and 2.3.

2.1.1 Limitations

An increasing number of applications are being installed in mobile devices. As the number of applications increases and their functionality becomes more advanced, they require greater processing power and data storage. The core hardware components of a mobile device include a processor, memory, storage, and a battery. The major problems regarding the integration of these parts into a small device are the small size restriction and heat problems. The processing power and amount of storage keep increasing as technologies for producing these components rapidly advance. However, these will not be major concerns in the future. On the other hand, battery technologies are also advancing rapidly, but are still insufficient for the demand. Thus, there are still limitations in battery power.

The user interface (UI) components of a mobile device include the screen, speaker, and input peripherals, such as the keyboard and microphone. These parts are in close interaction with the user. Some of the limitations are brought about by the device itself, such as the small screen size. These limitations go beyond the problem of not having enough resources to run the application. These limitations pose specific challenges to mobile application developers. In the following discussions, we will investigate each of the core hardware and UI component limitations of a mobile device [2, 3].

Core Hardware Components

Processor

Simple communication processors were sufficient for early mobile phones. However, mobile devices today are designed as embedded systems, requiring more powerful application processors. There are different processor architectures for embedded systems, such as the ARM, NEC MIPS, Motorola/IBM PowerPC, and so forth. As an embedded device, the mobile device is an application-specific integrated circuit (ASIC) system, in which the processor is integrated into the IC design. Processor speed will affect the types of applications that will run on the device. Among those processor architectures, the ARM is the most widely used 16/32-bit instruction set embedded RISC processors in embedded and low-power applications. The ARM is the only architecture to support all of the major mobile operating systems (i.e., Symbian OS, Windows CE/Mobile, Embedded Linux, and Palm OS).

The ARM architecture includes the core 16/32-bit instruction set architectures (ISA), along with built-in extensions to support Java acceleration, security, and other Java features. The ARM ISA is constantly improving to meet the increasing demands of new applications. ARM covers a wide range of speed, running from 1 MHz to 1 GHz, and provides applications with hardware and software tools support. Most of all, the ARM is designed for low power consumption, measured in Mips per watt, which enables long battery life with the intelligent energy management (IEM) feature. IEM technology can reduce processor energy consumption by as much as 50%, making it attractive for mobile devices.

Storage (RAM, Flash Memory, and Microdrive Hard Disk)

Mobile devices hold the program and data in the RAM during execution, and use the flash memory or hard disk for nonvolatile persistent storage. Most types of RAM are *volatile*, which means that they lose data when the computer is powered down. Compared to desktop computers, mobile devices have a limited amount of storage. Both the OS and the application software require a combination of careful design and skillful programming.

Flash memory is a nonvolatile type of electronic storage. Flash memory offers random-access read operations, but has just a small number of cycles for erase-write operations. A carefully designed operation of the flash file system can solve this problem. The file system will write a new copy of the changed data to a fresh block, and remap the file pointers. The original block of data can be erased later. The Treo 650 smart phone even uses flash as its main memory during normal operations. It gives the illusion that RAM does not lose data when the power is down.

Compared to storage on a hard disk, flash storage is fast and requires much lower power. There are different formats of flash memory, including CompactFlash, Sony's Memory Stick (MS), Multimedia Card (MMC), and Secure Digital (SD). A newer generation of flash memory cards, such as the reduced size MMC (S-MMC) and the mini-SD, provide better performance with reduced size. Management functions, such as security and digital right management (DRM), are also incorporated into these new flash cards.

The microdrive, a miniature hard disk, is incorporated into mobile devices as large-sized permanent storage. The microdriver was developed and launched in 1999 by IBM. Other manufacturers include Hitachi and Seagate. The microdrive capacity was expanded to 10 GB by 2006. Flash memory and the microdrive are close competitors when it comes to capacity, performance, and price. Currently, microdrives are used mainly for mobile devices with large storage requirements, such as those devices storing MP3 and videos.

Battery Limitation and Energy Management

The term "power" refers to the energy consumed per time unit. A battery is used to store energy, not power, for mobile devices. For most fixed-duration tasks, such as audio or video playing, the energy required is directly proportional to the average power consumed. For this class of tasks, minimizing the power will also minimize energy. On the other hand, some computing-oriented tasks require a longer time on low speed processors, which implies lower power consumption. These tasks may be executed with less energy on high-speed processors.

Lithium-ion cells offer the highest capacity among currently available rechargeable batteries. Their capacity has improved by approximately 10% per year since their commercial introduction by Sony in 1991. This rate of improvement is leveling off, and major improvements in rechargeable batteries in the near future are unlikely. Color screens and additional features in mobile devices result in higher energy consumption, and consequently, reduced battery life. Many new display and communication technologies, and their components, are developed with energy/power saving in mind [4, 5].

Energy management has become an important design issue, and one of the greatest challenges for mobile devices. Most mobile operating systems implement three power states: run, idle, and standby (or suspend, sleep) states. Mobile device users can adjust the power settings to maximize battery life. A lower screen brightness and turn-off wireless mode also help to conserve energy.

User Interface Component Limitations

Limited Screen Space

The screen format and content of traditional desktop computers are controlled by the software developers or the Web designer, and can be personalized by the user. Device manufacturers' wireless operators mostly determine screen layouts on mobile devices, and users have little freedom to control these.

The screens of mobile devices are much smaller than those of desktop computers. Unnecessary graphic user interface (GUI) components, images, and graphics not only increase the processing time, but also occupy the limited display area. Within the limited area, text length is restricted, menus have few options, images must be small, and pop-up error messages are preferred. Application and Web designers have to carefully plan the screen layout. Most UIs in traditional application systems cannot scale down directly to a mobile device.

Web surfing is much more difficult when using mobile devices rather than desktop computers. To provide a comfortable user experience, designers should abide by the following guidelines for browsing design.

- Use hyperlinks to replace other GUI components whenever possible to save space.
- Use vertical scrolling as much as possible. Unless long items are involved, avoid horizontal scrolling.
- Organize the content hierarchically from a broad scope to a narrow scope.
- Arrange the hyperlinks from top to bottom by importance.
- Number the hyperlinks when user selection is required. This allows users to use the keypad for one-handed interaction.
- Use link navigation, but not direct input, for information searching whenever possible.

There are continuous efforts to automatically transform ordinary Web pages to mobile device displays. For example, Opera's *Small-Screen-Rendering* eliminates the need for horizontal scrolling. Due to the numerous types of GUI and Web page designs, in addition to the variety of screen sizes and types of mobile devices, room for further improvements still exists.

Input Mechanisms and Limitations

The input mechanism is one of the important aspects of a mobile device. The input facilities of mobile devices are limited, when compared to those of a desktop computer. Application designers must carefully analyze the user interaction requirement of each application. The common inputs available for mobile devices include the keypad, keyboard, pen-and-touch screen, and voice input.

The 12-digit keypad is typical for all telephones, allowing users to perform single-handed operation. It is effective for numerical input, but requires multiple keystrokes for text input. Users must press a key a number of times to enter a single character. The keyboard is the most efficient and easiest-to-use input mechanism, especially for Western languages. Still, multiple keystrokes are required to enter a single Oriental character.

Touch screens for pen-based input allows a user to enter data using a stylus. The pen-based input requires a two-handed operation. The most common and simplest pen-based input mechanism is the on-screen soft keyboard, which allows users to enter data by pressing the virtual keys on the screen. Some devices support written character recognition by interpreting the letters entered on the screen.

Many mobile phones support the voice command feature, which compares the voice command with a recorded message, and then executes the corresponding action (e.g., dialing). The voice command feature is a simplified speech recognition application. For natural speech recognition and translation, more powerful processors are required.

Unreliable Wireless Environment

Earlier network protocols were designed mainly for the wired environment. Due to fading and interference, wireless networks are inherently unreliable. Wired networks drop packets due to congestion, while wireless networks lose packets mainly due to a high error rate. The end-to-end connection between the mobile device and the wired network is often dropped and needs to be reconnected. Mobile devices can deal with this problem through the following approaches.

- Transmit small amounts of data as much as possible.
- Transmit only the changed data.
- Compress the data prior to transmission.
- Monitor the status (e.g., connection dropping, bandwidth) during the transmission.
- Use an efficient connection reestablishment mechanism.
- Restart the download at the exact place where it was stopped.

The current proliferation of mobile devices has resulted in a large diversity of designs, each optimized for a specific application. Some of the approaches mentioned above have been incorporated into wireless transport protocols. Some still require the attention of device manufacturers and application developers.

We have shown that the user interface and environmental limitations of mobile devices are inevitable. Next, we will analyze the functional enhancements of mobile devices, and discuss the trends of convergence and evolution.

2.1.2 Evolution and Integration

As more technologies are integrated into a mobile device, the mobile device becomes more complicated. Modern mobile device platforms allow users and developers to manage and create new applications. In the evolution of the mobile device, the most significant improvements have been: (1) the separation of the device hardware and the operating systems, (2) the transformation from a proprietary OS to an open OS, and (3) the creation of the mobile application platform. The reference architecture for mobile devices is shown in Figure 2.1. The bottom part of the figure is the hardware layer, which consists of a processor, storage, peripherals, and communication modules. The core hardware components are introduced in Section 2.1.1. This section will focus on recent peripheral and other enhancements.

The mobile OS kernel is optimized for the processor architecture and application platform. The application system interface layer provides a set of software development kits (SDK) and tools that have been developed by operating systems or by third parties, and are made available to all applications. For example, the communication middleware provides access to the underlying communications hardware module, which establishes a connection for phone/voice service, such as dialing a call.

The application layer is the actual GUI and applications available to the user on a mobile device. It determines the look and feel of the mobile device. The mobile operation systems include the two middle layers (the OS kernel and the application system interface layers), as well as a part of the application layer. The mobile OS provides only a part of the application, and leaves the rest to third party developers. The open mobile terminal platform (OMTP) initiative was established

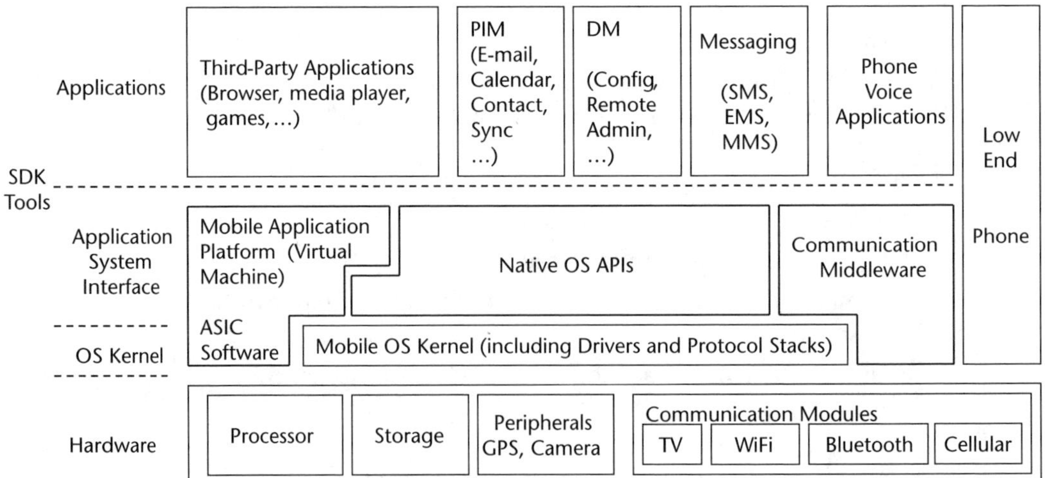

Figure 2.1 The mobile device system reference architecture.

in 2004 to define functional requirements for mobile phones by major mobile operators [3].

Modern mobile devices are descendent from PDAs, voice devices, and other handheld devices. From the standpoint of mobile phones, the enhancements start with a more powerful processor, larger storage, and a variety of communication support. In fact, the advancements of hardware technology have triggered the birth of the mobile OS and many other new applications. We can say that the trend of mobile device convergence can be encouraged from three different directions:

- *Communication capability enhancement*, including multiband/multimode phones, wireless local area network (WLAN) and wireless personal area network (WPAN) connectivity, and TV support;
- *Add-on peripherals*, including global positioning system (GPS), cameras, and games.
- *Add-on software*, including Portable Media Player (MP3 phones), voice over Internet (VoIP) on mobile devices, digital home phones for remote control, and mobile information management.

Communication Capability Enhancement

Multiband/Multimode Mobile Phone

A multiband (also known as a dual-band, triband, or quadband) mobile cellular phone can work on more than one radio frequency. For example, a GSM world phone supporting quadbands (850, 900, 1,800, and 1,900 MHz) can work on any GSM network. The multimode phone is designed to work on more than one wireless wide area network (WWAN) cellular technology, and is critical during the transition from 2G (e.g., GSM) to 3G (e.g., WCDMA) systems. Most 3G phones are actually dual-mode phones. The mobile network can also be integrated with fixed-line phone networks. Mobile users are now able to extend the fixed-line features of home and office telephones to mobile devices. The details of WWAN, also called cellular networking, are introduced in Chapter 7.

WLAN/WPAN Connectivity

The easiest way to set up a communication link between a mobile device and a host is through a cradle (or a physical wire) connection; however, by integrating the WLAN and WPAN into the cellular phone, mobile devices have more options to support data communication. A WiFi phone supports both cellular and IEEE 802.11 connections. The user can leverage the WLAN access points to get Internet access. Many enterprises have established WiFi networks, and integrating VoIP functionality has become a common practice. The details of WLAN are introduced in Chapter 6.

Bluetooth is a WPAN technology now found in many mobile devices. It has been used mainly to eliminate the need for wires between devices (e.g., between a mobile phone and a headset). Due to the advancement of WPAN technologies, such as near field communication (NFC) and radio frequency identification (RFID), new technologies are continuously being integrated into mobile devices. The details of WPAN are presented in Chapter 8.

TV Support

A television phone (e.g., Nokia N92) allows users to watch TV with its digital video broadcasting—handheld (DVB-H) receiver. The DVB-H is a handset version of the digital video broadcasting—terrestrial (DVB-T) television. The DVB-T has been adopted in many countries to provide HDTV and multichannel standard definition televisions to users. As compared to the traditional TV, the DVB-H with built-in cellular connectivity is more likely to become a two-way medium with interactive capabilities. However, the DVB-H is still in its infancy, and pilot trials are now underway in major cities (e.g., Helsinki, Paris, Pittsburgh, and Taipei). Another competing mobile TV system, the digital multimedia broadcasting (DMB), is in use in South Korea.

Add-On Peripherals

Camera Phone

All major manufacturers now produce cellular phones with a built-in camera. CMOS and CCD image sensors are both used in a camera phone. The former has the advantage of low power consumption. Many of the camera phone models have zoom lenses and removable digital media. Some newer camera phones are also videophones, and can transmit videos and video calls, a standard feature of the 3G phone. As the technology improves, camera phones rival digital cameras in terms of functionality and price. At the end of 2005, the highest resolution available for a camera phone was at 8 megapixels. Furthermore, camera phones are already outselling digital cameras.

Camera phones are used in many applications, such as pointing, authentication, and storage devices. These applications are used in the fields of crime prevention, journalism, and so forth. With additional software, the cameras in mobile phones can be turned into personal barcode scanners. After scanning and transmitting the barcode, consumers may receive a coupon, or be redirected to a Web site. On the other hand, camera phones are potential threats to security and privacy. Because of this, Saudi Arabia has banned the sale of camera phones, and the South Korean government requires that all camera phones make a clear and distinct sound whenever a picture is taken.

Mobile Game Device

Computer games have been one of the driving forces behind IT innovations. Each game console, such as the Sony PS3, Microsoft Xbox 360, and Nintendo Revolution, integrates many innovations. To speed up the game development process, special software platforms and technologies aimed at game development are being proposed. For example, the Microsoft XNA (Xbox Next-generation Architecture) focuses on game development through team collaboration, using pluggable components and the XNA Studio, which brings together Windows and Xbox software development tools [6].

A mobile game console is a video mobile device that integrates the control, the screen, and the speakers into a single unit. Mobile game consoles also converge with PMP and smart phones. Mobile games originally were single-player games, but with add-on communication capabilities, players can now connect with other players.

The Mobile Massively Multiplayer Online Game (MMMOG, or 3MOG), which enables hundreds or thousands of players to simultaneously interact in a game, connected via mobile devices, has quickly gained popularity among users.

The Nintendo DS and Sony PlayStation Portable (PSP) are mobile game devices with PMP and WiFi features. The Nokia N-Gage is the first game device from the telecommunication industry.

- *Nintendo DS*—Nintendo has dominated the handheld game market since the release of Game Boy in 1989. Later, Game Boy Advance (GBA) introduced the concept of connectivity, using the Game Boy as a controller for the Nintendo GameCube. The latest Nintendo DS was released in 2004. It featured dual screens and IEEE 802.11WiFi connectivity. The lower screen allows touch sensitivity that can provide interactive activities. The DS has a separate port for the loading of GBA games in a single player mode.
- *Sony PSP*—Sony's PSP was released in 2003. It featured USB connectivity and 802.11 WiFi support. PSP is able to play music and movies from Sony's proprietary Universal Media Disc (UMD), or Memory Stick.
- *Nokia N-Gage*—The Nokia N-Gage was first released in 2003. It was designed as a combination smart phone and gaming device. The first generation N-Gage, however, has not been commercially popular, due to its physical design and layout. The *N-Gage QD* released later has featured revisions in the design.

Two popular platforms for mobile gaming, the Java Micro Edition (Java ME) and the Morphun, are presented in Section 2.3.

GPS Phone and Assisted Global Positioning System

The U.S. Federal Communications Commission's enhanced 911 (e911) requires all cellular phone carriers to provide the ability to trace calls to a location. The geographical location of a mobile device can be known through GPS, which uses a satellite to determine location. GPS devices require a clear line-of-sight to at least four GPS satellites, and enough processing power to transmit data. Assisted GPS (A-GPS) uses cell towers to receive the signal from satellites, and an assistance server to reduce the time needed to find the location, and is commonly associated with location-based services (LBS) over cellular networks. With A-GPS, cellular phones do not need to be GPS-enabled. There are also many GPS-enabled cellular phones, such as the Motorola i88s, which can act as standalone GPS receivers [7].

The tracking service allows users to view and monitor location in real time on a map. The user may be the actual owner of the mobile device or some other person using the device. The owner of the device to be tracked must give permission to enable the device. Map and aerial photo products can be purchased or licensed from other companies. The LBS receives data from the tracking service and makes it accessible to the user. Many location-based applications have been recently developed. For example, GeoFencing is a location-based service that sends alerts automatically when the mobile device is removed from the home or office, to avoid unnecessary tracking.

Add-On Software

Portable Media Player and MP3 Phone

A portable media player (PMP) is a device that stores, organizes, and plays digital media files. Although the PMP is more commonly referred to as an MP3 player, PMPs often play many other media file formats, such as MP3, Windows Media Audio (WMA), MPEG4 Audio (M4A), and MPEG4 video. Apple's iPod is the most successful PMP with limited PDA functionalities. Apple's iPods also feature games.

There are two main types of PMP: a flash-based PMP, and a hard drive–based PMP. Flash-based PMPs are generally solid-state small-storage devices that are integrated into USB key drives. Hard drive–based PMPs, on the other hand, have larger storage, capable of storing thousands of songs or hours of videos. An MP3 phone is a phone with a PMP capability. It also requires a media player software. Popular media players include RealPlayer (for Nokia N91), and iTunes (for Motorola ROKR E1). The other standard features of an MP3 phone include an FM radio, stereo speakers, and Bluetooth/USB connectivity.

VoIP on Mobile Phones

VoIP allows users to bypass standard phone lines by using the Internet to make phone calls. It routes the voice conversations over any IP-based network. The voice data flows over a general-purpose packet-switched network (e.g., Internet), instead of traditional dedicated, circuit-switched voice transmission lines. Most VoIP services are based on a client-server model. Skype, the most popular of its kind, is a proprietary VoIP system based on a peer-to-peer (P2P) model. The Skype user directory is distributed among the nodes in the network. The network can scale very easily to large sizes, and each Skype user must have the Skype software running on his or her terminal.

The VoIP terminal can be any computing device with Internet protocol stack and connectivity support. The connection between the device and the Internet may either be wired or wireless. The wireless communication capability can be WiFi, WiMax, Bluetooth, or DECT. These mobile devices require more hardware resources than do traditional mobile phones, in order to support the Internet stack and additional peer operations.

Phones for Digital Home Coordination (Remote Control and Service Discovery)

Mobile devices are expected to interact with other devices in a networking environment through device coordination. Device coordination enables a device to discover services, as well as other remote control devices, in a network. Device coordination facilitates media sharing. For example, the audio and video stored on a mobile phone can be streamed to a living room TV. The mobile device with compatible device coordination capability eventually becomes the *universal media playback remote control* at the heart of the digital home. Two major techniques for device coordination, Jini and UPnP, are presented in Section 8.3. The Nokia N80 is the first UPnP-enabled phone.

Mobile Information Management

Mobile information management (MIM) is a standard feature of PDAs. It comprises personal information management (PIM) and device management (DM). PIM applications include e-mail, calendar, task lists, address books, and memo pads. DM was a task for users. As the functionalities and applications of a mobile device become more complicated, the tasks of DM increase.

Messaging Services and Applications

Messaging services have been with WWAN technology for a long time. These messaging services, from short message service (SMS), Enhanced Messaging Service (EMS), to Multimedia Messaging Service (MMS), have all functioned not only for message exchanges, but also have become application platforms. The details of messaging services are presented in Section 3.4.

In addition to traditional messaging services, the instant messengers (IM) over the Internet, including MSN, Yahoo!, AOL IM, Google Talk, Jabber, and ICQ, have become very popular in recent years. Most IM services offer a *presence aware-ness* feature, indicating whether friends in the contact list are currently online and available. Most IM applications also include the ability to set a status message, roughly analogous to the message on a telephone answering machine. Many mobile phones now include IM capabilities that are designed to hook into instant messaging presence service (IMPS) servers on a carrier's network. The chat client is accessed via the *My Presence* menu on the Nokia phone, *My Friends* on the Sony-Ericsson phone, and *IM* on the Motorola phone.

Keep Evolving? Yes. Continue Merging? Maybe Not.

It is very likely that the power of mobile devices will keep increasing in the near future. The trend of convergence is obvious, but it is not necessary to put all the functionalities into one device. The reason that users to not want a mobile device capable of solving all of their needs is that not all users have the same requirements. Most consumers do not like being involved in complicated operations. A personal-ized mobile device is more practical for both users and device manufacturers. We may see a variety of configurable mobile devices in the future market, in which each device is able to balance functionality and performance with cost.

Looking back at the progress of mobile devices in recent years, technologies for the *Quadruple Play* telecommunication service, which provides telephone, video-on-demand TV, Internet applications, and wireless communications services, are all available. The rapidly growing VoIP traffic has become a major threat to the profit of the telecommunication industry. The real problem then involves the estab-lishment of a business model in such a rapidly changing environment.

2.2   Mobile Operating Systems

The mobile OS market is vastly different from the desktop OS market. In the mobile world, there is no dominant OS. As mentioned in Section 2.1, the development of open mobile operating systems is one of the important factors that affect the emer-gence of recent mobile devices. Proprietary operating systems are available for manipulation only by the device manufacturers. For example, the research in motion (RIM) OS is a proprietary OS that is used only for RIM BlackBerry mobile devices. The use of proprietary operating systems is becoming rare, since it prevents large-scale development. Even for the proprietary RIM OS has worked together with Palm OS and Java for add-ons to extend the platforms of choice. To better serve device manufacturers and developers, open mobile operating systems must adapt to the limitations and trends that were introduced in Section 2.1, such as the limited IO capabilities and the needs of a wide variety of applications [8].

Figure 2.2 shows a mobile operation system structure. A hardware abstraction layer is between the hardware layer and the kernel layer. This abstraction layer simplifies the task of porting the OS on different processor architectures. The kernel is mainly responsible for resource management, which includes process/thread management, memory management, and persistent storage management. Device drivers are used to bridge the gap between the hardware and the software. A variety of native services are provided by the kernel. Applications are developed based on these services. Each OS comes with a different set of application program interfaces (APIs) and applications.

This section will introduce the common mobile OS structure and development issues, as well as case studies. First, we present general *architecture* issues (including multitasking, asynchronous calls, memory management, execute-in-place, and persistent data storage); and *application environment* issues (including standardized API and development, device emulator and testing, common features, and synchronization). Then, we present three widely used open operating systems (i.e., Symbian OS, Windows CE, and Embedded Linux).

Mobile Operating System Architecture

Multitasking and Asynchronous Calls
Multitasking allows more than one application to run at the same time. Due to the limited capability of their processors, early mobile systems did not often support

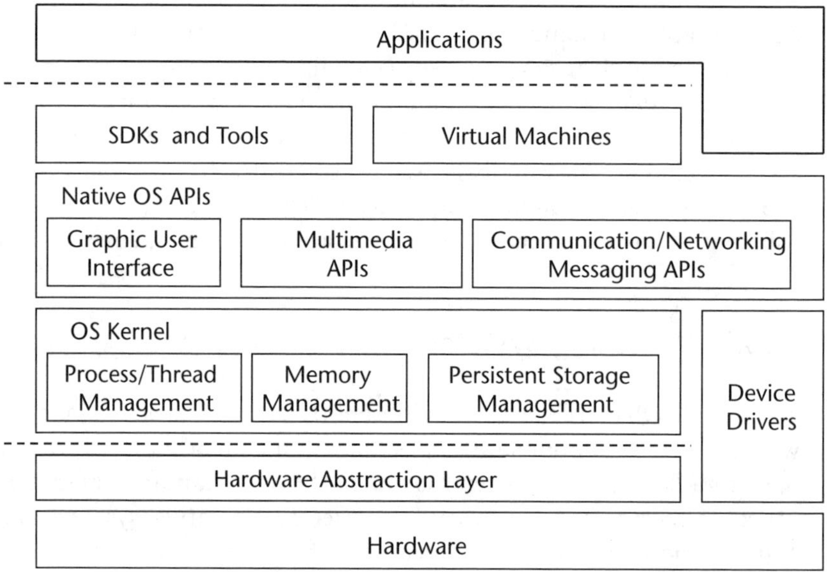

Figure 2.2   Mobile operating system structure.

multitasking. Recent mobile devices use more powerful processor architectures, such as ARM. Multitasking support became a common feature of mobile operating systems. It allows more than one service of the system and application task of the user to run simultaneously. For example, a user can read an e-mail and download a file at the same time. For a multitasking OS that supports multithreading, each process in the OS contains one or more *threads*. Communication between applications is facilitated by a lightweight interthread communications protocol.

Although multithreading applications are convenient to users, they involve additional programming complexity for developers who are writing true multithreaded programs with resource sharing and synchronization. Some systems provide an alternative application programming scheme to achieve the same effect as a preemptive multithreading application. This easier programming scheme is called *asynchronous function calls* [2, 9].

Using asynchronous function calls means that the message sender can do another task while waiting for the message receiver to complete a task. Note that the receiver can be any application on the same device or a separate server. When the receiver completes the task, it preemptively signals the sender. It is up to the message sender when it checks the response.

Memory Management

Although a large address space and demand paging are not essential for mobile devices, most modern OS kernels support *virtual memory* services. Other features of virtual memory services benefit mobile system operations. The virtual memory service provides a protected address space, shared virtual memory, and memory mapping. Each process receives its own virtual address space under a protected address space environment. The shared virtual memory support allows the code and multimedia frame buffer to be shared. A memory mapping object can be used for a shared object with a size larger than the physical frame. A memory mapping file also simplifies file operations.

eXecute-In-Place

For a desktop system, executable programs are loaded into the main memory before execution, in a process called copy-and-execute. Most mobile systems support eXecute-In-Place (XIP) instead. Under XIP, the executable of a program is stored in persistent storage, such as ROM and flash, and it is run directly from the location where it is stored. XIP reduces the amount of RAM required by an application. The shared system components stored in permanent storage can also be executed in-place.

XIP saves RAM, since it does copy programs and system code from the flash or ROM. However, XIP increases the flash/ROM requirements. Files and executable images are often stored in compressed form on the flash/ROM to save storage space. To support XIP, programs on flash cannot be compressed, which implies that more space is needed [1].

Persistent Data Storage (Flat-File Versus Relational Database)

All mobile operating systems have built-in persistent data storage, which is also called a database. These databases are best suited for simple data structures and operations, and suffer from poor performance in complex operations. These built-in databases often have limited programming language and synchronization support. Therefore, developers require a significant amount of effort to implement data synchronization. Mobile data databases can be classified into two types: a *flat-file database*, and a *relational database.*

A flat-file database is a simple database, in which all the information is stored in a plain text file. One line in the file stands for a database record. The main advantage of the flat-file database is ease of use. The user can easily add, delete, search, sort, and update records. A flat-file database does not support transaction, which means that data is either successfully written or not at all. Although transaction support is not critical for on-device local data processing, it is required for enterprise database applications that integrate with client mobile devices. Moreover, flat-file synchronization must be done by the developers.

A relational database is widely used in desktop and enterprise server systems. The relational databases of mobile operating systems support more advanced capabilities than do flat-file databases. Data access is often through a subset of SQL, and often comes with companion APIs. Using structured query language (SQL) to access data is more efficient than API in most cases, except for large amounts of data. A relational database also supports transaction, which saves the synchronization effort of developers.

Application Environment

Standardized API and Development

The application platform of an open mobile OS often provides a standard set of high-level language APIs and a development environment for both software execution and development. To further extend the usability of an application, cross-platform capabilities for more than one mobile OS are expected. A virtual machine (VM)–based application environment is aimed at this interoperability capability. Section 2.3 presents the cross-platform application execution and development environment.

Open standards can inspire more creative applications from the participation of third-party developers. However, a single type of mobile device may be delivered to a variety of customers with different installed applications and GUIs. Using specific add-on APIs and SDKs, customization can be preinstalled by device manufacturers or wireless operators.

One problem facing mobile application developers is the difficulty of keeping pace with the constantly changing technologies and environment. By the time a software product is released, newer versions of an OS or an application platform may already be in development.

Device Emulator and Testing

Most mobile operating systems have emulators for their supporting devices that are included with the development tool. The emulator provides a full target environment that runs on desktop platforms. The emulator is a good tool for learning APIs and testing an application throughout the development process. It can dramatically increase productivity. Deploying and testing the application on an emulator is much easier than on a real device.

On the other hand, an emulator is not the real device. Each device has different configurations, including different processors and hardware components. Deploying an application on different devices implies the possible change in the behavior of the application. Therefore, testing the physical device and communicating over the actual network are needed.

Common Features

Following the advancement of technologies, more features are supported by mobile operating systems. These features reflect the enhancements of hardware components and peripherals (see Section 2.1). The common applications and APIs supported include the following:

- *Communications*—Telephony (e.g., GPRS, WCDMA), Bluetooth, WiFi, InfraRed;
- *Networking*—OBEX, TCP/IP, WAP, Socket APIs;
- *Messaging*—SMS, EMS, MMS;
- *Security*—HTTPS, SSL;
- *Multimedia and Graphic User Interface*—2D/3D graphics, MP3, MPEG4;
- *Unicode and internationalization*;
- *Applications*—PIM (e.g., calendar, contact, e-mail), microbrowser, games;
- *Add-on VM Application environment*—Java Micro Edition, BREW.

These features and technologies are discussed in detail in subsequent sections and chapters.

Synchronization

Synchronization offers a mobile device the ability to connect to a network in order to update information. Through a cradle, a wireless connection, or over-the-air (OTA), a mobile user can synchronize data and applications between the device and a network. Basic proprietary conduits, such as backup and installation conduits, are often provided by the mobile operating system without extra charge.

The leading OS synchronization conduit software programs include Windows CE *ActiveSync*, *Symbian OS Connect*, and Palm OS *HotSync*. Many of the currently available commercial data synchronization solutions support ActiveSync and HotSync. Operating systems provide SDK [e.g., Palm Conduit Development Kit (CDK)], for their synchronization application development. For advanced custom conduits, such as enterprise data synchronization, users must purchase and integrate them with the applications.

2.2.1   Symbian OS

Symbian OS is a descendant of Psion's EPOC. Psion Software was founded by David Potter in 1980. Early EPOC releases were developed primarily for PDAs. In 1998, Symbian was formed as a joint venture of Nokia, Ericsson, Motorola,

Matsushita, and Psion, in order to explore the convergence between PDAs and mobile phones. As of Release 6, EPOC became known as Symbian OS. Nokia 9210 was the first Symbian OS phone. The Symbian OS has a generally maintained reasonable binary-backward compatibility. The latest Symbian OS is Version 9.1, which was announced in 2005. The Nokia N91 is the first Symbian OS 9.1 device on the market [10, 11].

Based on the Symbian OS, five different smart phone platforms were developed. Nokia developed three platforms—*Series 60*, *Series 80*, and *Series 90*. Nokia has licensed these platforms to many device manufacturers (e.g., Siemens, Panasonic, and Samsung), for high-end smart phones. NTT DoCoMo developed another platform for the Japanese market. The final platform, called *UIQ*, was developed by a subsidiary of Symbian. UIQ provides a customizable, pen-based user interface for mobile devices with large touch-sensitive color displays.

The Symbian OS is an operating system with much less dependence on peripherals. Peripherals are integrated with the kernel as individual system components. The kernel runs in the protected mode, and exports its services to applications via a user library. Each process, including the kernel, under the Symbian OS is run in its own protected address space. It is not possible for any application to overwrite any other application's address space. Thus, Symbian is reliable. Other unique features of the Symbian OS are introduced in the following sections.

Architecture

The Symbian OS is a full multitasking operating system that supports both multithreading and asynchronous calls. The asynchronous calls are often encapsulated as *active objects* in a Symbian OS. The active objects are used extensively in the Symbian OS for server-related operations.

Most mobile operating systems support a flat-file database. The Symbian OS provides a powerful relational database for persistent data storage. Furthermore, it makes the best use of a limited memory by means of XIP and the reuse of shared system components.

Dynamic Memory Allocation

A mobile device has a limited amount of main memory. Developers should always design applications with a small stack size, and should avoid allocating large objects into the memory. However, the Symbian OS supports an unlimited memory stack for dynamic memory allocation (e.g., dynamic buffer management for communication applications). The amount of memory available for dynamic allocation is restricted only by the hardware. Memory loss may occur for dynamic allocation. If an application forgets to release the allocated memory, it will be effectively unavailable to the rest of the system.

Framework Architecture
The framework architecture of Symbian OS allows software components to be added or updated to the phone while the phone is running, without rebooting. The

framework architecture can be used to add new system components, such as protocol stacks and device drivers.

UI Framework

Any user interface for a mobile application must be tailored specifically for its screen size and input methods. The Symbian OS UI framework defines an interface between the UI design and the framework. The UI framework design defines the look and feel of the user interface components, such as the test field, radio buttons, menu, and so forth. This framework allows the Symbian OS to support many different form factors. It also allows applications to run on multiple devices with minimal UI code rewriting [9, 12, 13].

Application Development

Symbian provides an SDK for application development. This kit includes a wide variety of APIs and an emulating environment. The Metrowerks' CodeWarrior for Symbian OS also provides a full range of C++ and Java-integrated development and debugging tools.

2.2.2   Windows CE/Mobile

The Microsoft Windows CE 1.0 operating system, which was released in 1995, was designed for handheld PCs. Windows CE 2.0, which was released in 1998, was aimed at the PDA market. Windows CE 3.0 came with Microsoft Office's support and built-in multimedia capabilities. Windows Mobile 2002, which is powered by Windows CE 3.0, is targeted specifically at Pocket PCs without keyboards. Pocket PCs and smart phones are two types of mobile devices that run these operating systems. Windows CEs are also run on Sega Dreamcast gaming devices [14].

Windows CE .NET 4.0 fits into the overall .NET strategy by allowing developers to use the full suite of Microsoft tools and Office products. *Windows Mobile 5*, which was based on *Windows CE 5.0* and the *.NET Compact Framework (CF)*, is the latest Microsoft mobile OS available at the time of this book's writing. Windows Mobile 5 marks the convergence of the Smartphone and Pocket PC editions of operating systems into one system that contains both phone and PDA capabilities. OEMs are allowed to configure the OS to their device requirements and designs [15].

Windows CE–based devices are required to have greater hardware capabilities. Extra computing power is required to run the Windows CE. A greater power requirement implies greater battery consumption. Windows CE devices typically must be recharged more frequently than other devices. Windows CE's architecture, environment, and features are discussed in the following sections.

Configurable System

Windows CE is a highly configurable OS. A Windows CE–based operating system for a specific device can be built by using a number of discrete modules. A module contains a collection of related API functions. Some modules are composed of components. Each component can contain a collection of API functions. By selecting only the required modules, the resources that are required by the target device, such as memory, can be minimized.

Program Execution

Windows CE supports multitasking and multithread process execution. The threads under Windows CE execute in one of two modes: user mode and kernel mode. The majority of threads, including applications and device drivers, execute under the user mode. A process under Windows CE is a 32-MB virtual address space. Windows CE supports up to 32 processes. When a user starts an application, an address space is created for the process. The executable code is mapped into this address space. Once the code is loaded, the system creates a new thread and starts its execution at the given entry point.

Windows CE executable files adhere to the portable executable (PE) standard. The PE header contains the chipset identities of the target device type. This prevents initiation of the executable on the wrong machine type. Following the PE standard, existing desktop-based tools can still be used to examine Windows CE executables.

Persistent Data Storage

Windows CE contains a built-in flat-file database. It supports timestamps as a data type. This allows the programmer to determine if the data has been altered by another application since its last use. However, a significant amount of effort is still required for the synchronization.

Windows CE's persistent storage is in the form of *object store*. The object store is considered to have three sections: the file system, the databases, and the registry. The object store is transaction-based.

Windows CE file system developers must pay attention to the following characteristics, which differ from those of desktop file systems.

- There is no current working directory, and fully qualified path names are used.
- Files are automatically compressed in the object store.
- There is no support for an overlapped IO.
- Installable file systems are not assigned a drive letter.

.NET Compact Framework

The .NET Compact Framework (.NET CF) is one of the two core technologies of Windows Mobile 5.0 for mobile client development. It allows developers to reuse existing programming skills and codes for all types of devices. The languages of choice for .NET CF development are C# and Visual Basic. The source code is compiled into a common language runtime (CLR) and deployed to the mobile device. The key .NET CF features include the following:

- *XML Web services*—to provide a standard way to support server-based functionality to a variety of client devices;
- *Corporate data access*—to access relational databases over a wireless connection with ADO.NET classes available on the client device;
- *High performance*—applications on .NET CF are executed as CLR native code;
- *Visual Studio .NET development tool integration.*

Mobile software codes on the Windows Mobile–based system can be one of the following three classes:

- *Native Code (C++)*—for high performance, direct hardware access, or a small footprint requirement;
- *Managed Code (C# or VB .NET with .the .NET CF)*—for business logic components, user interface, web services, and SQL Server-related applications;
- *Server-Side Code (ASP.NET Mobile Controls)*—for a wide variety of devices with a single codeset, and where there is guaranteed data bandwidth to the device.

Windows Mobile 5 Enhanced Features and Application Development

Windows Mobile 5 unifies the Pocket PC and Smartphone platforms by offering the following: common data services (i.e., SQL Server Mobile Edition), soft keys, persistent store, common installers, and application security models. Some of the new features include the following:

- The ability to be mounted as a USB mass storage device, bypassing ActiveSync;
- A taskbar that shows the start button, current time, volume, and connectivity status;
- Photo caller ID, picture, and video management supports.

Windows CE utilizes a large subset of approximately 10% of the full Win32 API. Windows Mobile 5 exposes OS features directly to application developers. It enables multimedia application development and other features by including new native and managed API sets, such as Direct3D Mobile, DirectDraw for graphic and game development, Windows Media Mobile for multimedia applications, Camera API, DRM, telephony, messaging, outlook mobile data, state and notification Broker, and more.

Windows Mobile 5 provides SDKs and Visual Studio 2005 for application development. The Visual Studio 2005 tool integrates the device development features from previous tools, such as Visual Studio .NET and eMbedded Visual C++, and adds new features, such as a new emulator, debugger, and data and UI designers. Visual Studio 2005 is a single tool for developing server, desktop, and mobile applications [16].

2.2.3 Embedded Linux (Mobile Linux)

Linux was written by Linus Torvalds and a group of volunteer programmers across the Internet. The *Linux kernel* and many system components are released under the GNU general public license (GPL), which dictates that the source code for those components, including modifications, must be made available under the GPL (or a compatible license) whenever a compiled version is distributed. The Linux kernel is constantly being updated. The latest stable version of the Linux kernel is 2.6 [17].

A *Linux distribution* bundles large quantities of application software with the core system, and provides more user-friendly installations and upgrades. Each Linux distributor determines the content of his or her Linux system, so the software actually included does vary. A complete GNU/Linux OS uses many packages from BSD UNIX, X window systems, GNU C Compiler, and so forth. The development language for Linux is C language. Due to Linux's openness, multitasking, multiusers, and multiplatform support, Linux has been widely used in the desktop and server area.

The Linux kernel and compiler are supported for all the popular 32-bit microprocessors, including ARM, Motorola/IBM PowerPC, and NEC MIPS, which are designed into embedded systems. Embedded Linux refers to the use of the Linux OS in embedded systems, such as mobile devices and other consumer electronic devices. There are many companies that sell embedded Linux solutions. These usually include a ported Linux kernel with cross-development tools, and sometimes with real-time extensions. A typical installation of an embedded Linux takes a small footprint (e.g., approximately 2 MB). Many device manufacturers, including Motorola, HP, Samsung, Sony, Sharp, NEC, and Panasonic, produce Linux-enabled mobile devices.

One key advantage of embedded Linux is that it can address a wider range of mobile device platforms. Other mobile operating systems are often targeted at high-end devices. Linux has the potential to become the leader for midrange feature phones. The Linux kernel for the embedded system supports true multitasking, virtual memory, shared libraries, execute-in-place file systems, flash management, and TCP/IP networking. The Linux persistent data storage and Nano-X window system are presented here [18].

Persistent Data Storage (File System)

The Linux kernel supports file systems for ROM/RAM residence. It also supports the XIP from ROM file systems. A compressed RAM disk image for system startup can be linked with the Linux kernel or be copied from a flash storage. Read-only compressed files can be partially decompressed, resulting in fast booting. Files for initial booting can be created and then discarded. Unnecessary codes are removed from the memory after execution.

GUI—Nano-X Window

The standard GUI used in all desktop Linux systems, X Window, requires a large amount of memory for execution, which is a disadvantage for mobile devices. Nano-X window, which used to be called Microwindows, is an open-source system aimed at producing GUI for small devices. The Linux kernel allows user applications to access the graphical display memory as a frame buffer. Thus, graphics applications do not need to know the underlying graphics hardware.

Microwindows requires memory in the range from 100 to 600 KB. Nano-X API allows applications to be built using a client/server protocol over a network or local socket. The client/server protocol can use shared memory for passing data between the client and the server. A compile time option allows all applications to be linked with the server for smaller environments.

Mobile Linux Standardization

Linux's openness and lack of royalties attract many device manufacturers and service providers. Most of the device manufacturers supporting Linux will have their own version of the OS. In addition, many commercial versions of embedded/mobile Linux are available. The major implementations of the Linux OS seem to be based on the *handhelds.org* kernel. Many companies, such as MontaVista's Mobilinux and TrollTech's Qtopia, offer Linux solutions for mobile devices. An industrial initiative, the embedded Linux consortium (ELC), has created a platform specification for embedded Linux. Many industrial consortiums have established Linux standardization. The Mobile Linux Initiative (MLI) of the Open-Source Development Lab (OSDL), and the Linux Phone Standard (LiPS) Forum are recently formed organizations that target mobile devices.

The OSDL, which was founded by major IT companies in 2000, is an industry consortium and resource lab for open-source developers. OSDL's MLI, which started in 2005, has been working on the improvement of Linux for mobile devices. MLI focuses on standard embedded Linux kernel-level issues and services, such as fast boot, memory footprint, multimedia framework, power management, radio interface, real time, security, and system-on-a-chip (SoC) data paths.

The LiPS Forum was launched in 2005 as a cross-industry consortium, and was chartered to turn embedded Linux into a plug-and-play mobile phone platform. LiPS's primary focus is defining the standards for the aspects of Linux that directly enable the deployment, development, and interoperability of applications and end-user services. Another main goal of LiPS is defining a platform that can be easily customized, so that wireless operators and device manufacturers can differentiate their products. This platform should provide all the services and applications for a complete mobile phone stack.

2.3 Mobile Application Platforms

Historically, mobile devices have been designed as closed systems. Applications under proprietary operating systems are developed either by wireless operators or by the device manufacturers themselves. Users are not allowed to download new applications, and have a very limited capability to personalize the devices. A mobile application platform provides an abstract layer over the native device for both application *development* and *execution*. Although most mobile OSs have their own application platforms (supporting SDKs and tools, as introduced in Section

2.2), developers often prefer to use an open, standard, and extendable platform that can produce applications that will work on all mobile devices.

Developing a mobile application platform allows users, mobile device manufacturers, and wireless operators to use a broad range of third-party applications. It can integrate unique features, such as GPS, voice, and cameras into the tools. Without worrying about the basic details of each peripheral, developers can focus on creating new functionalities for gaming, e-commerce, and other applications. Likewise, without worrying about application developments, device manufacturers are able to release new products faster. Applications can be deployed on any compliant device through OTA provisioning. In turn, wireless operators can take a percentage of the revenue for such OTA provisioning.

Most application platforms are designed to be independent of processor and operating systems. Developers do not have to repeat the development effort for different types of mobile devices. However, mobile devices do have large differences when it comes to their input/output and processing capabilities. In the real world, application testing is required for different types of mobile devices.

In this section, we introduce two widely used general-purpose platforms: Java and BREW. There are other platforms for specific types of applications. For example, the Opera platform is aimed at mobile Web applications [2, 3], and the *Morphun* platform is dedicated to mobile games [19].

2.3.1 Java Platform

Java technology plays an important role in Internet and Web software development. Java is both a programming language and a group of specialized platforms. Rather than running directly on the native operating system, Java programs are run on Java Virtual Machine (VM). The virtual machines make Java programs *machine independent*, which is one of the crucial requirements for modern mobile applications. To work with a wide range of devices and environments, Java supports four platforms: (1) Java Platform, Enterprise Edition (*Java EE*); (2) Java Platform, Standard Edition (*Java SE*); (3) Java Platform, Micro Edition (*Java ME*); and (4) *Java Card*. Note that the Java EE, Java SE, and Java ME formerly were called *J2EE*, *J2SE*, and *J2ME*, respectively.

Each Java platform is based on a VM to address applications *execution* and *development* to one particular target hardware environment. End users typically interface with the VM and the standard set of class libraries. Java SE provides an environment for core Java and desktop Java applications development, and is the basis for other Java platforms. Java EE defines the standard for developing component-based multitier enterprise applications. Java Card technology adapts the Java platform to enable smart cards and other intelligent devices with limited memory and processing capabilities. Java ME is comprised of a set of technologies and specifications targeted at consumer and embedded devices, such as mobile phones, PDAs, and set-top boxes. The architecture of Java platforms is shown in Figure 2.3. The Java ME architecture is also presented. The mobile information device profile (MIDP) sample program [20], the DoJa, and Java ME for i-mode are described in Chapter 4.

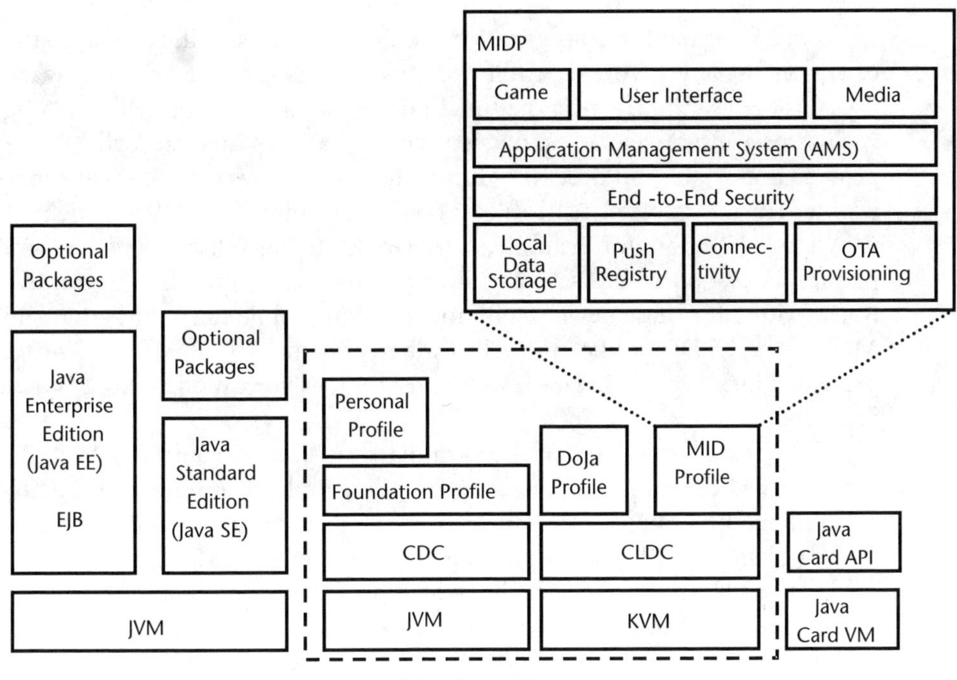

Figure 2.3   Java and Java ME MIDP platform structure.

Java ME Architecture

Java ME, which used to be called J2ME, enables Java applications to run on small, resource-limited embedded consumer devices. It adapts existing Java technology for handheld and embedded devices. Compatibility with Java SE is maintained wherever feasible. In fact, Java ME removes the parts of Java SE that are not applicable to constrained devices, such as abstract window toolkit (AWT), and other features. Two key components of Java ME are *configurations* and *profiles* [21, 22].

A *configuration* defines the basic Java ME runtime environment. This environment includes the *K virtual machine*, which is more limited than the VM used by Java SE, and a set of core classes derived primarily from Java SE. The key point is that each configuration is geared toward a specific family of devices with similar capabilities. Currently, two configurations are defined: the *connected device configuration* (CDC) and the *connected limited device configuration* (CLDC). Both target devices with network connectivity. CLDC targets the smaller devices: mobile phones and PDAs. CDC, on the other hand, targets devices that are less restricted in size, such as set-top boxes. The line between CDC and CLDC is not distinct. CDC is a superset of CLDC, which includes all of the classes defined by CLDC.

CLDC is for *limited* devices—devices that have severe limits on their computational power and battery life. CLDC devices are required to have some kind of network connection (thus, the term *connected device*), although it might be an intermittent, slow-speed connection. The CLDC supports Java language, except for the following differences: no floating point support, no object finalization, no finalization, no weak references, no support for JNI, and no thread groups and daemon threads.

A *profile* extends a configuration, adding domain-specific classes to the core set of classes. In other words, profiles provide classes that are geared toward specific uses of devices, and provide the functionality missing from the base configuration, such as user interface classes, persistence mechanisms, and so forth. While profiles provide important and necessary functionality, not every device will support every profile. The most complete profile of Java ME so far is the MIDP, which is based on the CLDC. In Japan, for example, NTT DoCoMo has released Java-enabled mobile phones based on the CLDC, but with their own proprietary DoJa profile. Applications written for these devices will not work on mobile telephones that support the MIDP. The CLDC connection API framework, the MIDP API, and programs are presented in Section 4.3. The Java ME technology for i-mode, DoJa, is described in Section 4.4.

The natural advantage of the Java platform comes from the VM design. However, the VM design produces extra compiling and execution overhead. Thus, Java was not the first choice for embedded system design. Processor designers realized this problem, and proposed solutions. For example, ARM developed a technology for Java acceleration called Jazelle, which provides a single-processor solution. The single processor solution allows developers the freedom to run Java codes, OS, and middleware on a single processor. Java byte codes are executed natively in the hardware.

Many tools are available for Java ME developers, including Sun's J2ME Wireless Toolkit, Metrowerk's CodeWarrior, Nokia's SDK for J2ME, Qualcomm's BREW SDK, BlackBerry Java Development Environment, Motorola's iDEN SDK for J2ME. Although Java ME is considered a platform-independent technology, the proprietary extensions created by mobile device manufacturers do generate some compatibility problems. Codes have to be tested before porting to another device. Among all possible modifications, the GUI adjustment is the part that requires the most attention.

2.3.2 BREW

BREW is an application development platform created by Qualcomm for mobile phones. Unlike Java, which is only aimed at application development, BREW provides an integrated solution, including applications development, device configuration, application distribution, and billing. The BREW platform is independent of the air interface. Handsets using other wireless technologies can be BREW-enabled, and the BREW environment can be used with other mobile operating systems, such as Symbian OS, Windows CE, and Palm OS. It also enables over-the-air provisioning of applications, and management of telephony functions in the devices [23, 24].

BREW manages all of the telephony functions on the device. Application developers can create advanced wireless applications without having to program to the device's system-level interface. A VM extension to BREW can actually be smaller than a VM that is natively integrated to a handset. In this latter implementation, phone-specific APIs must be written.

Currently, Java ME has a strong lead, in terms of the number of developers and potential market opportunity. It allows the JVM to be deployed to a BREW-enabled device, allowing Java ME and BREW to coexist within a single device. The Java ME

MIDP, with or without BREW, requires a significant amount of memory, so the device must have sufficient memory to accommodate an MIDP-compliant JVM. Device manufacturers and wireless operators usually support only one of these application platforms, due to cost and performance considerations.

Figure 2.4 shows the BREW architecture, and the relationship between the BREW API platform and the applications. The complete BREW application platform includes the following:

- The BREW SDK for application developers;
- The BREW client software and porting kit for device manufacturers;
- uiOne as the developing environment for BREW user interface;
- The BREW delivery system (BDS) controlled and managed by operators.

The BREW SDK

The BREW SDK provides general development and debugging tools, sample applications, and documents. C, C++, and assembly language are supported in BREW SDK. Provided that alternative VMs are installed, Java, extensible markup language (XML), and Flash can also be used. The SDK also includes a phone emulator that can be used for testing during the development process.

Client Software and Porting Kit

BREW is a thin client that sits between a software application and the ASIC software on the mobile device. Each SDK version is paired with a corresponding application execution environment (AEE) on the phone. BREW is much smaller than other application platforms or operating systems; its footprint is approximately 150 KB.

The BREW porting kit allows developers and publishers to easily port applications and content. Developers and publishers can port the *user interface* of a device directly on top of BREW. Device manufacturers and wireless operators can add new

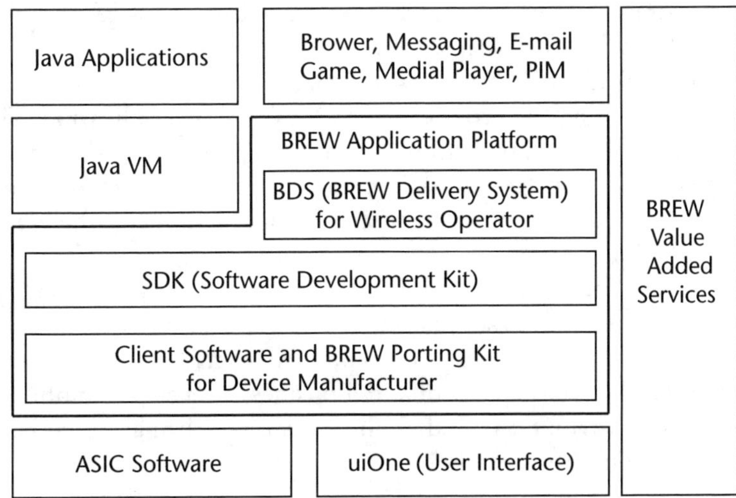

Figure 2.4　BREW architecture.

features or update their user interfaces directly over the air. Similarly, third parties can easily develop extensions to the BREW platform [25].

uiOne—BREW User Interface

BREW uiOne is a logical expansion of features and customization for mobile devices. It allows developers to more easily create a personalized and dynamic user interface. The main uiOne components include SDK; toolkit; application, resources, and templates (ART); and delivery system [24].

The *uiOne SDK* and *toolkit* help break UI development into manageable parts, and simplifies the customization and modification process. *uiOne ART* provides a global library of UI application resources and design templates, which is created by engineers and visual and interactive designers, for use by device manufacturers and operators. The *uiOne delivery system* handles UI theme downloads, ongoing casting of specific content to the UI, and bundled subscriber experience service packages. The delivery system enables ongoing "pushes" of content for UI processing.

BREW Delivery System

The BREW delivery system (BDS) is controlled and managed by operators. The main components of BDS include a download server, transaction manager, operator and developer extranets, and applications manager. The BDS integrates with the operator's existing billing and payment system. It provides an integrated architecture to enable BREW applications distribution.

The BDS allows operators to outsource their data solutions while still retaining most control. Operators can increase average revenue per user (ARPU) through secure applications downloads.

Both the developer and the wireless operator must digitally sign BREW applications. Developer-signed applications can only be executed on test-enabled handsets. The application must be submitted to Qualcomm for TRUE BREW Testing (TBT). After a pricing plan is negotiated between the developer and the wireless operator via a virtual marketplace, the application becomes available for download to general handsets. The application is then signed by the wireless operator to allow its execution on any BREW handset.

The BDS allows subscribers to shop, purchase, download, and install software over the air, using the browser or the MobileShop application. The subscriber sees only the catalog of applications that are supported on the handset. The wireless operator retains any retail markup and shares, which are 20% of the application's wholesale price with Qualcomm. The remaining 80% of the wholesale price flows to the developer.

Besides the BREW application platform, Qualcomm provides BREW value-added services to support operators, publishers, developers, and device manufacturers performing TBT; Qualcomm also signs BREW applications. BREW provides a development platform and a business process for mobile applications. The integrated distribution and billing features brings together the operators, developers, and users.

2.4 Summary

Mobile device technology is improving and changing at a rapid pace. Many powerful mobile devices integrated with new technologies have emerged in the market. The mobile device itself is now a complicated system. Open mobile operating systems and application platforms allow third party developers to participate in software development. Furthermore, users now have more freedom to control their mobile devices.

This chapter looked into mobile devices and operating systems from the perspectives of the user, the developer, the wireless operator, and the device manufacturer. We discussed the challenges facing the developers of software for mobile devices, including the restrictions imposed by limited hardware resources, different user input methods, and limited screen space. We also reviewed the features and architecture of three main mobile operating systems and two application platforms. Many of the technologies mentioned in this chapter will be discussed in detail in subsequent chapters.

References

[1]   Beaulieu, M., *Wireless Internet: Applications and Architecture*, Reading, MA: Addison-Wesley, 2002.

[2]   Mallick, M., *Mobile and Wireless Design Essentials*, New York: John Wiley & Sons, 2003.

[3]   OMA WAP specifications: http://www.openmobilealliance.org/tech/affiliates/wap/wapindex.html.

[4]   Viredaz, M. A., L. S. Brakmo, and W. R. Hamburgen, "Energy Management on Handheld Devices," *ACM Queue*, Vol. 1, No. 7, October 2003.

[5]   Casas, R., and O. Casas, "Battery Sensing for Energy-Aware System Design," *IEEE Computer*, November 2005, pp. 48–54.

[6]   Microsoft XNA Web site: http://www.microsoft.com/xna/.

[7]   *GPS World*: http://www.gpsworld.com/gpsworld/.

[8]   Johnston, C. J., and R. Evers, *Professional BlackBerry*, Hoboken, NJ: Wrox, 2005.

[9]   Hieda, S., et al., "Design of SMIL Browser Functionality in Mobile Phones," *IEICE Trans. on Communications*, Vol. E87-B, No. 2, February 2004, pp. 342–349.

[10] Yuan, M. J., and K. Sharp, *Developing Scalable Series 40 Applications: A Guide for Java Developers*, Reading, MA: Addison-Wesley, 2004.

[11] Symbian Ltd.: http://www.symbian.com/about/about.html.

[12] Sales, J., *Symbian OS Internals: Real-time Kernel Programming*, New York: John Wiley & Sons, 2005.

[13] Spence, E., *Rapid Mobile Enterprise Development for Symbian OS: An Introduction to OPL Application Design and Programming*, New York: John Wiley & Sons, 2005.

[14] Rich, J. R., *How to Do Everything with Your Smartphone, Windows Mobile Edition*, New York: McGraw-Hill Osborne Media, 2004.

[15] Boling, D., *Programming Microsoft Windows CE .Net*, 3rd ed., Redmond, WA: Microsoft Press, 2003.

[16] Windows Mobile 5: http://www.msdn.net/mobility/windowsmobile/howto/windowsmobile5/default.aspx.

[17] Yaghmour, K., *Building Embedded Linux Systems*, Sebastopol, CA: O'Reilly, 2003.

[18] Hollabaugh, C., *Embedded Linux: Hardware, Software, and Interfacing*, Reading, MA: Addison-Wesley, 2002.

[19] Developing Mophun games: http://www.mophun.com/developer/index.php.

[20] Mobile Information Device Profile (MIDP): http://java.sun.com/products/midp/index.jsp.

[21] Java 2 Platform, Micro Edition (J2ME): http://java.sun.com/j2me/index.jsp.

[22] Yuan, M. J., *Enterprise J2ME: Developing Mobile Java Applications*, Upper Saddle River, NJ: Prentice Hall, 2003.

[23] Rischpater, R., *Software Development for the QUALCOMM BREW Platform*, Berkeley, CA: Apress, 2003.

[24] Qualcomm BREW: http://brew.qualcomm.com/brew/en/about/about_brew.html.

[25] Barbagallo, R., *Wireless Game Development in C/C++ with BREW*, Plano, TX: Wordware Publishing, 2003.

CHAPTER 3

Wireless Application Protocols and Technology

The explosive growth of the Internet and mobile phones inspired the creation of the mobile Internet. The wireless application protocol (WAP) forum was founded to draft global specifications for mobile networks. WAP is positioned at the convergence of three rapidly evolving network technologies: the Internet, telecommunications, and wireless data.

WAP technology is expected to provide a long-term durable platform for mobile Internet system and application developers [1].

This chapter introduces WAP technology and protocols. Section 3.1 presents a short WAP background overview. Section 3.2 presents WAP architecture and protocols. The WAP proxy gateway and its push mechanism are also illustrated. Section 3.3 introduces the three main components in the WAP application environment (WAE): wireless markup language (WML), WMLScript, and user agent profile (UAProf).

Section 3.4 presents examples of WAP client software. This section focuses on messaging systems, in which short message service (SMS), multimedia messaging service (MMS), and their relationships with WAP are discussed. Section 3.5 discusses the future of WAP, and introduces the open mobile alliance (OMA), which is the organization that has continued WAP standardization since 2002.

3.1 Introduction to WAP

3.1.1 Brief History

WAP 1

In June 1997, Phone.com (formerly Unwired Planet; now Openwave), Ericsson, Motorola, and Nokia took the initiative to found the WAP Forum, with the aim of creating a standardized solution to the problem of providing Internet access from mobile devices. The basic idea of WAP technology is to use the existing digital wireless infrastructure to optimize information delivery between thin-client devices and the Internet. By 1999, the first WAP phones became available [2].

Until late 2000, WAP technology progressed at a steady pace. The membership of the WAP Forum was over 600, and doubled in 10 months. The members produced 99% of the handsets that carried the service to more than 300 million subscribers worldwide. Surveys also showed that there were 139 carriers deployed or in the final stages of testing at that time, and included 10,000 WAP sites from 95 countries, 7.8 million WAP-readable pages, and 50 million WAP-enabled handsets in circulation worldwide [3].

During the age of WAP 1 (i.e., WAP 1.0, WAP 1.1, and WAP 1.2) from 1997 to 2001 [1], most carriers deployed their WAP via the SMS of the global system for mobile (GSM) communications. Most of these systems ran at 9,600 bps. In essence, this means that WAP 1 is a slow circuit-switched system. In fact, the general packet radio service (GPRS) was a much better option for WAP, but the availability of GPRS and GPRS-compliant phones was still not ready at that time.

Instead of using HTML, which is popular in the Web environment, WAP 1 content is designed with an XML-based WML as its major format. As a consequence, a WAP device can only access newly written WML content, or pages that have been converted to WML. This greatly restricted the content development of WAP.

Between 2000 and 2002, most surveys showed that WAP users complained about the shortage of content and services, and the associated high prices. As many as 90% of corporate users who purchased WAP-enabled phones abandoned the data capabilities of the phones and used them only for voice communications. All surveys showed that mobile Internet communications during the period of WAP 1 was still at its infancy stage.

During that same period, the services provided by SMS (particularly straightforward text messaging), which was the bearer of most WAP 1, won greater popularity. The total number of chargeable SMS messages in the United Kingdom reached 1.2 billion in November 2001, and 1.5 billion in November 2002. The popularity of SMS was not limited to Europe, however. It was and still is a global trend. A total of 366 billion text messages were sent worldwide in 2002. SMS technology is presented in Section 3.4.1.

WAP 2

The WAP Forum announced the release of WAP 2.0 in June 2001. WAP 2.0 supports the standard Internet protocol suite on the client side. In particular, WAP 2.0 uses the XML-based XHTML for content presentation. It also optimizes the use of higher bandwidths and packet-based connections in wireless networks. WAP 2.0 technology represents an evolution to second generation mobile wireless Internet. To distinguish it from WAP 1, the WAP 2.0 release was named WAP 2. In June 2002, the WAP Forum was integrated into the OMA, and thereafter, all WAP standardization activities have been continued by the OMA [4].

Openwave announced the shipment of the 400 millionth handset using its WAP browser in August 2003. The company has shipped more than 45 varieties of WAP 2 phone models, with a total of 25 million units shipped worldwide. Openwave also noted the surprising growth in messaging client and WAP 2 product upgrades.

Not only did WAP phones and the browser sell well, but their actual usage also increased. According to the Mobile Data Association, the total number of WAP pages viewed in the United Kingdom during October 2003 totaled 897 million, across the four GSM networks. This daily average of 29 million compares to 9 million per day in August 2002. Ringtone and screensaver downloads dominated WAP usage. Other WAP-enabled services, such as news and sport services, also increased significantly [5].

The growth of WAP is driven not only by technology advancements (e.g., the introduction of WAP 2.0, colored handsets, GPRS), but also by the availability of a large amount of new content and services.

3.1.2   WAP Background

There are two major issues that designers of the wired Internet did not consider, for obvious reasons: (1) Internet access via handheld devices, and (2) the extension of the Internet protocol stack to wireless networks. Compared to desktop computers used for Internet access, handheld devices present a more constrained computing environment. Handheld devices tend to have (1) limited CPU and memory (ROM and RAM), (2) limited battery life, (3) smaller screens, and (4) limited input capability. Although the physical limitation of handheld devices is difficult to address, the enhancement of their CPU, memory, battery, and display performance have advanced rapidly. The cost of purchasing and using handheld devices keeps decreasing. Most importantly, wireless handheld devices are inherently mobile. This mobility introduces possibilities for services that are sensitive to such mobility, and can provide location-dependent information.

Wireless networks also present a more constrained communication environment compared to that of wired networks. Wireless data networks tend to have (1) a lower bandwidth, (2) longer latency, and (3) less connection stability and availability. Similar to the improvement in mobile device development, wireless data network technology has advanced rapidly [6].

To extend Internet technologies to handheld devices and wireless data networks, solutions must be found to fulfill the following basic requirements [7]:

- *Interoperability*—a device that is able to communicate with services from different providers in the mobile network;
- *Efficiency*—a of quality of service (QoS) suited to the behavior and characteristics of the mobile network;
- *Reliability*—a consistent and predictable platform for deploying services;
- *Security*—an ability to preserve the integrity of user data, and protect the devices and services from security problems.

To fulfill these requirements, WAP defines the system architecture and application development environment, which include the following important features:

- The ability to provide a layered, scaleable, and extensible architecture;
- The ability to optimize wireless transmission;
- The ability to leverage existing and evolving Internet standards;
- The ability to provide a Web-centric application model:
    - The ability to map well onto the existing Web and mobile phone interfaces, eliminating the need to reeducate end users;
    - The use of plain Web servers, and all existing server side Web development technologies (e.g., ASP.NET, PHP, and JSP);
- The ability to utilize XML as the foundation for presentation and processing;

• The ability to personalize the content and presentation of the device;

• The ease of using interoperable applications and communication.

3.2 Overview of WAP Architecture and Protocols

Accompanied by the changes in the wireless environment, WAP technology evolved from WAP 1 to WAP 2, and was finally integrated into the OMA. The basic architecture and principles of WAP are essentially unchanged. Since WAP 2 is backward-compatible with WAP 1, we use the general term WAP for both WAP 1 and WAP 2 in this chapter. This section introduces the WAP programming model, protocol stack, and service architecture. We also describe the operations of a proxy gateway, and how push can be implemented.

3.2.1 WAP Programming Model

The WAP programming model is closely aligned with the Web programming model. Most WAP operations are initiated by a client to request content from a server (i.e., *Pull* operation). Besides content pull, WAP extends the Web architecture by adding telephony support with wireless telephony applications (WTA), enabling a push model. Push indicates that a server can proactively send content to the client. As shown in Figure 3.1, the WAP programming model supports both content pull and push operations.

In WAP 1, a WAP gateway (often referred to as a WAP proxy gateway) is required to handle the protocol interworking between the client and the origin

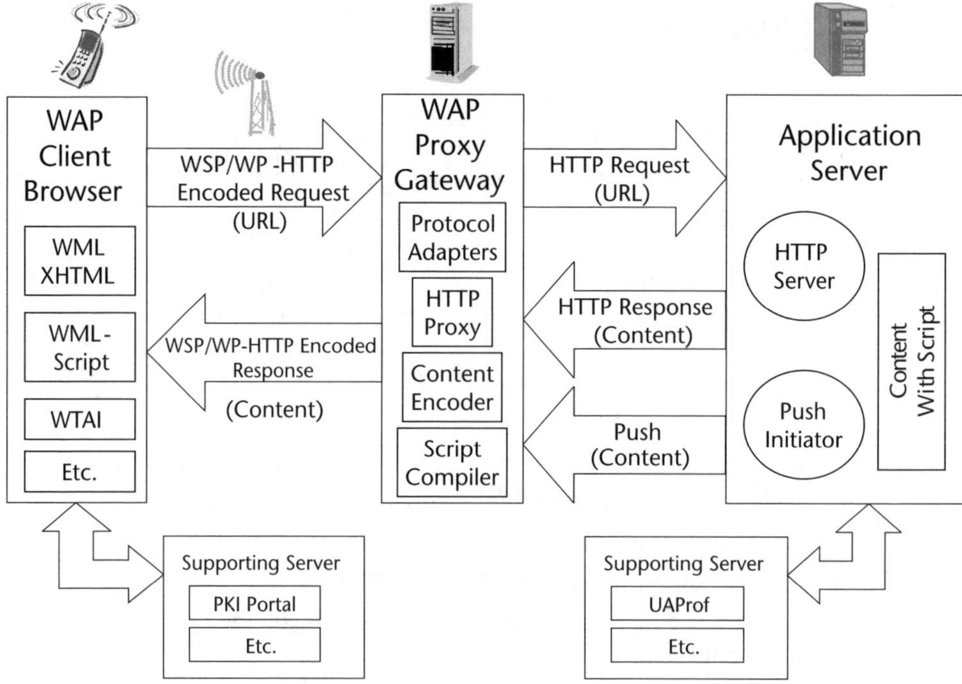

Figure 3.1 WAP programming model.

server. The WAP proxy gateway communicates with the client using the WAP protocols, but it communicates with the origin server using standard Internet protocols. WAP 2 does not need a WAP gateway to work on protocol translation, since the communication between the client and the origin server can be conducted by using the hypertext transfer protocol (HTTP) known as HTTP/1.1. However, deploying a WAP proxy gateway can optimize the communications process, and may offer mobile service enhancements, such as location, privacy, and presence-based services. In addition, a WAP proxy gateway is necessary to offer push functionality.

In addition to the WAP proxy gateway, some supporting servers may be used to serve WAP devices, proxies, and applications. These services are often specific in function, but are of general use to a wide variety of applications. The supporting servers include the following:

- *PKI Portal*—The PKI Portal allows devices to initiate the creation of new public key certificates.
- *UAProf Server*—The UAProf Server allows applications to retrieve the client capabilities and personal profiles (CC/PP) of user agents and individual users. More details on the UAProf are given in Section 3.3.3.

3.2.2   WAP Protocol Stack

WAP architecture provides a scaleable and extensible application development environment for mobile communication devices. This is achieved through a layered protocol stack. As shown in Figure 3.2, each layer provides a set of services to other services through a set of well-defined interfaces. Each of the layers of the architecture is accessible through the layers above it, as well as through other services and applications [7]. The application layer is defined by the WAE, which is described in Section 3.3.

Figure 3.2 shows the WAP 1 and WAP 2 stacks. An important part of WAP 2 is the support for Internet protocols to the mobile device. WAP 2 also contains support for the legacy 'WAP 1 Stack,' which is used over those networks that do not provide IP as well as that provided by low-bandwidth IP bearers. Both stacks are supported in WAP 2, and provide similar services to the application environment.

WAP 1 Legacy Protocol Layers

The following WAP 1 protocols have been optimized for low-bandwidth bearer networks with a relatively long latency.

- *Wireless Session Protocol* (WSP)—WSP provides HTTP/1.1 functionality and incorporates new features, such as long-lived sessions and suspend/resume functions. It provides the upper-level application layer of WAP with a consistent interface for both connection-oriented and connectionless service.
- *Wireless Transaction Protocol* (WTP)—WTP has been defined as a lightweight transaction protocol that is suitable for implementation in "thin" clients (mobile devices). WTP operates efficiently over wireless datagram networks, and relieves the upper layer from retransmissions and acknowledgments. It has no explicit connection setup or tear-down phases.

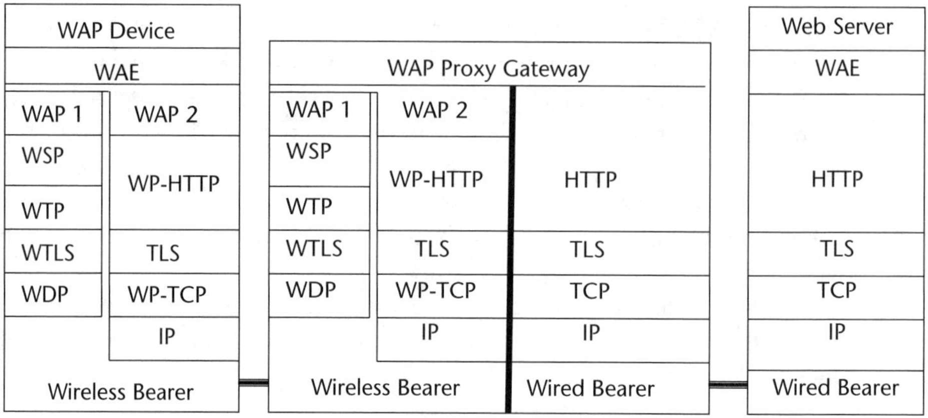

Wireless Profiled HTTP (WP-HTTP) : supports message body compression of responses
Wireless Profiled TCP (WP-TCP) : optimized for wireless environments

Figure 3.2   WAP 1 and WAP 2 protocol stacks with the proxy gateway.

- *Wireless Transport Layer Security* (WTLS)—WTLS is designed to provide privacy, data integrity, and authentication between two communicating applications. It provides an interface for managing (e.g., creating and terminating) secure connections. It provides functionality similar to that of TLS 1.0 (i.e., SSL).
- *Wireless Datagram Protocol* (WDP)—WDP is a general datagram service that can function independently of the services of the wireless network.

WAP 2 Protocol Layers for Networks Supporting IP

WAP 2 introduced the Internet protocols into the client side of the WAP environment. This support has been motivated by the emergence of high-speed wireless networks (e.g., GPRS) that directly provide IP support to wireless devices.

- *Wireless Profiled HTTP* (WP-HTTP)—WP-HTTP is a profile of HTTP for the wireless environment, which is fully interoperable with HTTP/1.1. The interaction between the WAP device and WAP proxy/WAP server is based on the HTTP request/response. WP-HTTP supports message-body compression of responses and the establishment of secure tunnels.
- *Transport Layer Security* (TLS)—A wireless profile of the TLS protocol permits interoperability for secure transactions.
- *Wireless Profiled TCP* (WP-TCP)—WP-TCP provides connection-oriented services. It is optimized for wireless environments, and is fully interoperable with standard TCP implementations. To simplify the implementation, WP-TCP makes four minor modifications: (1) a fixed 64-KB window, (2) no slow start, (3) a maximum MTU of 1,500 bytes, and (4) a slightly different retransmission algorithm [8].

Figure 3.2 shows how WAP 1 and WAP 2 stacks operate between WAP devices and the WAP proxy gateway. In the figure, the WAP proxy gateway interconnects the services offered by WSP/WP-HTTP to the HTTP protocol to permit access to data on the wired Internet.

Bearer Networks

Protocols have either been designed or selected to operate over a variety of different bearer services, including short message, circuit-switched data, and packet data. The bearers offer different levels of quality of service with respect to throughput, error rate, and delays. The protocols are designed to compensate for, or to tolerate these varying levels of service. Two common bearers for WAP, SMS and GPRS, are introduced in Section 3.4.1.

3.2.3   WAP Service Architecture

WAP architecture separates service interfaces from the protocols. This separation provides flexibility in handling the specification evolution and in selecting the appropriate protocol for a given context. Many of the services in the stack may be provided by more than one protocol. For example, either HTTP or WSP may provide the hypermedia transfer service. Figure 3.3 illustrates the WAP service architecture. The WAE-related services, such as user agent and content format, are introduced in Section 3.3.

Transfer Services

Transfer services provide structured information transfer between network elements. The transfer services include the following.

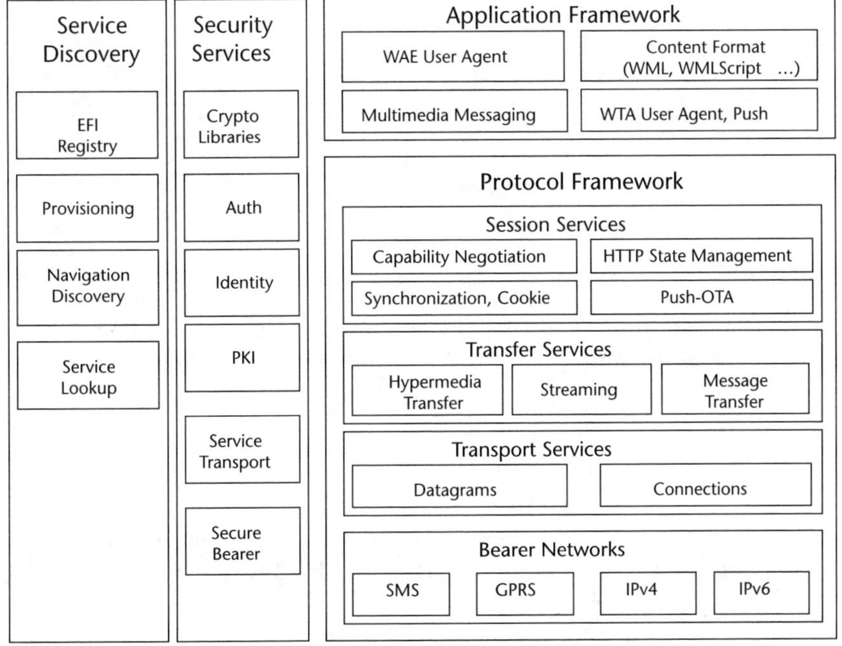

Figure 3.3   WAP service architecture.

- *Hypermedia Transfer*—Hypermedia transfer services provide for the transfer of self-describing hypermedia resources. The combination of WSP and WTP in WAP 1 provides the hypermedia transfer service over secure and nonsecure datagram transports. HTTP provides the hypermedia transfer service over secure and nonsecure connection-oriented transports.

- *Streaming*—Streaming services provide a means for transferring isochronous data, such as audio and video.

- *Message Transfer*—Message transfer services provide the means to transfer asynchronous multimedia messages. MMS encapsulation is a protocol used to transfer multimedia messages between WAP devices and MMS servers. An example of MMS message transfer is illustrated in Section 3.4.3.

Session Services

Session services provide for the establishment of a shared state between network elements that span multiple network requests or data transfers. For example, the push session represents that the WAP client device is ready and able to receive pushes from the push proxy gateway.

The session services include the following:

- *Capability Negotiation*—The WAP architecture includes specifications for describing, transmitting, and managing capabilities, as well as for preference information. This allows for customization of information and content returned by the origin server or pushed by the application.

- *Push-OTA*—The Push Over-the-Air (Push-OTA) session service provides for network-initiated transactions to be delivered to wireless devices that are intermittently able to receive data (e.g., devices with dynamically assigned addresses). The Push-OTA service operates over connection-oriented transport services and datagram transports [9]. The push operation sequence is illustrated in the latter part of this section.

- *Sync*—The Sync service provides for the synchronization of replicated data.

- *Cookies*—The Cookies service allows applications to store the hypermedia transfer state on the client or proxy.

Application Framework

The application framework provides an application environment based on a combination of World Wide Web (WWW) and mobile telephony technologies. Its primary objective is to establish an interoperable environment that will allow operators and service providers to build applications and services in an efficient and useful manner.

The application framework includes the following.

- *WAE User-Agent*—WAE is a microbrowser environment containing or allowing markup (including WML and XHTML), scripting, style-sheet languages, telephony services, and programming interfaces, all of which are optimized for use in handheld mobile terminals.

- *Wireless Telephony Application* (WTA)—This service provides tools for telephony processing (e.g., call forwarding) from within the application environment that traditionally supports data functionality [10].

- *Push*—The push service provides a general mechanism for the network to initiate the transmission of data to applications that are resident on WAP devices [11].

- *Multimedia Messaging*—Depending on the service model, MMS could permit a quick delivery paradigm (e.g., SMS) or a store-and-forward approach (e.g., e-mail), or could permit both modes to operate.

- *Content Formats*—The application framework includes support for a set of well-defined data formats, such as WML, WMLScript, images, audio, video, animation, phone book, and calendar information.

- *Pictogram*—This service permits the use of tiny images in a consistent fashion. Such images can be used to quickly convey concepts that permit efficient communication in a small amount of space. The 264 defined pictograms include animals, appliances, dress, emotion, food, human body, gender, maps, music, plants, sports, time, tools, vehicles, weapons, and weather. Notice that the WAP standard only names each pictogram; it does not give the actual bitmap image [12].

Security Services

Security forms a fundamental part of WAP architecture, and its services can be found in many layers of its architecture. In general, WAP offers privacy, authentication, and nonrepudiation security facilities. Examples of the security services include cryptographic libraries, HTTP client authentication, wireless identity module (WIM), WTLS and TLS handshaking, and PKI services.

Service Discovery

Service discovery forms a fundamental part of WAP architecture, and its services can be found in many layers of its architecture. Basic service discovery services include the following:

- *EFI*—The external functionality interface (EFI) specifies the EFI application interface (EFI AI) between WAE and components or entities with embedded applications that execute outside the defined WAE capabilities. This is analogous to providing a plug-in module for future growth and extendibility of supported WAP devices. EFI services are accessible through a WMLScript API and a markup API.

- *Persistent Storage Interface*—This capability specifies a standard set of storage services that are coupled with a well-defined interface for processing data on the wireless device. Data may be stored in the flash ROM of a mobile device.

- *Provisioning*—Provisioning service involves a standard approach (via OTA or SIM cards) for providing WAP clients with information that is needed to operate on the wireless networks. This permits the network operator to manage the devices on its network using a common set of tools.

- *Navigation Discovery*—This service allows a device to discover new network services (e.g., secure pull proxies) during navigation.
- *Service Lookup*—The service lookup service provides for the discovery of a service's parameters through a directory lookup by name. One example of this is the domain name system (DNS).

The WAP layered architecture builds upon an extensible set of protocols. External applications may directly access the various services. This allows the WAP stack to be used for applications and services not currently specified by WAP. Many applications, such as e-mail, calendar, phone book, notepad, and Yellow Pages, can be developed with WAP.

3.2.4 From Web Proxy to WAP Proxy Gateway

Accompanying the rapid growth of the Internet, insufficient network bandwidth and long latencies have become more critical concerns. Web proxy (i.e., HTTP proxy) is the most effective and widely adopted mechanism to shorten the response time. The WAP proxy gateway utilizes the Web proxy concept to provide efficient wireless access to the Internet. A proxy gateway makes requests on behalf of the client, and plays the role of both server and client.

Figure 3.4 shows the request/response sequence chain between a client and a Web server through an HTTP proxy. The sequence consists of the following six steps, corresponding to the six labels in the figure:

1. The user agent (i.e., client system) sends a request to the proxy and asks for a particular resource (e.g., a page).

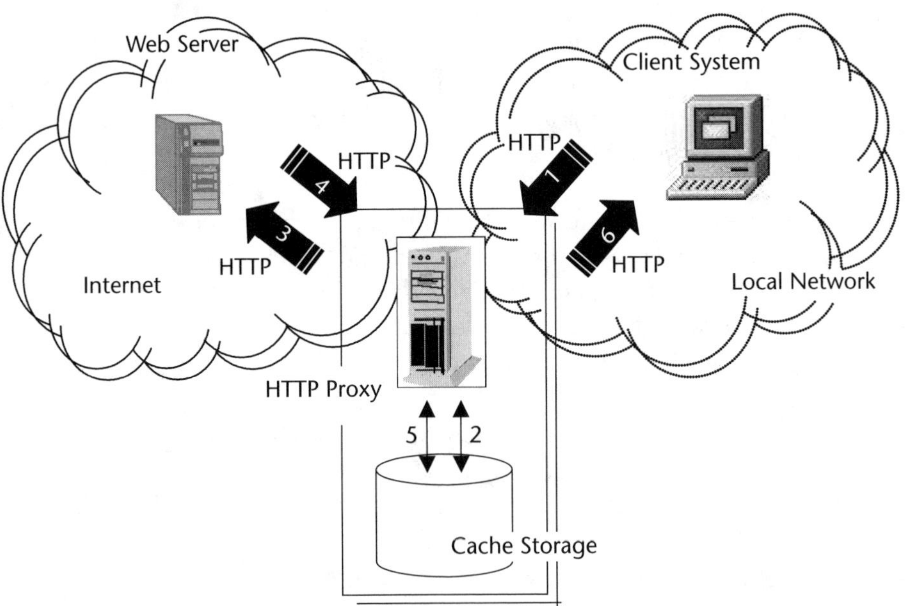

Figure 3.4 HTTP proxy operation sequence.

2. After receiving and interpreting the request, the proxy checks if the resource was cached in the disk. It is called a *Hit* when the resource is found and is sufficiently fresh. Then the proxy sends a response to the user agent. This operation shortens the request/response chain. Otherwise, a *Miss* situation occurs, and the sequences continue to the next step.
3. The proxy rewrites all or part of the request, and forwards the reformatted message to the origin server.
4. The Web server receives the request and replies with a suitable response.
5. After receiving the response sent by the origin server, the proxy would cache the message, if it is cacheable.
6. Lastly, the response is sent to the user agent.

WAP utilizes a proxy gateway technology to enhance the performance and optimize the connection between the wireless domain and the Web server [13]. The WAP proxy gateway may provide a variety of functions, including the following.

- *Caching Proxy*—A caching proxy can improve perceived performance and network utilization by maintaining a cache of frequently accessed resources. Caching is the most important function of a Web proxy, but it is optional for the WAP proxy gateway.
- *Protocol Gateway*—The protocol gateway translates requests from a WAP protocol stack to the Web protocols (HTTP and TCP/IP). The gateway also performs DNS lookups of the servers named in the request URLs.
- *Content Encoders and Decoders*—The content encoders can be used to translate WAP content into a compact format that allows for better utilization of the underlying link, due to its reduced size.
- *UAProf Management*—User agent profiles describing client capabilities and personal preferences are composed and presented to the applications. The UAProf is described in Section 3.3.3.

In general, a WAP proxy gateway is expected to complete three tasks: (1) header translation, (2) push operation, and (3) content compiling. The header translation allows the client system to access the Internet via a different protocol. The push operation allows the server to send the right information to the client, which will be illustrated in detail in the next section. The content compiling will compact the data for low-bandwidth support, which is not a part of this book's scope.

Header translation is the first task to be done in the WAP gateway. In WAP 1, the WSP headers are defined in binary format, but the HTTP headers are in string type. In order to enable a mobile handset to access the IP network, the translation of the coded WSP and traditional HTTP requests/replies is needed.

The operation sequence chain between a WAP mobile phone client system and a Web server through a WAP 1 gateway is shown in Figure 3.5. The seven steps in the operation sequence are shown as labels in the figure.

1. The user agent (i.e., client system) sends a URL request to a WAP gateway following the WSP protocol.

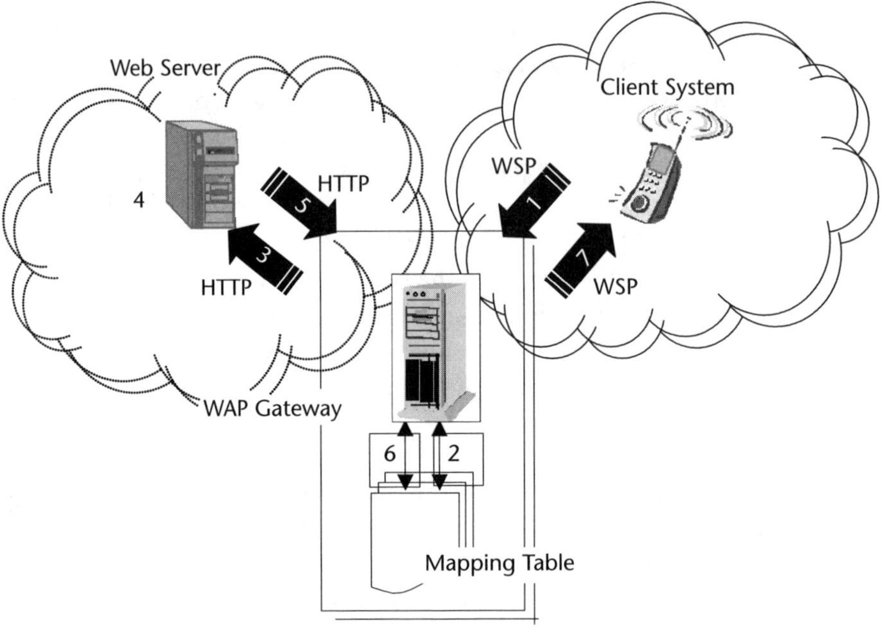

Figure 3.5   WAP 1 gateway operation sequence.

2. The WAP gateway decodes the requested message, and translates the request line and request header (in binary format) into HTTP format through a mapping table.

3. The WAP gateway creates a connection to the Web server and sends an HTTP request to it.

4. The Web server processes the HTTP request. If the URL refers to a static file, the Web server fetches the file to respond with an HTTP reply header; if the URL refers to a script application, the Web server starts the application.

5. The Web server returns an HTTP reply message that contains data or result from the script application.

6. The WAP gateway encodes the received reply message, and translates the well-known and formatted HTTP reply line and header to WSP with a binary format using the mapping table.

7. The WAP gateway creates a connection, and a WSP response containing the WML is returned to the client system.

As soon as the WAP 1 gateway receives the WSP request, it is processed in two steps: (1) request line decoding, and (2) request header decoding. The request line parser gets the transaction ID (TID), protocol data unit (PDU) type, and uniform resource identifier (URI) of the request packet. It then translates the PDU type to a string format by using a defined mapping table. The request header parser gets the well-known header values and translates them to the string type header's names through a defined mapping table. According to the values, the parser calls the corresponding method to get the string type header's values. The mapping table mentioned above can be found in the WSP specification.

The WAP 1 gateway also encodes the reply message in two steps: (1) response status line encoding, and (2) response header encoding. The response status line parser gets the version, response status code, and reason of the reply packet. It then encodes this information into a binary format through a mapping table. The response header parser gets the string type header's names and translates them to the well-known header values through the defined mapping table. According to these values, the parser calls the corresponding method to get the binary format header's values [14].

Table 3.1 shows an example of the header translation. A WSP-coded request header is sent to the WAP gateway. The first section of Table 3.1 shows the

Table 3.1   AP Gateway Header Translation Example

WSP Coded Data

01 40 25 68 74 74 70 3a 2f 2f 77 65 63 6f 2e 63 73 69 65 2e 66 6a 75 2e 65 64 75 2e 74 77 2f 69 6e 64 65 78 2e 77 6d 6c 81 ea 81 84 80 94 80 95 80 a1 80 9d a9 4e 6f 6b 69 61 2d 57 41 50 2d 54 6f 6f 6c 6b 69 74 2f 31 2e 33 62 65 74 61 00

Decoded HTTP Data

TID=1
Type=GET
URL=http://weco.csie.fju.edu.tw/index.wml
Accept-Charset:Somali ,Amharic
Accept:application/vnd.wap.wmlc, application/vnd.wap.wmlscriptc, image/vnd.wap.wbmp, image/gif
User-Agent:Nokia-WAP-Toolkit/1.3beta

WSP/HTTP Mapping

WSP Code	WSP Value (Hex)	Filed Name	HTTP Value
01	0x01	Transaction ID	1
40	0x40	PDU Type	GET
25	0x25	URI Length	37
(Following 37 bytes) 68 74 74 70 3a 2f 2f 77 65 63 6f 2e 63 73 69 65 2e 66 6a 75 2e 65 64 75 2e 74 77 2f 69 6e 64 65 78 2e 77 6d 6c		URI	http://weco.csie.fju.edu.tw/index.html
81	0x01	Header Name	Accpet-Charset
ea	0x6A	Header Value	Somail
81	0x01	Header Name	Accpet-Charset
84	0x04	Header Value	Amharic
80	0x00	Header Name	Accept
94	0x14	Header Value	application/vnd.wap.wmlc
80	0x00	Header Name	Accept
95	0x15	Header Value	application/vnd.wap.wmlscriptc
80	0x00	Header Name	Accept
a1	0x21	Header Value	image/vnd.wap.wbmp
80	0x00	Header Name	Accept
9d	0x1D	Header Value	image/gif
a9	0x19	Header Name	User-Agent
4e 6f 6b 69 61 2d 57 41 50 2d 54 6f 6f 6c 6b 69 74 2f 31 2e 33 62 65 74 61 00		Header Value	Nokia-WAP-Toolkit/1.3beta.

WSP-coded request. The request decode module of the WAP gateway then parses the data and translates them to an HTTP value. The second section of Table 3.1 shows the decoded HTTP request. The third section of Table 3.1 describes the mapping details.

3.2.5　WAP Push Mechanism

The WAP push framework introduces a mechanism in which the server transmits information via a push proxy gateway (PPG) to a client without a previous user action. Push functionality is especially relevant to real-time applications that send notifications to users (e.g., text messaging, stock prices, and traffic update alerts). Without push functionality, these types of applications would require the devices to poll application servers for new information or status.

A push operation in WAP occurs when the push initiator transmits content to a user agent. The push initiator is on the Internet, and the WAP client is in the wireless domain. If the user agent and the push initiator share no protocol, then the PPG acts as an intermediate translating server. Thus, the push initiator contacts the PPG from the Internet, delivering content for the client using Internet protocols. The PPG forwards the pushed content to the wireless domain, and the content is then transmitted over the air in the mobile network to the client.

In addition to providing simple proxy gateway services, the PPG has the capability of notifying the push initiator about the final outcome of the push operation. The PPG may also provide the push initiator with client capability lookup services, enabling a push initiator to select the optimal type of content for sending to the client.

The encoded content is pushed from the push initiator to the PPG, following the push access protocol (PAP), and from the PPG to the client following the Push-OTA Protocol. PAP uses XML messages that may be tunneled through HTTP.

A user agent of the mobile client in the WAE handles the pushed content based on its media type. Each media type specifies both the data structure and its semantics. The user agent is responsible for interpreting the content according to the rules specified for each media type. A user agent should not take any other actions other than discarding the content or placing it in cache, if neither the definition of the media type itself nor the user agent specifies any particular push behavior for that media type.

An example of a push operation is presented in Figure 3.6. The PAP and Push-OTA protocols define the required elements of the push operation through the PPG. The 10 steps in the push operation sequences are the following.

1. A push initiator submits a push message to a PPG. The push message is a multipart type that contains an XML control entity and a content entity, and may contain an RDF format capability entity that is defined in the user agent profile.
2. The PPG determines whether to accept or reject the received push message. If the PAP push message element is not valid with respect to its document type definition (DTD), then the PPG must reject it.
3. The PPG reports the acceptance or rejection result in the response, which is an XML document.

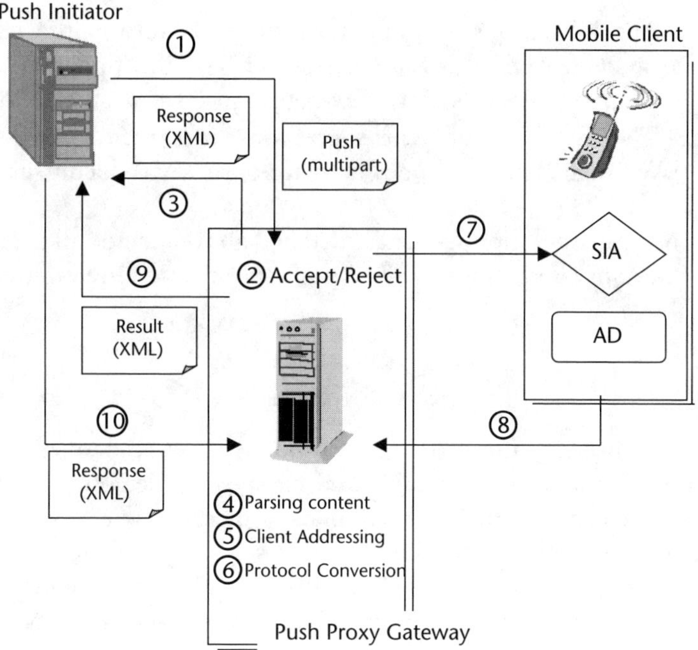

Figure 3.6   Push operations with PPG.

4. The PPG parses the control information of the push message to determine where and how it should be delivered.

5. A push initiator must identify the client addresses for submission. A client address is composed of a client specifier and a PPG specifier. The client specifier is used to find a target client, and determine if the content is either user-defined identifiers (e.g., e-mail address) or a device address (e.g., phone number). The PPG specifier permits a push message to be routed through the proxy, and the content is a site of the PPG. In this step, the PPG parses the client addresses for delivery.

6. The PPG converts the protocols between the Internet and the WAP domain.

7. If there is no error, the PPG must deliver either the confirmed or unconfirmed push primitives. In a confirmed push, an active WSP session that only the client can create is required. This situation is also called a connection-oriented push. If the PPG gets a connection-oriented push message to a client, and there are no active sessions to that client, the PPG cannot deliver the push message. In order to solve this problem, the PPG sends a request that contains the necessary information for creating a session to a session initiation application (SIA) in the client side. The SIA determines whether or not to establish a session. Push delivery may also be performed without the use of sessions in an unconfirmed push.

8. When a client receives a pushed message, the application dispatcher (AD) in the client parses the push message header, in order to determine its destination application and to run it. If the push delivery method is a confirmed push, then the client responds with a primitive to acknowledge receiving the PPG's push content.

9. The PPG sends a result notification to inform the push initiator of the outcome of the push submission. This notification would report whether the push message was sent, delivered, expired, cancelled, or if an error occurred.

10. The push initiator returns a response message to the result notification message. These two messages are both XML documents.

The push operation sequences detailed above include all the required operations. Some optional operations in WAP push specification are omitted for simplicity.

3.3 WAE

A WAE represents the WAP application layer. WAE defines the structure in which the various forms of executable and nonexecutable content interact. Markup language specifications for the WAP microbrowser in wireless devices are critical to a content developer. The markup languages specified include (1) the XHTML mobile profile markup language (XHTML-MP) for WAP 2 content, and (2) WML to support the legacy WAP 1 content. A full backward compatibility support for the original WML (WML 1) applications is provided in the WAE for WAP 2. These markup languages provide the display services for wireless devices [15].

The accompanying client-side scripting language, WMLScript, provides for additional intelligence and control over presentation. To improve the efficiency of transmission and client implementation for the handling of both WML and WMLScript, WAE supports the tokenization of WML 1 and the compilation of WMLScript before the gateway sends the content to the device.

This section introduces the WAE. The characteristics of WML, XHTML-MP, and WMLScript are illustrated here. We will also introduce the UAProf. The examples of WAP client software development will be presented in the mobile client software chapter (Chapter 4).

3.3.1 WML

WML is based on XML. Although WML is deprecated in WAE 2, WML 1 is included in WAP 2 to ensure backward compatibility. WML is designed with the constraints of small narrowband devices in mind.

WML has the following specific features:

- WML supports deck-and-card architecture. A deck is a transmission unit consisting of multiple cards; a card is a user interaction unit.
- WML includes navigation and event-handling models.
- WML includes support for managing the user agent state through the use of variables and navigational history [16].

XHTML Basic and XHTML Mobile Profile

The XHTML Basic Profile is a subset of XHTML recommended by the World Wide Web Consortium (W3C). It is an XML-based markup language intended for small devices. It does not support some complicated components, such as style sheets, scripts, and frames, which are common for normal Web pages. However, most of the XHTML standard tags are preserved. They are grouped into 11 modules. Some are required and some are optional [17].

WAP 2 supports XHTML Basic. Since NTT DoCoMo (provider of i-mode) has also agreed to support XHTML Basic, Web designers can use this format to design pages displayable on desktop as well as mobile devices. These decisions will end the markup language format wars that impede the growth of the mobile Internet industry. Where there is overlap between WML 1 and XHTML, preference has been given to the XHTML expression of a feature.

The XHTML Mobile Profile (XHTML-MP) is the basic markup language for the WAE in WAP 2, which extends the Basic profile of XHTML, as defined by W3C. This core was designed to be extensible, and WAE takes advantage of this capability by defining additional markup features for enhanced functionality. XHTML-MP is extensible, permitting additional language elements to be added as needed. A pure XHTML basic document is also a valid XHTML Mobile Profile document. The XHTML Mobile Profile browser would indicate if it has accepted both XHTML Basic and XHTML Mobile Profile documents. XHTML-MP should be used for new applications [18].

WML 2

WML 2 is built on top of XHTML-MP, with additional extension modules that provide the WML-specific features. Thus, WML 2 provides convergence with the existing Internet standard, as well as backward compatibility with WML 1.

- A pure XHTML Mobile Profile document is also a valid WML 2 document.
- A WML 2 browser would indicate if it has accepted the XHTML Basic, the XHTML Mobile Profile, and the WML 2 documents.

WML 2 is deprecated in WAE 2, and should not be used to create new applications. It should only be sent to a WAE user agent as the result of WML 1 transformation. Backward compatibility is provided through either native support for both languages (WML and XHTML-MP), or through a defined transformation operation of WML 1 to WML 2. The transformation process provides for the conversion from WML 1 to XHTML-MP, as well as support for features specific to WML 1. WAP 2 provides a transformation model using extensible stylesheet language transformation (XSLT), which permits documents defined in WML 1 language to be converted to WML 2 code.

Cascading Style Sheets Mobile Profile

The WAE in WAP 2 also enhances the presentation of the content by supporting style sheets. Based on the Cascading Style Sheets (CSS) Mobile Profile from the W3C, WAP support covers both the inline and external style sheets that are commonly supported by most Web browsers. This extended subset is used to specify document presentation in conjunction with WML 2 and the XHTML Mobile Profile. This subset is further optimized for small devices that have small displays and limited computational resources.

Some features in WAP 1 can be achieved through both WML proprietary elements and attributes and XHTML elements and attributes, using CSS properties. Content authors are encouraged to use XHTML, if possible, to leverage the convergence of WAP and the standard Internet.

3.3.2 WMLScript

WMLScript is a lightweight procedural scripting language. It enhances the standard browsing and presentation facilities of WML with behavioral capabilities, support of more advanced UI behavior, and client intelligence. It also provides a convenient mechanism to access the device and its peripherals, and reduces the need for round trips to the origin server [19].

WMLScript is a client-only scripting platform used in combination with WML to provide client-side procedural logic. It can be used for the following functions: (1) to validate user input, (2) to generate message boxes and dialog boxes locally, (3) to view error messages and confirmations faster, and (4) to access facilities of the user agent. The key WMLScript features include the following.

- *ECMAScript (or JavaScript)-based Scripting Language*—WMLScript is a loosely extended subset of ECMAScript. It refines ECMAScript for the narrowband environment.
- *Procedural Logic*—WMLScript adds the power of procedural logic to WAE.
- *Event-based*—WMLScript may be invoked in response to certain user or environmental events.
- *Compiled Implementation*—To deal with better utilization of narrowband wireless channels and the limited memory on handheld devices, WMLScript is compiled into a bytecode on the server, or WAP gateway, before it is sent to the client browser. A WAE user agent executes WMLScript in bytecode format. Where a user agent requires WMLScript bytecode, WAP proxies should provide the compilation function.
- *Integrated into WAE*—WMLScript is fully integrated with the WML browser. This allows authors to construct their services using both technologies, through the most appropriate solution for the task at hand. It has access to the WML state model, and can set and obtain the WML variables. This enables a variety of functionality (e.g., validation of user input collected by a WML card).

The basic syntax features of WMLScript are similar to JavaScript: case-sensitivity; statements ending with a semicolon (;); literal character stringing enclosed within double or single quotes; and so forth. WMLScript is a weakly-typed language; that is, no type-checking is done at compile-time or run-time, and no variable types are explicitly declared. Internally, the following data types are supported: Boolean, Integer, Floating-point, String, and Invalid. The programmer does not need to specify the type of any variable. New variables are declared using the var keyword (i.e., var x;). WMLScript will automatically attempt to convert between the different types of variables as needed. The operators and expressions supported by WMLScript are virtually identical to those of the JavaScript programming language, including: assignment operations, arithmetic operations, logical operations, string operations, comparison operations, and array operations.

Furthermore, WMLScript supports three categories of functions, including:

- *Local Script Functions*—script functions defined inside the same script that also includes the calling expression;
- *External Script Functions*—script functions defined in another script not containing the calling expression;
- *Standard Library Functions*—functions defined in a library that is part of the WAE specification.

WMLScript defines several standard libraries, including a language library, a string library, a browser library, a floating point library, and a dialog library.

The high reliance of WMLScript on WML implies that WMLScript's capabilities can only be fully utilized in conjunction with WML. To use WMLScript with XHTML-MP is not as direct as with WML. Programs written in WMLScript cannot be called directly from an XHTML-MP. However, developers can call a WML from an XHTML-MP, and WML can then call WMLScript.

3.3.3 UAProf

As more WAP-enabled devices are designed, the assumption of device homogeneity is no longer valid. The UAProf service provides a mechanism for describing the capabilities of clients and the preferences of users to an application server. It supports the client-server transaction model by sending client and user information to servers with the request. This information permits the servers to adapt their content accordingly when preparing its response [20].

The UAProf captures classes of device information for content formatting by origin servers, WAP proxy gateways, and other interim servers. These classes include the hardware and software characteristics of the device, as well as information about the network to which the device is connected. WAP 2 supports the end-to-end flow of a UAProf, also referred to as capability and preference information (CPI). CPI includes information about the following:

- *HardwarePlatform*—screen size, color and image capabilities, manufacturer, and so forth;
- *SoftwarePlatform*—operating system vendor and version, list of audio and video encoders, and so forth;
- *BrowserUA*—browser manufacturer and version, mark-up language and versions supported, scripting languages supported, and so forth;
- *NetworkCharacteristics*—bearer characteristics, such as latency and reliability, and so forth.
- *WapCharacteristics*—WMLScript libraries, WAP version, WML document size, and so forth.

The CPI is transmitted and maintained using designated hypermedia transfer service (i.e., HTTP or WSP) headers. The hypermedia transfer service uses Profile and Profile-Diff headers to convey the CPI. A Profile header contains URL(s), where each URL references an externally accessible CPI document. The Profile-Diff header contains a CPI document. Multiple Profile and Profile-Diff headers may be cached by a WAP proxy and/or included with a request.

When transmitted using WSP, this information is initially conveyed when a WSP session is established with a compliant WAP proxy gateway. The WAP proxy gateway caches the CPI and applies it on all requests during the lifetime of the WSP session. The WAP proxy gateway is also responsible for translating HTTP responses into appropriate WSP responses (as described in Section 3.2.4) for delivery over the wireless network to the requesting client device. In forwarding these responses, the WAP proxy gateway must also forward any CPI usage headers provided by the origin server and/or any intermediate HTTP proxies.

As a request travels over the network from the client device to the origin server, each network element may optionally add additional profile information to the transmitted CPI. These additions may provide information available solely to that particular network element. Alternatively, this information may override the capabilities exposed by the client, particularly in cases where that network element is capable of performing in-band content transformations to meet the capability requirements of the requesting client device.

Origin servers, gateways, and proxies can use the CPI to ensure that the user receives content that is particularly tailored for the environment in which it will be presented. Moreover, this specification permits the origin server to select and deliver services that are appropriate to the capabilities of the requesting client. Finally, it is expected that this specification will be used to enhance content personalization based on user preferences.

The user agent profile is defined to comply with the W3C standards for CC/PP. The user agent profile uses a resource description framework (RDF) schema and vocabulary, as defined in the CC/PP model, to define a robust, extensible framework for describing and transmitting CPI [21]. See Section 4.1.4 for RDF.

3.4 WAP and Messaging Services

Most of WAP 1 implementations choose SMS as their bearer network. SMS evolved from a GSM service to a new business that contributes considerable traffic. SMS has been one of the most successful services in mobile telecommunications. As accompanying network bandwidth has increased, the delivery of multimedia content by MMS has also become feasible. This section introduces messaging services, and their relations to WAP. A scenario of sending MMS messages with the assistance of WAP and SMS is presented in Section 3.4.3.

3.4.1 SMS

Developed in 1991, SMS was thought of as an application of service, almost included in the first GSM cell phones. SMS is a mechanism to deliver short messages over the mobile networks. It is a store-and-forward way of transmitting messages to and from mobiles. The text-only message from the transmitting mobile is stored in a central short message center (SMC), which then forwards the message to the destination mobile. If the recipient is not available, the short message is stored and sent later. Each short message can be no longer than 160 characters. These characters can be alphanumeric text messages, or binary nontext short messages [22].

SMS supports a delivery confirmation (i.e., return receipt) mechanism, which means that the sender can get a small message notification if the short message was delivered to the intended recipient. Since SMS uses the signaling channel as opposed to dedicated channels, SMS messages can be transmitted simultaneously with voice, data, and fax services. It also supports roaming, which means that a user can send short messages to any other GSM mobile user around the world. SMS is a universal mobile data service.

As technology and networks evolved, a variety of value-added services based on SMS were introduced, including interactive banking, various information services, integration with Internet-based applications, e-mail, paging, and so forth. SMS not only provides a useful mechanism for a host of innovative services over mobile networks, but also acts as a point of entry for new data services (e.g., WAP over SMS) in mobile networks.

The success of SMS has certainly been possible mainly due to its substantial simplicity, availability, and low cost. However, limitations have usually been associated with its simplicity. Aside from the limitation of plain text content and 160-character message size, SMS has some other inherent disadvantages.

- SMS protocol data units, as defined in GSM 03.40, are not efficient.

- SMS uses the slow signaling channel, which is also used for many other purposes in GSM. Ultimately, SMS has a low data rate and long latency.

- The store-and-forward nature of SMS is useful in many applications, but is accompanied by large overheads for the case of WAP over SMS. Even for a simple request from a WAP client device, SMS center resources must be used to store and forward the request and the corresponding response during the information transfer.

Enhanced Messaging Service

Enhanced messaging service (EMS) is based on the existing SMS. EMS is a mechanism by which a user can send a comparatively richer message, combining text, simple melodies, pictures, and animations, to an EMS-compliant handset. It extends the user data header (UDH) in SMS, which makes it possible to include binary information in the message header, and needs no upgrade to the network infrastructure. However, the handsets need to be EMS-compliant. In EMS, there are 10 different predefined sounds. EMS is delivered via SMS, which means that it can be very costly. It also uses the limited control channel capacity in a GSM network.

WAP over GPRS

Although most WAP 1 implementations select SMS as the bearer network, GPRS is considered to be a better choice. Unlike SMS, GPRS provides a real-time data bearer. The users always have a virtual connection to the network. The GPRS is a packet-based bearer that was introduced on many GSM and TDMA mobile networks, beginning in 2000. GPRS is immediate (i.e., no dial-up connection); relatively fast (up to 177.2 Kbps); and it supports virtual connectivity, allowing relevant information to be sent from the network as it is generated. It makes efficient use of radio and network resources, and is key to the evolution toward 3G products. Thus, GPRS is also considered as 2.5G technology [4]. More details of GPRS and 3G technologies are given in Sections 7.2 and 7.3.

3.4.2 MMS

Whereas SMS is considered a bearer for WAP, MMS is itself a bearer-independent service for the mobile environment. MMS standards have been designed by several standardization bodies, including the Third Generation Partnership Project (3GPP) and the WAP Forum (now integrated into OMA) [23].

Multimedia messaging allows a variety of message elements to be sent to a user. These can contain text, sounds, images, and video. Users can compose their own messages, receive rich content messages from content providers, and forward them. However, multimedia messages can only be sent to the owners of compatible terminals.

The MMS standard does not specify the maximum size of a message. This is to ensure future interoperability, and to avoid the 160-character limitation of SMS. The size of any MMS message is therefore an implementation issue. Maximum size may be dependent on operator preferences; for instance, a standardized message size is more convenient for billing purposes. Most manufacturers, concerned with the amount of memory on devices, presently limit the message between 30 and 100 Kb in size.

MMS primarily targets phone-to-phone communication. Therefore, the possibility will always exist that the message cannot be delivered when the receiving phone is switched off, has an empty battery, or has poor network coverage. Thus, a network element, called the multimedia messaging service center (MMSC), is needed to store undelivered MMS messages until the receiving phones can be reached. In addition, the MMSC hosts a number of interfaces to connecting networks, and an application programming interface (API) to enable delivery of value-added services, such as network interconnection to e-mail. Nokia and several other manufacturers produce MMSCs.

3.4.3 MMS Works Together with WAP and SMS

As introduced earlier, SMS is not an adequate bearer for WAP because of the weight of the protocol. However, the fact that SMS is suitable for delivering short text messages has remained unchanged. It provides a good platform for delivering control information for MMS. For phones without the capability to receive an MMS, the receiving terminal is identified as a non-MMS phone, in which case, the message is stored on a Web page instead of being sent to the phone. An SMS is then sent to the non-MMS phone, with the address of the Web page where the message can be retrieved.

MMS also uses many services of WAP, particularly the lower level WAP transport mechanisms that are optimized for operation over the GSM radio interface. MMS also uses WSP's push mechanism to transparently notify users of the receipt of a new message. Besides the MMS phones for originator and receiver, other network elements include: the WAP/HTTP push proxy gateway, the MMSC, and the SMC. Figure 3.7 shows an example of an MMS message delivery in a four-step sequence, as described here.

1. The originator sends an MMS message with the WSP or HTTP Get method. The message is processed by the push proxy gateway, then delivered and stored in the MMSC.
2. The MMS center pushes the SMS message to the receiver, and gives a notification of the arrival of the MMS message. The message is pushed via the push proxy gateway, and sent out from the SMC.
3. The receiver retrieves the MMS message with the WSP or HTTP Get method.
4. The MMSC pushes an SMS acknowledge message to the sender. The message is pushed via the push proxy gateway, and sent out from the SMC.

3.5 Future of WAP—OMA

3.5.1 Converging to WAP 2

During the period between 2000 and 2002, there was a general feeling by most experts that WAP would not survive. Following the release of WAP 2.0, most experts think that WAP technology now will become a critical player in the industry.

To be fair, WAP 1 was probably a little ahead of its time, with regard to bearer availability and BW and the protocols used. When WAP 1.0 first started, XML was

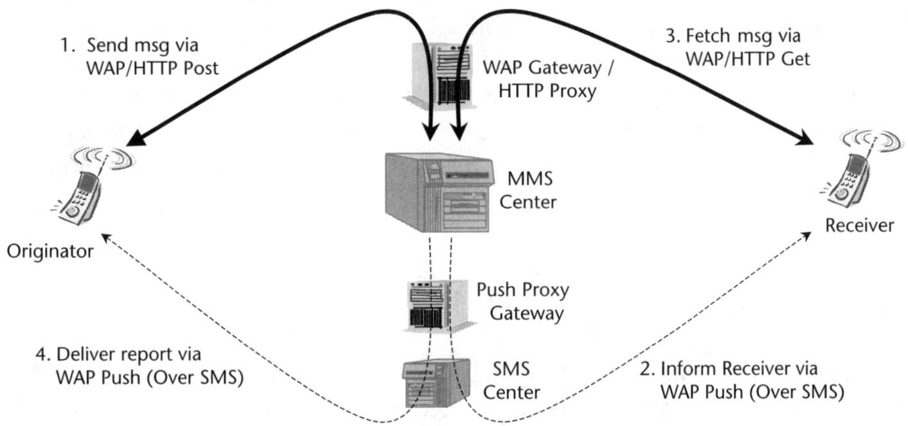

Figure 3.7 MMS message delivery and network element.

hardly known outside the W3C. The press reported WAP's launch as "WAP does not use HTML and HTTP." In fact, WAP defined WML with XML, and the new HTML (i.e., XHTML) is defined based on XML. On the protocol part, WSP headers could easily map one-to-one to HTTP headers by the gateway. Once the damage by the initial press reports was done, it was hard to repair, and WAP 1.0 never caught on.

Retaining backward compatibility is a major problem when changing any specification. Because WAP 1 is based on XML, it is possible to retain backward compatibility using XML namespaces. It is also possible to use the CC/PP mechanisms to minimize the issues with backward compatibility. Although WAP 1 was not a success in the market, it provided a solid foundation for WAP 2.

Shortly after WAP 1 was released, a strong competitor of WAP was announced in February 1999. The i-mode is a similar technology developed by the Japanese company NTT DoCoMo. Up to late 2002, most subscribers of voice-oriented service by PDC (Japan's cellular mobile system) also subscribed to the i-mode service. Starting in 2002, DoCoMo made investments and created partnerships with worldwide telecommunications operators. Despite its massive success in Japan, it is not clear if it will catch on in the United States and Europe. From a technical point of view, both i-mode and WAP 2 are converging. For example, both technologies agree to adopt the open XHTML Basic Profile as content format. The main difference is on the operational policy. More details on i-mode are introduced in Chapter 4.

Furthermore, the rapid growth of the 802.11 Wireless LAN also plays a major role in the future development of WAP. WAP 2 on GPRS is supposed to run at 384 Kbps, far better than the 9,600 bps of most WAP 1. However, WAP 2 is still much slower than the 11 to 54 Mbps offered by the 802.11 standard. Although the coverage of 802.11 is spotty and limited to certain areas of cities, more and more businesses, schools, and organizations are deciding to install base stations. The WiFi Alliance provides a Web site for WAP phone users to discover public hotspots. Businesses use 802.11 logos to attract customers. Thus, the development of wireless LAN-WAN devices has become a hot topic in the communications field. These dual-mode devices will initially seek the availability of wireless LAN, and if this is not available, then the devices will connect to wireless WAN. Thus, users may wirelessly connect to the Internet without device or software switching.

As more businesses and organizations become involved in the wireless industry, some technologies may be specified, though inconsistently, through separate specification organizations. In June 2002, the WAP Forum was merged into the OMA. The tasks previously handled within the WAP Forum continued within the new OMA. The formation and progress of the OMA marks an important step in the history of the mobile services industry. By integrating several specification organizations into one, information and resources are shared, work processes are streamlined, and overlapping efforts are reduced.

3.5.2 OMA

The WAP Forum and the Open Mobile Architecture initiative formed the foundation for the OMA in June 2002. At that time, there were nearly 200 members, including the mobile operators, device and network vendors, information technology companies, and content and service providers. The organizations integrated into the OMA include: the WAP Forum, the Location Interoperability Forum (LIF), the SyncML Initiative, the Multimedia Messaging Interoperability Process (MMS-IOP), the Wireless Village, the Mobile Gaming Interoperability Forum (MGIF), and the Mobile Wireless Internet Forum (MWIF). This consolidation promotes end-to-end interoperability across different devices, geographies, service providers, operators, and networks, and further supports the OMA's market and user requirements focus to guide the specification work [24].

The OMA technical plenary is responsible for technical specifications. At the time of this book's writing, there are 14 technical working groups and 2 committees of the technical plenary. Selected OMA working groups and their relations to merged organizations are shown in Table 3.2.

Table 3.2 Selected OMA Working Groups and Committees

Working Group and Committee	Description	Organizations for Previous Work
Architecture	Responsible for defining the overall OMA architecture, enabling specification work in work groups and assuring, through review, adherence of specification work to OMA architecture.	
Browser and content (BAC)	Responsible for the specification of application technologies used in the OMA architecture. Subgroups: Download + DRM, Browser and Content Environment (MAE), Push, Standard Transcoding Interface (STI), UAProf	WAP Forum W3C
Data Synchronization	Continues development of specifications for data synchronization, and the development of other similar specifications, including but not limited to SyncML technology.	SyncML Initiative
Device management	Defines management protocols and mechanisms that enable robust management of the life cycle of the device and its applications over a variety of bearers.	SyncML Initiative WAP Forum
Games services	Responsible for developing interoperability specifications, APIs, and protocols for network enabled gaming; as well as enabling game developers to develop and deploy mobile games to efficiently interoperate with OMA platforms and enable cost reduction for game developers, game platform owners, and service providers.	MGIF
Interoperability (IOP)	Acts as a center of excellence to identify, specify and maintain the required processes, policies, and test programs for ensuring interoperability for OMA-specified enablers and end-to-end services.	
Location	Continues the work originated in the LIF, developing specifications to ensure interoperability of mobile location services on an end-to-end basis.	LIF
Messaging	Responsible for the specification of messaging and related enabling technologies, with the goal of specifying a set of basic messaging features that may be used to enable specific messaging paradigms.	MMS-IOP Wireless Village
Presence and availability	Specifying the service enablers to permit the deployment of interoperable mobile presence and availability services. "Presence and availability services" enable applications to exchange dynamic information (e.g., status, location, and capabilities) about resources (e.g., users and devices).	Wireless Village

The OMA derives service enabler architectures and open enabler interfaces that are independent of the underlying wireless networks and platforms. It creates interoperable mobile data service enablers that work across devices, service providers, operators, networks, and geographies. Toward that end, the OMA will develop test specifications, encourage third party tool development, and conduct test activities that will allow vendors to test their implementations.

There are four key principles on which the OMA is founded. These principles encourage competition through innovation and differentiation, while ensuring the interoperability of new and existing mobile services across the entire value chain.

- Services are based on open standards, protocols, and interfaces, and are not locked to proprietary technologies.
- The applications layer is bearer-independent.
- The architecture framework and service enablers are independent of operating systems.
- Applications and platforms are interoperable, providing seamless geographic and intergenerational roaming.

Following these key principles, the OMA adopted a specification release process that begins with defining prioritized market requirements and use cases, and ends with OMA interoperability releases that enable end-to-end, interoperable mobile services. The OMA specification release program is completed in three phases:

- *Candidate Enabler Release:* Approved specification for a single OMA service enabler;
- *Approved Enabler Release:* To ensure that products and services supporting a single OMA service enabler work together seamlessly and end-to-end across different devices and networks;
- *OMA Interoperability Release:* To ensure that a thorough device and network interoperability testing is completed across multiple OMA service enablers.

One legal framework enables true interoperability testing across technologies and standards. This enables a much faster time to bring a product to market, and reduces overall cost and complexity in the industry.

3.6 Summary

"Accessing the Internet by phone" in the late twentieth century meant connecting to the Internet using a desktop through public telephone connections. Due to the efforts by the WAP Forum (and continued by OMA), Internet technology has extended to handheld mobile devices and wireless networks.

This chapter presented the WAP architecture and protocols. WAP protocols are largely based on Internet technologies. The role and operations of the WAP proxy gateway and push mechanism are discussed. WAP provides a general application environment as well as an optimized wireless network environment. The WAE includes: WML, WMLScript, UAProf, and so forth. Examples of WAP client software are left for Chapter 4.

SMS is one of the most successful services in the history of the telecommunications industry. SMS has also been the most commonly used bearer network of WAP 1. MMS is another messaging service that requires assistance from the WAP push mechanism. All of these technologies are destined to coexist for many years to come. There will be a great deal of change and new technologies emerging into the mobile Internet area in the future, but with WAP's principle of flexibility and interoperability, smooth evolution is expected.

References

[1]   Chidambaram, N., (ed.), *Professional WAP*, 1st ed., Hoboken, NJ: Wrox Press, 2001.
[2]   OMA WAP specifications, http://www.openmobilealliance.org/tech/affiliates/wap/wapindex.html.
[3]   "Latest Mobile, GSM, Global, Handset, Base Station, & Regional Cellular Statistics," http://www.mobileoffice.co.za/stats/stats-main.htm.
[4]   Beaulieu, M., *Wireless Internet: Applications and Architecture*, Reading, MA: Addison-Wesley, 2002.
[5]   The Mobile Data Association (MDA), http://www.mda-mobiledata.org/resource/press/index.asp.
[6]   Scn Education, B.V., (ed.), *Mobile Networking with WAP: The Ultimate Guide to the Efficient Use of Wireless Application Protocol*, Wiesbaden, Germany: Gwv Vieweg Verlag, 2004.
[7]   Wireless Application Protocol Architecture Specification, WAP-210-WAPArch-20010712-a, http://www.openmobilealliance.org/tech/affiliates/wap/wapindex.html.
[8]   Wireless Profiled TCP Specification, WAP-225-TCP-20010331-a.
[9]   Push OTA Protocol Specification, WAP-235-PushOTA-20010425-a.
[10] Wireless Telephony Application Specification, WAP-266-WTA-20010908-a.
[11] Push Architectural Overview, WAP-250-PushArchOverview-20010703-a.
[12] WAP Pictogram Specification, WAP-213-WAPInterPic-20010406-a.
[13] Push Proxy Gateway Service Specification, WAP-249-PPGService-20010713-a.
[14] Mei, H., Y. M. Wen, and W. J. Lin, "Turning a HTTP Proxy Server to WAP Gateway," *INET2000—10th Annual Internet Society Conference*, July 2000, http://www.isoc.org/inet2000/cdproceedings/3b/3b_1.htm.
[15] Wireless Application Environment Specification, WAP-236-WAESpec-20020207-a.
[16] Wireless Markup Language version 1.3 Specification, WAP-191-WML-20000219-a.
[17] Wireless Markup Language version 2 Specification, WAP-238-WML-20010911-a.
[18] XHTML Mobile Profile Specification, WAP-277-XHTMLMP-20011029-a.
[19] WMLScript Language Specification, WAP-193-WMLScript-20001025-a.
[20] Reynolds, F., et al., "Composite Capability/Preference Profiles (CC/PP): A User Side Framework for Content Negotiation," *W3C Note*, July 1999, http://www.w3.org/1999/07/NOTE-CCPP-19990727/.
[21] User Agent Profiling Specification, WAP-248-UAProf-20011020-a.
[22] Gupta, P., "Short Message Service: What, How and Where?" http://www.wirelessdevnet.com/channels/sms/features/sms.html.
[23] Multimedia Messaging Service Architecture Overview, WAP-205-MMSArchOverview-20010425-a.
[24] Open Mobile Alliance Technical Plenary Structure. http://www.openmobilealliance.org/about_OMA/orgchart.html.

CHAPTER 4

Technologies for Mobile Client Software

The client/server architecture has been the most popular model for software infrastructure since the 1980s. A client is defined as a requester of services, and a server is defined as a provider of services. In recent years, HTTP services (e.g., Web services) have grown rapidly. They now dominate Internet traffic [1]. When the Internet extends to the mobile wireless world, the details of the server are unchanged. Services provided on the wired Internet, such as browsing, searching, dynamic content generation, and messaging, are still available from a mobile client. Mobile Web services are developed and stored on the server, then passed to the client. Other applications (e.g., mobile browser and tiny games) are installed and executed on client devices [2, 3]. To deal with the narrow bandwidth and limited capability of client devices, the content presentation and development techniques are tailored particularly for mobile wireless software.

The most visible piece of software on a mobile client is the mobile microbrowser. There is a wide range of features and technologies that may be integrated into a mobile browser. These include markup language support, Java support, push, message service support, multimedia support, and security support. Among these technologies, push and message services are described in Chapter 3. Wireless multimedia support is described in Chapter 5, and security issues are covered in Chapters 11 and 12.

This chapter is aimed at mobile client software development, and is organized as follows. XML is used extensively in data representation and exchanging. In Section 4.1, we introduce the syntax rules of XML, and the technologies that support XML, such as DTD/Schema for validating and CSS/XSL for styling. Section 4.2 focuses on mobile Web content generation. XHTML Basic, XHTML-MP, and the issues of mobile page generation are described. The Java ME platform and MIDP program development are illustrated in Section 4.3. The i-mode infrastructure and server side software technologies are introduced in Section 4.4. Section 4.5 summarizes the key points and discusses the future for integration issues.

4.1 XML-Based Technologies

Extensible markup language (XML) is a simple, flexible text format derived from standard generalized markup language (SGML). XML was designed to meet the challenges of large-scale electronic publishing. An XML document is considered *self-describing*, since XML documents are clearly structured. Tags surround every piece of data in an XML document. In recent years, XML has played a key role in the exchange of a wide variety of data on the wired and wireless Internet [4, 5].

Many computer companies are working on integrating their products using XML. Microsoft's word processors, spreadsheet applications, and databases are likely to read each other's data in XML format. It is expected that most, if not all, future applications will exchange their data in XML. Data can be exchanged between incompatible systems with XML. In the real world, computer systems and databases contain data in incompatible formats. One of the most time-consuming challenges for developers has been to exchange data between such systems over the Internet. Converting the data to XML can greatly reduce this complexity and create data that can be read by many different types of applications.

XML enables document authors to describe data of any type. It can also create new standard markup languages (i.e., XML applications). Many of the XML-based markup languages are used for wireless mobile Internet development (e.g., SMIL, SyncML, Schema, XSLT, WML, and XHTML-MP).

Figure 4.1 shows the XML family of technologies. XML-based markup languages are also called XML applications. To better utilize these applications, technologies for validation (e.g., DTD, Schema); processing (e.g., XPath, XSLT, SAX/DOM); and styling (e.g., CSS, XSLT, XSL-FO) are developed. The XML syntax is introduced first, and the associated technologies are illustrated in the rest of the chapter.

4.1.1 XML Syntax

An XML document is just plain text with the addition of some tags enclosed in angle brackets. An XML *element* is everything from (and including) the element's start tag, to (and including) its end tag. XML elements are related as parents and children. Just like HTML, an element can have element content, mixed content, simple content, or empty content. XML elements can have *attributes* in the start tag. Attributes are used to provide additional information about the elements [4, 6].

A *well-formed* XML document is a document that conforms to the XML syntax rules. Since XML is used as a metalanguage, there are syntax rules to observe. Conforming to XML syntax rules is necessary for well-formed XML documents. The syntax rules of XML include:

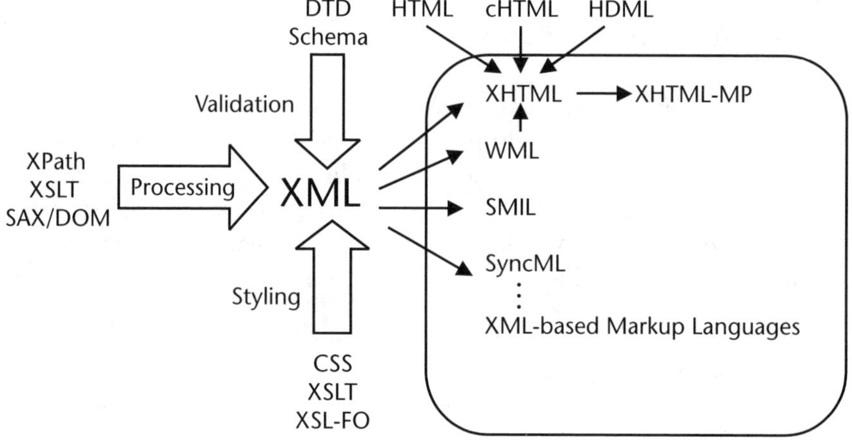

Figure 4.1 XML-related technologies.

- All XML documents must begin with the XML declaration.
- All XML elements must have a closing tag.
- XML tags are case-sensitive.
- All XML elements must be properly nested.
- All XML documents must have a unique root element.
- Names of elements can contain letters, numbers, and other characters, but not spaces. Names must not start with a number, punctuation character, or the reserved letters xml (or XML, Xml, . . .).
- Attribute values must always be quoted.
- With XML, white space is preserved.
- With XML, carriage return/linefeed is converted to linefeed.
- XML entities must be used for special characters.
- Comments in XML are written as <!— This is a comment —>.

Data can be stored in either elements or attributes. Although there are no specific rules as to the location of stored data, some developers have suggested avoiding the use of attributes for data. In general, data is preferably stored in child elements. In comparison with using elements, storing data in attributes may encounter the following difficulties:

- Attributes cannot contain multiple values.
- Attributes are not easily expandable.
- Attributes cannot describe structures.
- Attributes are more difficult to manipulate and validate.

If attributes are used as containers for data, documents are difficult to read and maintain. Thus, attributes are used to provide information that is not relevant to the data itself. In general, metadata (i.e., data about data) is stored as an attribute, and data is stored as an element.

One exception in which an attribute is used for data is the *ID reference* attribute type. ID references, a tokenized attribute type, can be used to access XML elements in much the same way as the NAME or ID attributes in HTML. The ID is just a unique identifier to identify the different notes in the XML file, and not a part of the note data.

Software that can handle plain text can also handle XML. XML elements are visible in a simple text editor. To process the XML document, developers may either write programs themselves, or use the XSL family of techniques (see Section 4.1.3).

4.1.2   XML Validation: Document Type Definition/Schema

An XML document with correct syntax is called a well-formed XML document. Even if documents are well formed, they may still contain errors, and those errors can have serious consequences. When data is sent from a sender to a receiver, it is essential that both parties have the same expectations about the content. These errors are analogous to the logical errors in the software. To deal with these errors, validating techniques are required. XML documents can have a reference to a document type definition (DTD) or a schema [7]. With a DTD and schema, the sender can describe the data in a way that the receiver will understand. DTD and schema technologies are used to build the XML document templates, and to validate such documents.

XML DTD

The purpose of a DTD is to define the legal building blocks of an XML document. It defines the document structure. A DTD can be declared within an XML document, or as an external reference. Each XML file can carry a description of its own format with it. Independent groups of people can agree to use a common DTD for interchanging data. An application can use a standard DTD to verify whether the data received is valid or not [4, 5].

DTD was inherited from HTML. The syntax rules of DTD are defined using extended Backus-Naur form (EBNF) grammar. Since DTD is not XML-based, many developers in the XML community feel that DTD is not flexible enough to meet the needs for processing.

From a DTD's point of view, all XML documents are made up from the following simple building blocks: Elements, Tags, Attributes, Entities, PCDATA, and CDATA. A DTD defines the order and relationship of these building blocks. Next, a simple XML document, named "note.xml," contains elements with PCDATA.

```
<!-- An XML Document: note.xml -->

<?xml version="1.0"?>
<note>
  <from>Emily</from>
  <to>Wesley</to>
  <heading>Reminder</heading>
  <body>Hello XML!</body>
</note>
```

The DTD file here, called "note.dtd," defines the elements of the XML document above:

```
<!ELEMENT note (from, to, heading, body)>
<!ELEMENT from (#PCDATA)>
<!ELEMENT to (#PCDATA)>
<!ELEMENT heading (#PCDATA)>
<!ELEMENT body (#PCDATA)>
```

A valid XML document is a well-formed XML document, which also conforms to the rules of a DTD/schema.

XML Schema

The XML schema is an XML-based alternative to DTD. Similar to a DTD, an XML schema describes the structure of an XML document. The XML schema language is also referred to as XML schema definition (XSD). Besides defining the elements, attributes, and their relationship in an XML document, an XML schema:

- Defines data types for elements and attributes;
- Defines default and fixed values for elements and attributes.

XML schema is not only XML-based, it also has the following advantages:

- XML schema is extensible.
- XML schema is richer than DTD.
- XML schema supports data types, including: string, numeric, date.
- XML schema supports namespaces.

It is believed that XML schemas will be used in most Web applications as replacements for DTDs in the future.

One of the greatest strengths of the XML schema is its support for data types, making it is easier to describe and validate data. For example, the date 2005-06-07 might be interpreted in some countries as June 7th, or in some other countries as July 6th, but an XML element with a date type ensures a mutual understanding of the content, because the XML data type date requires the format CCYY-MM-DD.

The following is a simple XML schema file, named "note.xsd," which defines the elements of the XML document above ("note.xml"):

```
<!-- The XML Schema for note.xml: note.xsd -->

<?xml version="1.0"?>
<xsd:schema xmlns:xs=http://www.w3.org/2001/XMLSchema>
 <xsd:element name="note">
  <xsd:complexType>
   <xsd:sequence>
     <xsd:element name="from" type="xs:string"/>
     <xsd:element name="to" type="xs:string"/>
     <xsd:element name="heading" type="xs:string"/>
     <xsd:element name="body" type="xs:string"/>
   </xsd:sequence>
  </xsd:complexType>
 </xsd:element>
</xsd:schema>
```

The note is a complex type element because it contains other elements. The other elements (to, from, heading, body) are *simple types* element because they do not contain other elements.

The XML schema was originally proposed by Microsoft, but became an official World Wide Web Consortium (W3C) recommendation in May 2001. Other XML schemas are also defined. Among those, RELAX NG is an XML schema based on RELAX and TREX. A RELAX NG schema specifies a pattern for the structure and content of an XML document. A RELAX NG schema thus identifies a class of XML documents consisting of those documents that match the pattern [8]. Yet another schema language, called Schematron, differs from most other schema languages, since it is a rule-based language that uses path expressions instead of grammar. A Schematron schema makes assertions applied to a specific context within the document. If the assertion fails, a diagnostic message that is supplied by the author of the schema can be displayed. One advantage of a rule-based approach is that, in many cases, modifying the desired constraint, written in plain English, can easily create the Schematron rule [9]. Both RELAX and Schematron are being standardized to become part of ISO document schema definition languages (DSDLs) [10].

4.1.3   XML Styling: Cascading Style Sheets and Extensible Stylesheet Language

A validated XML document is still a document in text format. To process or present the XML document, styling and processing techniques are required. A cascading style sheets (CSS) and extensible stylesheet language (XSL) are used widely to show an XML document in a desired style. SAX and DOM APIs are standards for processing XML document with programs.

CSS

A CSS is a style sheet language that allows users to attach styles (e.g., font, color, spacing) to structured documents (e.g., HTML documents, XML applications). Style sheets describe how documents are to be presented. By separating the presentation style of documents from the content of documents, CSS simplifies Web authoring and site maintenance [4, 6].

CSS is a breakthrough in Web design because it allows developers to control the style and layout of multiple Web pages. Web developers can define a style for each element and apply it to many Web pages. To make a global change, once the style sheet is changed, all elements in the Web pages are updated automatically. Style sheets allow style information to be specified in many ways. Styles can be specified inside a single HTML element, inside the [head] element of an HTML page, or in an external CSS file. Even multiple external style sheets can be referenced inside a single HTML document. In general, the styles will "cascade" into a new "virtual" style sheet through the following precedence rules, where the first has the highest priority:

1. Inline style inside an HTML tag;
2. Internal style sheet (<style> element inside the <head> tag);
3. External style sheet;
4. Browser default.

An inline style will override every style declared inside the <head> tag, in an external style sheet, and in a browser (a default value). In general, external CSSs can be shared across multiple documents. However, an external CSS document causes a document rebuild and a relayout. A page with inline style is rendered faster than a page with the CSS in other forms.

For mobile pages, WAP/Wireless CSS (WCSS) is tailored for WML/XHTML. The W3C CSS Mobile Profile (CSS-MP) specifies a conformance profile for mobile devices, identifying a minimum set of properties, values, selectors, and cascading rules [11]. The CSS-MP is very similar to the CSS. It supports media-specific style sheets, so that authors may tailor the presentation of their documents to various handheld devices.

XSL

The W3C started to develop XSL because there was a need for an XML-based stylesheet language [12]. XSL is considered to be the preferred stylesheet technique for XML, and is far more sophisticated than CSS. It consists of three parts:

- XSL transformations (XSLT) is a language for transforming an XML document from one style to another.
- XPath is a language for locating parts of an XML document.
- XSL-FO is a language for formatting XML documents.

XSLT is the most important part of the XSL standards. It is used to transform an XML document into another XML document (e.g., XHTML). It can also remove or add elements into the output file. It can rearrange and sort elements, test and make decisions about which elements to display, and a much more. A common way to describe the transformation process is to say that XSLT transforms an XML *source tree* into an XML *result tree*.

XSLT uses XPath to define matching patterns for transformations. XPath is a string-based language of expressions, but not written in XML. It is a set of syntax rules for locating specific parts of an XML document. It also defines node types, and uses location path expressions to specify how to navigate an XPath tree. XPath string library functions can be used to specify the matching patterns in an XML document. Without XPath, XSLT documents cannot be created.

An XSLT document is an XML document with a root element *stylesheet*. The namespace for an XSLT document is http://www.w3.org/1999/XSL/Transform. To match an output with an input, templates are defined in an XSLT document. The <xsl:template> element contains rules to apply when a specified node is matched. The *match* attribute is used to associate the template with an XML element. The match attribute can also be used to define a template for a whole branch of the XML document (e.g., match="/" defines the whole document). When a corresponding input is found, the contents of the template are executed.

A simple transformation example is shown here. The original XML document, "note.xml," is the input source to the XSLT document, "note.xsl." The output XML document is "note-t,xml."

```
<!-- Input document : note.xml -->

<?xml version="1.0"?>
<?xml:stylesheet type = "text/xsl" href = "note.xsl"?>
<note>
   <from>Emily</from>
   <to>Wesley</to>
   <heading>Reminder</heading>
   <body> Hello! DTD/Schema.</body>
</note>

<!--The XSLT file : trans.xsl -->

<?xml version = "1.0"?>
<xsl:stylesheet version = "1.0"
  xmlns:xsl = "http://www.w3.org/1999/XSL/Transform">
```

```
<xsl:template match = "/">
  <xsl:apply-templates/>
</xsl:template>

<xsl:template match = "note">
  <xsl:element name = "{@to}"">

    <xsl:attribute name = "from">
       <xsl:value-of select = "from"/>
    </xsl:attribute>

    <title>
       <xsl:value-of select = "heading"/>
    </title>

    <content>
       <xsl:value-of select = "body"/>
    </content>

  </xsl:element>
 </xsl:template>
</xsl:stylesheet>

<!-- Result : note-t.xml -->
<?xml version="1.0"?>

<Wesley from = "Emily">

  <title>Reminder</title>
  <content>Hello! DTD/Schema.</content>

</Wesley>
```

Using XSLT to transform the XML content to fit various mobile devices seems an ideal solution for the mobile Web. Nonetheless, there are unavoidable hidden overheads. Each time a device makes a request, the XSLT processor (built within the browser) must transform the content once. Although a caching mechanism can reduce the number of conversions, problems for new or heavily loaded applications are still significant. Another potential problem is the number of XSLT documents required. If a developer has to manually create XSLTs for each page on each device, huge efforts are required for development and testing. Automated processes and tools are expected in the future.

Document Object Model/Simple API for XML APIs

Programs written as standard APIs can also process XML documents. W3C provides a standard API interface called document object model (DOM) for building and processing the XML tree in memory. Each *node* represents an element, attribute, or another characteristic. An alternative to DOM is called the simple API for XML (SAX) [4, 5]. SAX-based parsers invoke methods when a markup (e.g., a start tag, end tag, and so forth) is encountered. With this event-based model, the SAX-based parser creates no tree structure. Thus, parsing an XML document with SAX results in better performance and less memory overhead than if the document were parsed with DOM. In fact, many DOM parsers use a SAX parser to retrieve data from a document for building the DOM tree. However, programmers may feel that it is easier to traverse and manipulate XML documents using the DOM tree structure. As a result, SAX parsers are typically used for reading XML documents that will not be modified.

4.1.4   XML-Based Markup Languages

As a metalanguage, XML-based standard markup languages cover a wide variety of areas. For example:

- MathML is developed to describe mathematical notations and expressions in an XML-like manner.
- The chemical markup language (CML) is an XML-based language used to represent molecular and chemical information.
- The geography markup language (GML) is developed by the OpenGIS Consortium (OGC) to describe geographic information.
- The scalable vector graphics (SVG) markup language describes two-dimensional vector graphics. The Macromedia Flash Lite (for mobile device)s supports a simplified version of SVG, called SVG Tiny (SVG-T).
- The extensible 3D (X3D) language is the result of the combined efforts of the W3C and the Web 3D Consortium to extend VRML with XML.
- Electronic business using extensible markup language (ebXML) provides a standard infrastructure for global electronic business that enables medium-to-large businesses to exchange information.

These markup languages may be used as standard formats in many applications [5, 6].

In the area of wireless mobile software, SyncML, synchronized multimedia integration language (SMIL), and composite capabilities/preferences profile (CC/PP) are three of the XML-based applications. SyncML's purpose is to allow the synchronization of data; for instance, to synchronize an event list between a mobile device and a desktop computer [13]. Many of the personal information management (PIM) (e.g., e-mail, calendar, contact list, memo pad) software vendors have incorporated SyncML support into their products.

The SMIL enables simple authoring of interactive audiovisual presentations. SMIL is typically used for "rich media" or multimedia presentations that integrate streaming audio and video with images, text, or any other media type. The syntax and processing of SMIL are presented in Section 5.1.3. Here, we introduce the user agent profile (UAProf), the CC/PP, and resource description framework (RDF).

UAProf, CC/PP, RDF

The UAProf, as introduced in Section 3.3.3, is a WAP service that provides a mechanism for describing the capabilities of clients and the preferences of users to an application server. By knowing the information display capabilities of a terminal, the server can create a display that is optimized for that terminal. Including profile information with the request minimizes the number of transactions needed to optimize such information. Furthermore, this information can be cached in a proxy or retrieved from a repository that the device manufacturer maintains. This minimizes the amount of information transmitted over the air, and speeds up information access. Designers can create pages or page templates to be used with application servers, and display them in formats that are adapted to user devices.

The UAProf is defined to comply with the W3C standards for the CC/PP. The CC/PP is defined in an XML application called the RDF, which enables users to connect a property to an object. The CC/PP is an application of the RDF, and are, in effect, written in XML.

The RDF provides the means for exchanging modular and interoperable metadata across different communities. Instead of specifying for each community, RDF provides the ability for those communities to define metadata elements to fulfill their own needs, and do so within a certain framework. For encoding and exchange of metadata, RDF uses XML as a common syntax [14].

4.2   Mobile Web Client

When Web applications for desktop computers were being developed, HTML had been the only markup language for years. HTML was designed to display data. Some markup languages were specifically developed for wireless mobile devices, such as the handheld device markup language (HDML), the wireless markup language (WML), and the compact HTML (cHTML). HDML was developed by Unwired Planet (now Openwave) in 1997. WML is a successor to HDML, and also the first XML-based markup language for mobile devices. WML is also deprecated in WAP 2. cHTML is a subset of HTML for the i-mode service in Japan. Creating unified solutions has always been a challenge in the mobile industry. WAP 2 (see Chapter 3) defined the infrastructure and application execution environment of the mobile Web. In this section, techniques and issues of content presentation on the mobile client are introduced.

Many people considered XML to be the successor of HTML, but this is not correct. XHTML is the XML version of HTML. As we learned from Section 4.1, XML was designed to describe data, and HTML was designed to display data. Separation of data and presentation is important for browsers on mobile devices with limited resources. For both the wired and the wireless mobile Web, the trend of markup languages is moving toward XHTML.

While HDML, cHTML, and WML were designed to work with *some* mobile devices, WAP 2 defined a strict subset of XHTML called the XHTML Mobile Profile. The profile is in turn a superset of the W3C XHTML Basic Profile. These solutions are made to be trusted with *all* displays and hosts.

4.2.1   XHTML Basic

Although XHTML's elements are almost identical to the elements of HTML, some notable differences exist. All XHTML tags must be in lower case. All XHTML start tags must have a corresponding end tag. Empty tags (i.e., tags that do not have closing tags) must be terminated using the forward slash character (/). Therefore, the

 tag in HTML becomes the <br/> tag in XHTML. Since XHTML is required to be well-formed, the job of browsers becomes easier.

XHTML Basic is the mobile version of XHTML from W3C. Not all the elements of XHTML are required for limited display and memory [15]. XHTML Basic makes use of XHTML Modularization, and builds a document type that consists of a minimal set of modules from XHTML [16]. Combining these with images, forms, basic tables, and object support, XHTML Basic provides an XHTML document type that is implementation-efficient for small devices. The resulting document type is much more versatile compared to the WML of WAP 1.

XHTML Modularization is a decomposition of XHTML into a collection of abstract modules that provide specific types of functionality. These abstract modules are implemented using the XML DTD/Schemas. These modules may then be combined with each other and with other modules to create an XHTML subset and extension document types that qualify as members of the XHTML family of document types.

XHTML Basic offers almost all the standard features available with mobile device-specific subsets of HTML, such as cHTML and WML. These features include support for basic text formatting, forms, hyperlinks, basic tables, images, and metainformation. Essentially, XHTML Basic is XHTML minus some capabilities, such as frames, which are not very relevant for mobile devices with limited display capabilities. Other notable XHTML features that are not included in XHTML Basic are the style element (although an external style sheet is still linkable), script elements (optional), and events elements. Only basic tables and forms are supported in XHTML Basic. Core modules of XHTML Basic are shown in Table 4.1.

Tables are widely used on desktop Web pages. XHTML Basic supports only the basic table. Most mobile browsers have limited or nonexistent table support. The obvious solution to this problem is to stop using tables, or just replace a table with an ordered list. If a table is required, table support can be detected by using the *getDeviceCapability ('table_support')* method.

Table 4.1 Core Modules of XHTML Basic

Module	*Element*
Structure	body, head, html, title
Text	abbr, acronym, address, blockquote, br, cite, code, dfn, div, em, h1, h2, h3, h4, h5, h6, kbd, p, pre, q, samp, span, strong, var
Hypertext	a
List	dl, dt, dd, ol, ul, li
Basic forms	form, input, label, select, option, textarea
Basic tables	caption, table, td, th, tr
Image	img
Object	object, param
Metainformation	meta
Link	link
Base	base

4.2.2 The XHTML Mobile Profile

XHTML Mobile Profile (XHTML-MP) is defined as the authoring language for
WAP 2. It is designed for resource-constrained Web clients that do not support full
XHTML features. It builds on top of XHTML Basic, and adds a few more elements
and attributes from XHTML. These include some additional presentation elements,
along with support for internal style sheets. Like XHTML Basic, XHTML-MP is a
strict subset of XHTML [17].

Figure 4.2 shows the output of a simple XHTMP-MP with CSS example. In
"MP.xhtml" here, the page is declared as an XML document. Then, DOCTYPE is
set to XHTML Mobile 1.0 and associated to the DTD located on WAP Forum's
server. It is also possible to use proprietary extensions to implement XHTML-MP,
and the corresponding DTD should be specified. The rest of the XHTML-MP document consists of a simple well-formed XHTML page.

```
<!-- MP.xhtml -->

<?xml version="1.0" encoding="utf-8"?>
<!DOCTYPE html PUBLIC "-//WAPFORUM//DTD XHTML Mobile 1.0//EN"
    "http://www.wapforum.org/DTD/xhtml-mobile10.dtd" >

<html xmlns="http://www.w3.org/1999/xhtml">
 <head>
   <title>XHTML Mobile Profile Document</title>
   <link rel="stylesheet" href="style.css" type="text/css"/>
   <!--
   <style> document-wide styles can be put here </style>
   -- >
 </head>
 <body>
   <h1 class="b"> Welcome </h1>
   <p class="big">
    Hello! XHTML-MP.
   </p>
   <p class="u">
           XHTML- Mobile profile.
   </p>
```

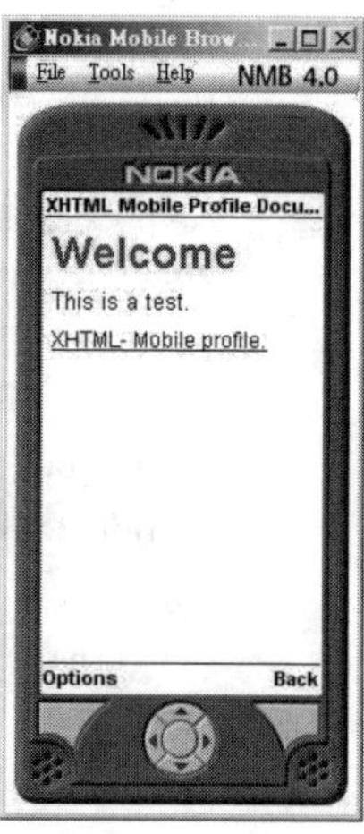

Figure 4.2 Output of the XHTML/CSS example.

```
  </body>
  </html>
```

CSS provides the presentation elements to the XHTML-MP documents. XHTML-MP supports the mobile version of CSS (i.e., WCSS or CSS-MP). The style sheet is embedded into the page using the LINK tag:

```
<link rel="stylesheet" href="style.css" type="text/css"/>
```

There are alternative ways of embedding style sheets into XHTML-MP pages. Although inline style is not defined in XHTML-MP, most vendors support inline style. The following is an example of wireless CSS code that can be applied to the XHTML-MP page. Here, styles for HTML-MP elements are defined.

```
/* styles.css */

td,th {border-width:1px; border-color:black; border-style:solid;
text-align:center}
h1 {color:red}
.b {font-weight:bold}
.u {text-decoration:underline}
.big {font-size:larger}
.i {font-style:italic}
.small {font-size:smaller}
```

Due to device limitations, mobile clients usually cannot cache files like regular Web browsers. Each time a user accesses a page, the CSS file will be downloaded into the device once.

XHTML-MP stands for a major step toward an interoperable mobile world. However, most of the XHTML-MP–supported browsers include their vendor's specific features. One intuitive solution to achieve a crossplatform is by using the core of the XTML-MP, which is XHTML Basic. Theoretically, XHTML Basic pages are supported on all mobile browsers. An alternative solution is to use a wireless universal resource file (WURFL) and server side program to detect mobile devices. The WURFL project is derived from the UAProf of W3C. WURFL uses an ambitious XML configuration file, tracking the capabilities of other mobile devices. It will match the user agent within the XML file, check the resources available to the targeted device, and return the result.

Transform for Compatibility

Not all new WAP 2 devices support WML. To support backward compatibility, a transformation model from WML to XHTML-MP is defined in WAE. Some features in WML for resource-limited devices are not included in XHTML-MP. Among those features, the *card* (as the page in XHTML) and *event* mechanisms are not transformable to XHTML-MP. As to the *markup language variables*, *timer*, and *softkey* in WML, there are matched operations in XHTML-MP to reach the same effects.

- *Markup language variables* locally store user data across the application. Embedded variables are useful for minimizing the need of the user to reenter data that has been entered previously. In the XHTML-MP, server side applications can prefill forms with the data that users have entered earlier.

- *Timers* are useful in activating the time-controlled operation. To simulate timers in the XHTML-MP, use META tags with http-equiv="refresh" to instruct the browser to request a different document after a specified time. For example: *<meta http-equiv="refresh" content="30; url=(http://www.weco.net/tim. html"/>*

- *Softkeys* allow users to trigger obvious operations with just one click. The only thing similar to softkeys in the XHTML-MP is the accesskey attribute, which lets a numeric keyboard create shortcuts to hyperlinks and form elements. Some vendors (e.g., Openwave) still support softkeys for their browsers using CSS.

Both WML and XHTML-MP are XML-based. Notwithstanding the direction of transformation, XSLT is the preferred technique to define the matching process.

4.3   Java Mobile Client

As described in Section 2.3, Java is a critical technology for recent Internet and Web software development. The Java Platform, Micro Edition (Java ME, also called J2ME) is a set of technologies and specifications targeted toward consumer and embedded devices. The architecture of Java platforms is shown in Figure 2.3. The Java ME connected limited device configuration (CLDC) connection framework and the mobile information device profile (MIDP) program are presented in this section. The Java ME technology for i-mode, DoJa, is described in Section 4.4.

4.3.1   CLDC Connection Framework

The mobile clients are referred to as small mobile devices in this book. The CLDC of Java ME includes small devices with network connections, such as mobile phones and PDAs [18, 19]. The connection scheme of the CLDC is referred to as the generic connection framework (GCF).

The GCF of CLDC abstracts the concepts of files, sockets, HTTP requests, and other input/output mechanisms, into a set of classes that is simpler than the classes defined by Java SE. To a programmer, the GCF provides the same functionality as the classes from the java.io and java.net packages, but without requiring specific capabilities from a device. With the GCF, all communication is abstracted through a set of well-defined interfaces. Instead of creating a specific class of communication objects, such as java.io.File or java.net.Socket, the application asks the GCF to create a connection that uses a specific protocol. The protocol is passed in as part of a universal resource identifier (URI). The GCF then determines whether or not the implementation supports that protocol, and returns an appropriate interface if it does. The application then uses this interface to interact with the implementation in sending or receiving data.

The GCF classes are defined as part of the javax.microedition.io package. An application uses one of the static Connector.open methods to obtain an object that implements the Connection interface or one of its subinterfaces. The application then uses the methods defined by the interface to read and/or write data. Most of the interfaces work on stream-based data, and therefore expose input or output streams. There are six subinterfaces of Connection defined. The *InputConnection* and *OutputConnection* interfaces are for one-way stream connections. The *StreamConnection* interface is for two-way stream connections; it extends both *InputConnection* and *OutputConnection*. The *ContentConnection* interface extends *StreamConnection* with methods for determining information about the content itself, such as its type and length. The *DatagramConnection* interface is for sending and receiving packet data. Finally, the *StreamConnectionNotifier* interface is for implementing server-side connections, in which the application must wait for a client to connect. The interface hierarchy of the GCF is shown in Figure 4.3.

4.3.2   MIDP Development

The most complete profile of Java ME so far is the MIDP, which is based on the CLDC. The MIDP is designed for mobile phones and other small devices with limited resources. It is built on the CLDC, and includes APIs for creating applications (with the javax.microedition.midlet package), building user interfaces (with the javax.microedition.lcdui package), and working with persistent storage (with the javax.microedition.rms package) [20–22]. Network APIs are included with the CLDC.

All applications for the MID Profile must be derived from a special class, MIDlet. The MIDlet class manages the life cycle of the application. It is located in the javax.microedition.midlet package. A MIDlet can exist in one of the four states: loaded, active, paused, or destroyed. Figure 4.4 gives an overview of the MIDlet life cycle. When a MIDlet is loaded into the device and the constructor is called, it is in the loaded state. This can happen at any time before the program manager starts the application by calling the *startApp()* method. After the *startApp()* is called, the MIDlet is in the active state, until the program manager calls *pauseApp()* or *destroyApp()*. All state change callback methods should terminate quickly, because the state is not changed completely before the method returns.

In the *pauseApp()* method, applications should stop animations and release resources that are not needed while the application is paused. This action avoids resource conflicts with the application running in the foreground. It also avoids

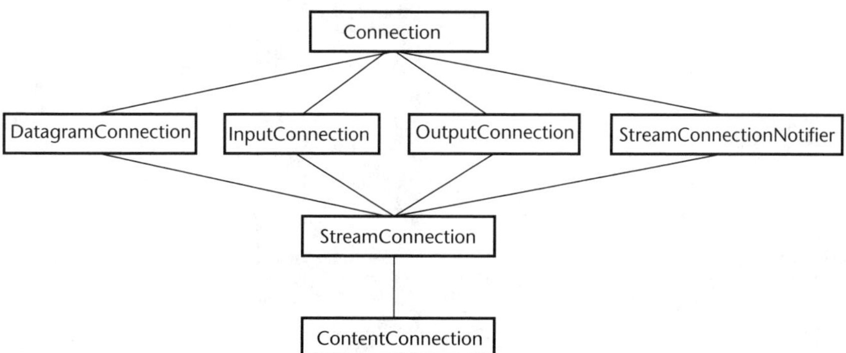

Figure 4.3   The CLDC GCF connection interface hierarchy.

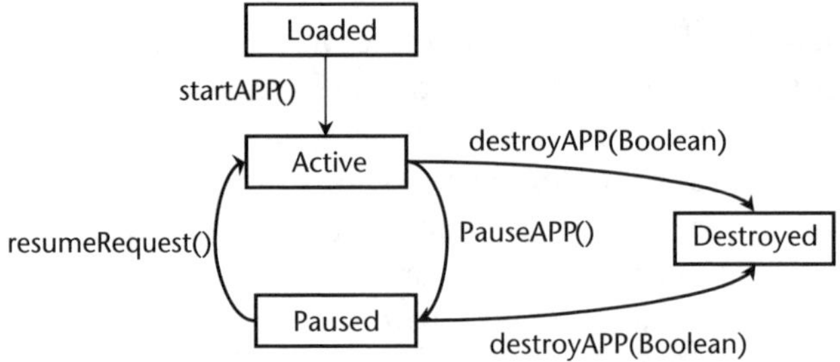

Figure 4.4   Life cycle of a CLDC MIDlet.

unnecessary battery consumption. The *destroyApp()* method provides an unconditional destroy parameter. The MIDlet can resume activity by calling *resumeRequest()*. If the MIDlet decides to go to the paused state, it should notify the application manager by calling *notifyPaused()*. In order to terminate, a MIDlet can call *notifyDestroyed()*.

The MIDlets are stored in a MIDlet suite (*jar* file). The MIDP device runs application management software (AMS), and AMS runs the MIDlet from the MIDlet suite. The AMS keeps MIDlet information in an application descriptor file (*jad* extension).

The following is a simple MIDlet example, whose output is shown in Figure 4.5.

```
import javax.microedition.midlet.*;
import javax.microedition.lcdui.*;

public class TestMIDlet extends MIDlet {

  private Display display;
  private TextBox textbox;

  public TestMIDlet() {
    textbox = new TextBox("TEXT EDITOR", "Welcome to J2me ", 50,
    TextField.ANY);
    display = Display.getdisplay(this);
  }
```

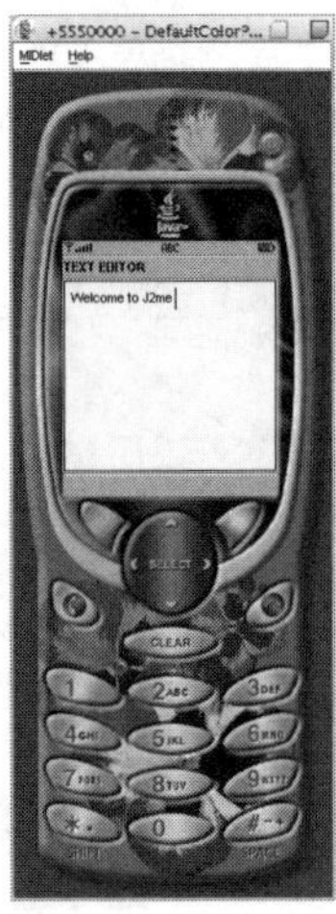

Figure 4.5   Output of a simple CLDC MIDlet.

```
public void startApp() {
  display.setCurrent(textbox);
}

protected void destroyApp(Boolean unconditional) {
}

protected void pauseApp() (
}

}
```

The MIDP user interface API is divided into two parts: high-level and low-level API. Figure 4.6 shows an overview of the MIDP graphic user interface (GUI) classes and their inheritance hierarchy. The high-level *Screens* and the low-level class *Canvas* have the common base class *Displayable*. All subclasses of *Displayable* fill the whole screen of the device. Subclasses of *Displayable* can be shown on the device using the *setCurrent()* method of the *Display* object. The display hardware of a MIDlet can then be accessed by calling the static method *getDisplay()*, where the MIDlet itself is given as a parameter. In the TestMIDlet example, this step is performed in the lines:

```
Display display = Display.getDisplay (this);

...

display.setCurrent (testbox);
```

The high-level API provides input elements, such as text fields, choices, and gauges. In contrast to the Java SE Abstract Windows Toolkit (AWT), the high-level components cannot be freely positioned or nested. There are only two fixed levels: *Screens* and *Items*. The *Items* can be placed in a *Form*, which is a specialized *Screen*.

The low-level API allows full control of the MID display at the pixel level. For this purpose, the lcdui package contains a special kind of screen called *Canvas*. The *Canvas* itself does not provide any drawing methods, but it does provide a *paint()* callback method, similar to the *paint()* method in AWT components. Whenever the program manager determines that it is necessary to draw the content of the screen,

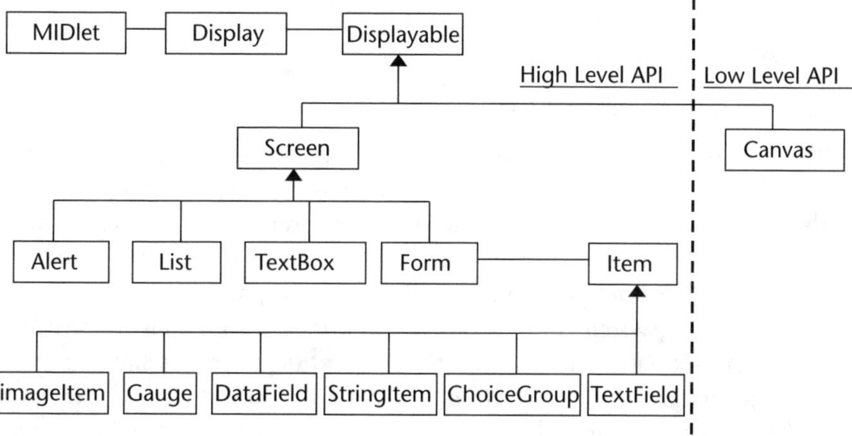

Figure 4.6   The MIDP GUI class hierarchies.

the *paint()* callback method of *Canvas* is called. The only parameter of the *paint()* method is a *Graphics* object.

With the GCF of CLDC and the GUI of the MIDP, many applications have developed and run on mobile devices.

4.4 i-mode

While the WAP Forum was striving to work out an open standard for wireless Web, Japan's NTT DoCoMo launched the information-mode (i-mode) wireless Internet service. The commuting time spent on trains and subways has become the most popular time for wireless Internet services in Japan. In 2002, i-mode had more than 35 million Japanese subscribers, who could access over 40,000 special i-mode Web sites. Compared with WAP, i-mode is not only a technology, but is also a business model. In Japan, the customers buy mobile phones already included with phone service. The i-mode service is available only to users who have signed up for NTT DoCoMo's phone service.

Outside Japan, i-mode has extended to other parts of Asia and Europe. In 2004, i-mode services were available in Taiwan, Germany, the Netherlands, Belgium, France, Spain, Italy, and Greece. NTT DoCoMo has also launched the world's first commercial 3G wireless service. On top of a wideband code-division multiple access (WCDMA) packet transmission network, video and audio files with the MPEG4 format could be sent and received by mobile phone [23].

However, i-mode and WAP have not been compatible. The markup language in i-mode is based on HTML, whereas WAP is based on the XML format. The protocols are also different. To make life easier for developers, NTT DoCoMo had been working with the WAP Forum to develop the WAP 2. The future versions of i-mode will eventually evolve to WAP 2. As introduced in Chapter 3, the core technologies in WAP 2 are XHTML-MP, HTTP, and wireless profiled TCP/IP. The i-mode protocol stack is similar to that of the WAP 2 stack. The XHTML-MP is closer to the i-mode cHTML than to the WAP 1 WML standard, while the wireless profiled TCP/IP is based on research performed by NTT DoCoMo. NTT DoCoMo is now a major sponsor of the OMA, which is the successor to the WAP Forum. In the following sections, we introduce the i-mode business model, infrastructure, and client development system.

4.4.1 i-mode Infrastructure

The i-mode network consists of two separate systems: the existing circuit-switched mobile phone network, and a packet-switched network constructed specifically for the i-mode service. Voice traffic uses the circuit-switched voice network, and is billed by connection time; whereas i-mode data traffic uses the always-on packet-switched network, and is billed for each packet sent.

In general, wireless Internet services are used to fill short periods of time when the user is not occupied with anything else. Quick and easy navigation is crucial. The content/service providers are linked to NTT DoCoMo's i-mode service as either official or unofficial providers. Official content/service providers are validated and approved by NTT DoCoMo. They are directly connected to the i-mode server, are listed on the i-mode menu, and can easily be reached. These sites can charge for their content, and NTT DoCoMo's billing system handles their billing. The official service list is grouped as an Internet portal. The major categories on the official menu include e-mail, ringing tones, and games [24].

Instead of being accessed through menu item selection, Unofficial content/services are accessed by typing a URL. The i-mode server enables Internet access from i-mode phones by relaying communications between the NTT DoCoMo packet network and the open Internet network. The architecture of the i-mode data network is presented in Figure 4.7. Handsets talk over the air to a gateway. The protocol conversion in the gateway translates the wireless-profiled TCP to a regular TCP. The gateway has a high-speed connection to the i-mode server, which is connected to all the i-mode services. When the user selects an official service, the request is sent to the i-mode server, which then either retrieves the content from a cache or asks the official service/content provider to respond. All other requests go directly to the Internet.

4.4.2 i-mode Client Technology

On top of the TCP/IP transport service and HTTP, a variety of standards are used for the i-mode service. Pictures are supported in GIF format, ringtone downloads are SMF-based, and security is supported via SSL. Besides the cHTML-based mobile Web, Java ME–based applications can be downloaded over HTTP, stored, and run on some phones.

The official mobile Web markup language of the i-mode is i-mode–compatible HTML (iHTML), which is cHTML with i-mode–specific extensions included. The cHTML is a combination of tags from the HTML specifications with some "mobile" extensions. Although cHTML was submitted to the W3C as a markup language for small information appliances including mobile phones, W3C has not endorsed it. cHTML is still considered as a proprietary product in the industry. DoCoMo itself also considered cHTML as a transitional markup language. XHTML is replacing the i-mode services implemented by the cHTML.

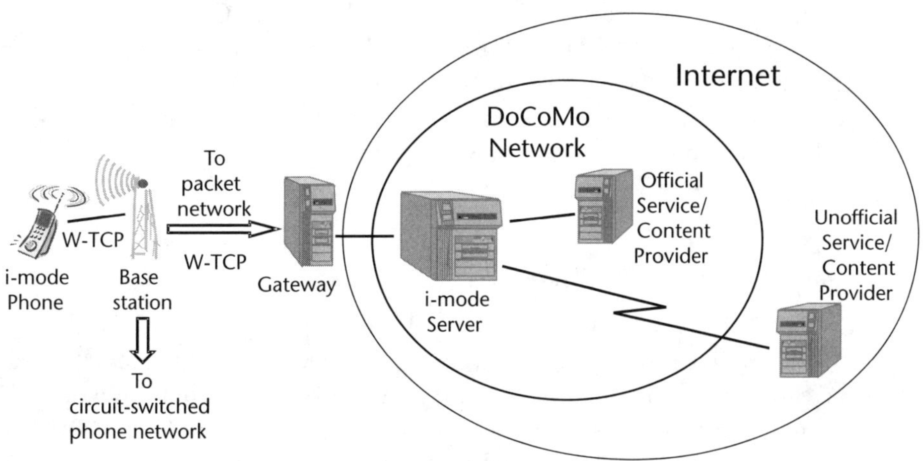

Figure 4.7 The i-mode data network architecture.

The cHTML is defined so that all the basic operations and site navigation can be done by using a combination of four buttons: Forward, Backward, Select, and Back/Stop (Return to the previous page). Because the cHTML is a subset of HTML, most of the basic HTML tags are allowed, including <html>, <head>, <title>, <body>, <hn >, <center>, <ul>, <ol>, <menù>, <li>, <br>, <p>, <hr>, <img>, <form>, and <input>. The <a> tag is allowed for linking to other pages, but with the additional scheme *tel* for dialing telephone numbers. The *tel* is a telephone URL analogous to *mailto*. The cHTML leaves out coding for tables, image map, multiple character fonts and styles, background, frames, and cascading style sheets. These parts are excluded, due to the low bandwidth and limited screen size of handheld devices. It also does not support JPEG images, because they take too much time to decompress.

In addition to the mobile Web, i-mode also supports services implemented using Java. The iAppli service of i-mode is based on the Java ME CLDC, with added functions unique to i-mode (e.g., GUI, communications, Japanese language processing). iAppli is targeted toward standalone applications, such as games, weather charts, and stock charts. Java applications may then be either client-only, or client/server-based. For safety, an application is not allowed to directly dial out or call other applications.

The structure of the i-mode Java Application Environment consists of three layers of APIs: (1) Java ME/CLDC API; (2) subpackages io, net, util, lang, and ui; and (3) manufacturer API extensions. For simplicity, network connections are made to look like I/O streams (e.g., GCF as defined by the CLDC specifications). Client/server applications are written by using methods to get the input and output stream of a connection. The extensions include APIs for low-level graphics controls, a scratch pad for storing application data (accessed using the GCF), HTTP and HTTPS support, and components for defining user interface elements. All iAppli programs are downloaded over the HTTP.

iAppli extends the *com.nttdocomo.ui.iApplication* class, and implements the *start()* method. It also must use an application descriptor, with a different file extension (.jam instead of .jad). The profile designed for iAppli, called DoJa, is incompatible with Sun's MIDP specifications (as shown in Figure 4.3) [25].

One last handy feature of the i-mode is called *emoji*. Although the Japanese language has tens of thousands of *kanji* (i.e., the name of Chinese characters in the Japanese language), NTT DoCoMo invented more than 160 new *emoji*. These *emoji* are similar to the pictograms of WAP 2 (see Section 3.2.3), which can quickly convey concepts that permit efficient communication in a small amount of space. They include symbols for the animals, birthday, holidays, broken heart, kiss, mood, sleepy, and so forth.

4.4.3 Server-Side Technologies

End users of the wireless Internet see only mobile browsers. Servers for wireless mobile applications are mostly, if not all, connected to the wired network. Although the client-side capability of WAP and i-mode are somewhat limited, the server-side technologies to generate dynamic content are the same as those for Web-based software. Originally, the Web was designed for convenient retrieval and presentation of static remote content. Because of the rapid growth of the Web, dynamic content generation, such as customized information searching, database retrieving, and news reporting, became a major part of the Web. Unfortunately, the Web's core technologies (e.g., HTML and HTTP) were not designed for dynamic content and software execution [26, 27].

Common gateway interface (CGI) technology was first introduced for dynamic content creation in the Web. Then, a series of technologies were developed, including Java Servlet and JavaServer Pages (JSP) from Sun; Active Server Pages (ASP) and ASP.NET from Microsoft; and open source software PHP and Cold Fusion from Macromedia.

Programs can also be executed on either server or client. For example, Java applets and Microsoft's ActiveX controls are downloaded from the server, and then executed on the client. With Java Script and VB Script, developers may have their code executed either on the server or client. Because codes executed on the server cannot be viewed from the browser, they are considered to be more secure. Properly distributing the tasks between the client and the server could reduce the network communication overhead. In general, the server-side technologies may:

- Dynamically create or modify any content of a Web page;
- Respond to user queries or data submitted from HTML forms;
- Access any server-side data or databases, and return the results to a browser.

CGI

CGI is a simple interface for running external programs (also called gateways or CGI programs) under a Web server in a platform-independent manner. CGI specifies how data is sent through the server to the external program and returned back to the browser. A CGI external program is executed in real time, so that it can produce dynamic and interactive Web pages. Although the CGI programs may be written in any language, Perl has been the most popular language due to its strong string-processing ability. However, the major problems of using CGI are poor scalability and performance. A standalone process is required for each request received by the Web server. If there is a large number of simultaneous requests to the same CGI program, then the memory and processing overhead become intolerable for the Web server.

Java Servlets

Servlets are Java technology's first answer to CGI programming. As the CGI programs, servlet programs also run on a Web server and generate dynamic pages. However, servlets have the following advantages over CGI [28]:

- *Less overhead.* A lightweight Java thread handles each request to a servlet program. The thread-per-request model is more efficient than the process-per-request of traditional CGI.
- *Easy development.* Servlets have an extensive infrastructure for HTML form and HTTP processing, such as HTTP headers parsing, cookies handling, sessions tracking, and so forth.

- *Better performance.* Servlets can talk directly to the Web server. Operations that are not possible in CGI, such as connection pools and information sharing, are built and optimized within the servlet engine.
- *Portability.* Because servlets are part of the Java platform, they inherit the cross-platform advantage. Since servlets are supported on almost every major Web server, servlets written for one server can run virtually unchanged on all other servers.

Servlets follow the syntax of Java. For Java programmers, the servlet API itself is straightforward. To deploy a servlet application requires further work. The developer must plan the directory structure of the context root, and prepare a deployment descriptor, generally XML-based.

JSP

JSP is an extension of Java servlet technology. JSP enables developers to separate programming logic from page design through the use of components that are culled from the page itself. Servlets generate the entire page via the Java program, even though most of it is always the same. It is more convenient to write regular HTML, rather than to have println statements that generate the HTML. With JSP, developers can create dynamic pages by putting Java codes inside the HTML pages. The Web server executes the code before the page is returned to the browser. By separating the display characteristics from the content, development tasks by Web page designers and Java programmers are separated [29].

JSP programs are executed in the JSP container, which is part of the Web server. The key components of JSP include directives, actions, scriptlets, and tag libraries. Directives are messages sent to a JSP container. Programmers can use directives to specify page settings, content that must be included from other resources, and custom-tag libraries used in the JSP. JSP tags that encapsulate functionality predefine actions. Scriptlets, also called "scripting elements," enable programmers to insert Java code in JSPs. Custom-tag libraries encapsulate complex functionality for use in JSPs.

Other than the syntax and deployment, JSP and Java servlets are nearly the same. During the first request of each JSP program, the JSP container translates that JSP into a servlet. If the JSP source has not changed, then the servlet can be executed immediately for later requests.

ASP

In 1997, before the creation of Java Servlet and JSP, Microsoft introduced their server-side scripting technology, ASP. Similar to JSP, ASP also combined a markup language, scripting, and server component into an ASP file (with the file extension .asp). Scripts in an ASP file were executed on Microsoft's Web server (e.g., IIS). Although the Java Script (called JScript in ASP) was supported in ASP, VBScript was more popular for its easy integration with Microsoft's tools and systems.

When a browser requests an ASP file, IIS passes the request to the ASP engine. The ASP engine then reads the ASP file, line by line, and executes the scripts in the file. Finally, the ASP file is returned to the browser as plain HTML. The advantages of using ASP instead of the conditional CGI and Perl are better performance and easier development.

.NET Framework and ASP.NET

ASP.NET is a part of the Microsoft .NET (dotnet) Framework. The .NET Framework provides a common environment for building, deploying, and running Web applications and Web services. The .NET Framework contains a common language runtime (CLR), which is similar to the Java VM in the Java platform, and common class libraries of standard services. Due to the CLR, .NET Framework is language-neutral; that is, it supports languages such as C#, C++, Visual Basic, and so forth. Because the application development is part of .NET, Microsoft's Visual Studio.NET is integrated into the .NET Framework as a common development environment [30].

.NET is an Internet- and Web-based infrastructure, which combines an operating system, a Web sever, a development environment, and an office tool set. .NET uses XML for data exchange. The new Web services standards (e.g., SOAP, WSDL, and UDDI) are also included in .NET.

Microsoft intends to not only dominate the desktop operating systems with .NET, but also to extend the infrastructure toward Internet-enabled devices, such as PDAs and mobile phones, using the Microsoft Mobile Internet Toolkit (.NET Mobile). The .NET Mobile is a set of server-side Web forms controls to build applications for wireless mobile devices. These controls produce different output for different devices by generating WML, XHTML, or cHTML.

ASP.NET is considered to be the next generation ASP, but the ASP.NET file (with file extension .aspx) is not fully backward-compatible with ASP. Compared to the original ASP, ASP.NET supports a large set of new controls and XML-based components, which in turn provide better performance and interoperability. Compared to other server-side technologies, ASP.NET is not just for server-side programs development and execution. It is also part of the .NET Framework product suite. The relationship between ASP.NET and .NET is similar to the relationship between JSP and Java EE.

Hypertext Preprocessor

The hypertext preprocessor (PHP) is a widely used open source general-purpose server-side scripting technique. Similar to other CGI-based technologies, PHP allows developers to create dynamic Web pages by putting script code inside HTML pages. The Web server executes the code before the page is returned to the browser. PHP supports many databases (e.g., MySQL, Informix, Oracle, Sybase, and so forth). However, the Linux, Apache, MySQL, and PHP (LAMP) architecture has become popular for inexpensive Web application development. PHP files have the file extensions .php, .php3, or .phtml. They are not as complicated as JSP and the ASP.NET [31].

Technical and business debates are ongoing for choosing between .NET and Java EE programs. In general, the .NET is a well-integrated product suite for

Microsoft users, and the Java EE is an industry-supported open platform. PHP provides a nonenterprise solution that is compatible with almost all servers.

Based on more powerful mobile devices and standardized XML-based communications, rich client applications become possible. New development techniques, such as AJAX and Flash Lite, are proposed for rich mobile clients [32, 33]. If the XML-based service oriented architecture (SOA) is adopted, then the server-side development technology will be independent from the client-side technologies. Thus, a true open environment for mobile applications is expected in the future.

4.5 Summary

Mobile software development encompasses a wide range of technologies. The server platforms used for mobile software are very similar to desktop Web-based applications servers. The standard format for describing and exchanging data in the mobile Internet is XML. XML-based technologies are widely used for mobile software development and data exchange. Two of the most important technologies for mobile client software are the XHTML-based Mobile Web technology, and Mobile Java MIDP technology.

The XHTML-based Mobile Web and Java MIDP should be viewed as complementing rather than competing technologies. Similar to the integration of Java applets into HTML in the desktop Web, MIDlets can be integrated into an XHTML-MP page. The XHTML-MP page can then be called from a mobile browser, and the embedded MIDlet gets installed on the device. The first generation MIDlets are not currently being provisioned over the air. In order to facilitate over-the-air provisioning, the handset must allow the user to enter a URL for a MIDlet suite. An XHTML-MP–supported mobile browser can be the best choice for this environment. An alternative is to use message services (e.g., SMS, MMS) to send the MIDlets to mobile devices. It is believed that mobile clients supporting WAP, Java, and messaging services will dominate the market in the near future.

References

[1]   Tanenbaum, A. S., *Computer Networks,* 4th ed., Upper Saddle River, NJ: Prentice Hall, 2003.
[2]   Mallick, M., *Mobile and Wireless Design Essentials*, New York: John Wiley & Sons, 2003.
[3]   Deitel, H. M., et al., *Wireless Internet & Mobile Business: How to Program*, Upper Saddle River, NJ: Prentice Hall, 2002.
[4]   W3Schools Online Web Tutorials, http://www.w3schools.com/default.asp.
[5]   Deitel, H. M., *XML How to Program*, Upper Saddle River, NJ: Prentice Hall, 2000.
[6]   Elliotte, R. H., *XML Bible*, 2nd ed., New York: John Wiley & Sons, 2001.
[7]   XML Schema, http://www.w3c.org/XML/Schema.
[8]   RELAX NG home page, http://relaxng.org/.
[9]   Academia Sinica Computing Centre's Schematron home page, http://xml.ascc.net/resource/schematron/.
[10] DSDL home page, http://dsdl.org/.
[11] CSS Mobile Profile 1.0, http://www.w3.org/TR/css-mobile.
[12] The Extensible Stylesheet Language Family (XSL), http://www.w3c.org/Style/XSL/.

[13]   OMA Technical Section—Affiliates—SyncML, http://www.openmobilealliance.org/tech/
       affiliates/syncml/syncmlindex.html.

[14]   Resource Description Framework (RDF)—W3C Semantic Web Activity, http://www.w3.
       org/RDF/.

[15]   XHTML Basic, http://www.w3.org/TR/xhtml-basic/.

[16]   XHTML Modularization, http://www.w3.org/MarkUp/modularization.

[17]   XHTML Mobile Profile Specifications, http://www.wapforum.org/tech/documents/WAP-
       277-XHTMLMP-20011029-a.pdf.

[18]   Java 2 Platform, Micro Edition (J2ME), http://java.sun.com/j2me/index.jsp.

[19]   Yuan, M. J., *Enterprise J2ME: Developing Mobile Java Applications*, Upper Saddle River,
       NJ: Prentice Hall, 2003.

[20]   Mobile Information Device Profile (MIDP), http://java.sun.com/products/midp/index.jsp.

[21]   Riggs, R., et al., *Programming Wireless Devices with the Java(TM) 2 Platform (Micro
       Edition)*, June 2001.

[22]   Bloch, C., and A. Wagner, *MIDP 2.0 Style Guide for the Java 2 Platform, Micro Edition*,
       Addison-Wesley, 2003.

[23]   Wallace, P., et al., *i-Mode Developer's Guide*, Reading, MA: Addison-Wesley, 2002.

[24]   How to Create an i-Mode Site Version 1.1, http://www.nttdocomo.com/corebiz/imode/
       why/tech.html.

[25]   DoJa 1.5 Overseas Edition DoJa Java Content Developer's Guide Release 1.2, 2002, NTT
       DoCoMo, Inc., http://www.nttdocomo.com/corebiz/imode/why/tech.html.

[26]   XHTML Mobile Profile and CSS Reference, http://developer.openwave.com/
       documentation/xhtml_mp_css_reference/front.html.

[27]   Jing, J., A. Helal, and A. Elmagarmid, "Client-Server Computing in Mobile
       Environments," *ACM Computing Surveys*, Vol. 31, No. 2, June 1999, pp. 117–157.

[28]   Java Servlet Technology, http://java.sun.com/products/servlet/.

[29]   JavaServer Pages Technology, http://java.sun.com/products/jsp/.

[30]   Microsoft .NET, http://www.microsoft.com/net.

[31]   PHP Hypertext Preprocessor, http://www.php.net.

[32]   Enabling AJAX Applications on Mobile Phones, http://www.opera.com/products/mobile/
       platform/.

[33]   Macromedia Flash Lite, http://www.macromedia.com/software/flashlite/.

CHAPTER 5

Wireless Multimedia Technologies

The mobile Internet has brought dramatic changes to both individuals and businesses. Individuals and businesses are communicating through e-mail and instant messages, and engaging in mobile commerce (m-commerce). M-commerce allows goods and services to be bought and sold electronically through mobile devices, and provides exchanges of electronic information on transactions. The information from these transactions further facilitates the study of focused groups of consumers. Merchants can address the needs of target customers more closely to their profiles and behaviors. Along with broadband connections to mobile phones, the convergence of video, voice, and data is imminent. The new era of mobile service (M-Service) has arrived, which combines the power of streaming video on the Internet with the mobility of cell phones. Users can watch news, TV programming, and sporting events with the click of a button. These services will be provided anywhere that users roam with their cell phones. In a broad sense, M-Service consists of different types of mobile services, such as chatting, distance learning, and video streaming services. Chatting and distance learning services can be combined with streaming video to create value-added services. For instance, adding video to a chatting service creates powerful videophone services. Similarly, live lectures can be delivered to a cell phone. These new services are creating new opportunities for wireless companies, service providers, and users. New broadband wireless networks are making the delivery of these mobile multimedia services ever more possible. The driving force behind M-Service is the rapidly increasing number of cell phone users. According to the market research firm Market Intelligence Center (MIC), the M-Service market has huge growth potential [1]. There were 2 billion cell phone users worldwide in 2005 and the number is expected to reach more than 3 billion by 2010.

One of the prime examples of multimedia streaming services in wireless networks is mobile streaming service (MSS), which delivers live media, TV programming, or other types of videos on demand. By definition, MSS is a suite of software programs that is used to implement and manage wireless networked multimedia content and services, which viewers access through their handsets [2].

MSS offers traditional TV programming, and adds new services, such as video-on-demand (VOD). In addition to VOD, there are new services on the horizon, which include interactive program guide (IPG), enhanced TV programming, TV-based Web access, and videophone services. In particular, VOD is a new service resulting from the rise of broadband wireless networks such as 3G networks, which deliver videos on demand over the bidirectional network. For instance, viewers can choose between multiple channels or seek further information while watching a video. As the VOD services progress to include viewer interactions, viewers are active rather than passive participants. Service providers and viewers both benefit from the interactive features, with higher revenues and more choices, respectively. There is a clear trend toward an increase in the amount of services, including Web access and online transactions.

In this chapter, we discuss infrastructure issues to support mobile streaming services or interactive features within the broadband wireless network, as wireless multimedia technology progresses.

5.1 Multimedia Streaming

In simple terms, multimedia streaming describes the techniques of transferring more than one medium (usually audio and video multiplexed into one stream) over the network. Since most video and audio files are large, they are streamed, rather than downloaded. In other words, media files are divided into individual frames, which are then further separated into a series of small network packets and transferred to the destination over the network. Once the packets are delivered to the destination, they are reassembled into individual frames, with each frame displayed in real time. Unlike downloading, streaming does not require the entire content of a file to be stored on the target device. Table 5.1 lists types of streaming media.

Mobile streaming products enable visual communication and rich multimedia experiences anywhere and at any time. The idea of "anywhere" implies that the technology for these products must be portable, and operate under low power and mobile conditions (e.g., on a bus, in a train, in a car). "Anytime" means that users will decide when to use these products, based on their schedule.

In Internet-based multimedia streaming, the enterprise market is one of the important areas of growth. The same is true for the mobile services market. Multimedia applications are the tools of choice for enterprises to better communicate with the concerned parties. Many businesses provide their customers with multimedia streaming in the high-speed wireless network. Let us examine some of the reasons why business and individual cell phone users may want to adopt multimedia streaming. Mobile streaming has the following distinguishing characteristics over Internet-based multimedia streaming to PCs.

- *Ease of access:* Mobile/wireless devices can be used anywhere and at any time.
- *Reasonable costs:* As wireless devices get cheaper and the wireless network provides higher bandwidth, it is cheaper to deliver multimedia contents to wireless devices rather than to PCs.

Table 5.1   Types of Streaming Media

Types	Description
Video	Streaming video to mobile phones. The most common data forms of streaming that include audio and video.
Flash animation	Streaming animation. This often includes audio and images multiplexed into one stream.
Presentation	Streaming text such as Microsoft PowerPoint files. Texts are presented with user control.

- *Ease of personalization:* Mobile devices, such as cell phones, have a unique identity for each device, making it is easier to keep track of individual users.

With the increase in mobility of multimedia services, the wireless industry is rapidly transitioning from voice-oriented services to video-based applications. Some of these video-based applications are combined with instant messaging and location-based services. The industry needs to define robust and open standards in order to ensure the interoperability of different devices. This is an important step to ensure the success of multimedia mobile services. The authors expect an industry-wide standard in the near future. For example, in the area of interactive presentations that is the essential component of future mobile services, there is an open standard called synchronized multimedia integration language (SMIL). This standard is used to guarantee the interoperability of interactive multimedia presentations, regardless of the platform used (e.g., cell phones, PCs, or set-top boxes).

5.1.1 Interactive Mobile Services

Interactive mobile services (IMS) combines any type of service with interactive features. For example, viewers can participate in a game show, or request more information about products. If viewers identify themselves in terms of interest areas for programming content, subsequent contents would be tailored for the users. It also allows viewers to participate in live broadcast programs. For example, viewers can solve problems together with contestants in a game show and receive an instant feedback, or can choose different camera angles while watching a sporting event. Some IMS solutions provide an end-to-end software platform for electronic commerce transactions over the broadband wireless network. For example, while watching a commercial, viewers would also browse different models of a product, or request additional information related to it. A viewer can purchase a product while watching TV. Background software handles the transactions and ensures that the inquiries are routed to the right merchants. It essentially combines the power of the broadband wireless network (3G) with interactive viewing and m-commerce. IMS, combined with industry standards such as SMIL, must provide those new services seamlessly with the traditional programs.

5.1.2 Extensible Markup Language

Extensible Markup Language (XML) was an official recommendation by the W3C in early 1998 [3]. XML addresses the issues of proprietary solutions that have been tried in the past to enhance HTML. XML covers the functionality of HTML and other document types. XML documents are made up of logical units called *elements*. An XML parser is used to read XML documents, and provide access to their contents and structures. Applications use an XML parser to access different elements inside an XML file. XML fixes some of the problems in HTML. For instance, XML addresses the dead links, slow searches, and static pages of HTML. As opposed to HTML, XML: (1) is used for creating markup languages, in which users define their own tags; (2) is a subset of SGML; (3) describes the structure of a document; and (4) is usually human readable.

5.1.3   SMIL

SMIL became an official recommendation by the W3C in mid-1998 [4]. SMIL documents are XML documents. SMIL has its own document type definition (DTD) in the XML syntax [3]. Using DTD, SMIL defines unique tags to be used in an SMIL file. One can make use of SMIL-specific tags to specify the synchronization between audio, video, image, and text components in a multimedia presentation. Currently, Nokia supports SMIL in its cell phones for interactive multimedia presentations [5].

5.1.4   Mobile Streaming Architecture

The core of mobile multimedia streaming is a video server. In addition to the video service, it is often the integration of other services, such as m-commerce. The mobile streaming architecture provides an infrastructure to support these varying services in a unified setup. It must be flexible enough to support different running environments. Figure 5.1 shows the architecture of a wireless streaming system. It consists of two major components. They are the backend server and the client platform. The backend server includes the creation of contents and the integration of service applications with streaming videos. The client platform renders media contents and provides interactivity for users. Videos are streamed from a media server. Different types of media data, such as images and texts, are retrieved from a WAP/Web server.

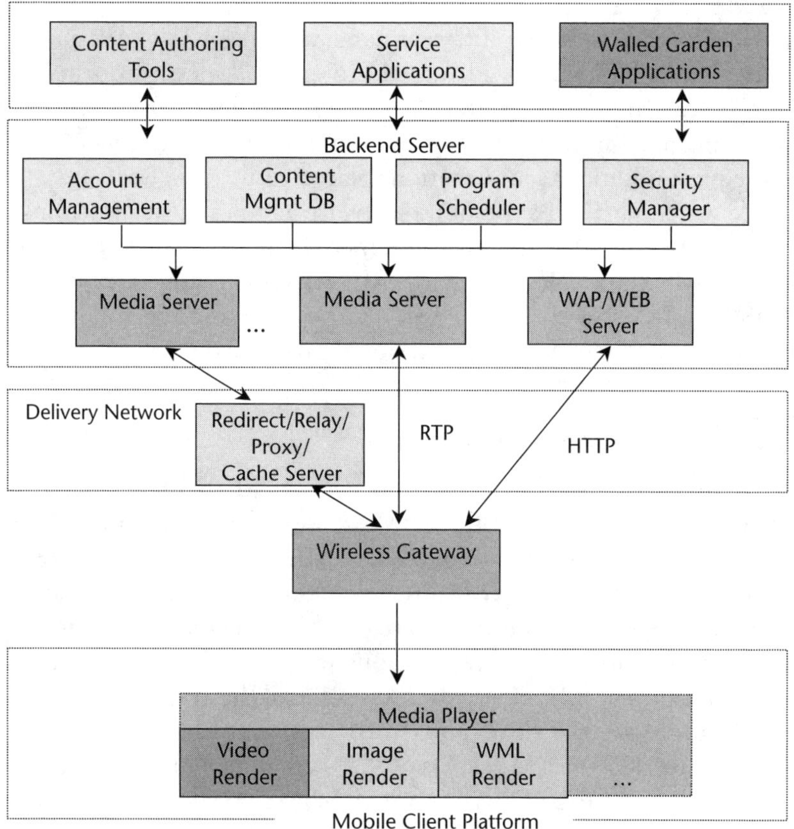

Figure 5.1   Mobile streaming architecture.

A media server is a software component that runs on a high-capacity server, because videos require high-capacity storage and network subsystems. It accepts the client's connection and delivers the requested stream, as well as other metadata.

On the network, videos use real-time protocol (RTP) through a proxy or cache server. The redirect server provides fault-tolerant connection management and load-balancing services across all members of media servers. It routes the streaming service connection to a member server with optimal availability and capacity. Initial client requests can be appropriately routed, based on current network and server loads. All the data traffic is routed to wireless gateways, where the data is transmitted to carriers' mobile networks. On the client side, the multimedia presentation framework supports interactive requests from viewers, and generates user interfaces to provide enhanced features from an XML file. It also interacts with the service applications. The media player displays the final contents. The client software may reside in cell phones or personal digital assistants (PDAs).

5.1.5 Backend Server

The wireless multimedia backend server mainly consists of four parts: the multimedia server, a WAP server, management software, and content-creation software.

Media Server

The media server provides rich media content delivery and a distribution platform that supports both unicast and multicast broadband streaming of live or prerecorded media. It consists of a streaming server engine, real-time streaming file systems, and network quality of service (QoS) functions. It supports both a local file system and fiber-based storage area networks for a high-capacity setup. A specialized media file system supports load balancing among storage and significantly increases the efficiency of the file system.

Some of the leading providers of server technology are Real Networks, Microsoft, and Apple, among others. Each of the developers provides a proprietary server solution. Table 5.2 lists features of these proprietary servers.

- *Real Networks:* Real Networks pioneered streaming media on the Internet in 1995. It is a global provider of network-delivered digital audio and video services. The company provides server solutions for building an internal content delivery network or outsource media delivery to a service provider [6].
- *Microsoft:* Another provider of mobile streaming server technology is Microsoft, founded in 1975. The Windows Media 9 series software development kit (SDK) provides the tools that developers need to build their own solutions on this platform [7]. Windows Media Server Format 9 series has core functionality of writing and reading files.
- *Apple:* QuickTime is Apple's standards-based software for developing, producing, and delivering audio and video over IP, wireless, and broadband networks [8].
- *Darwin:* Darwin Streaming Server is an open-source version of QuickTime Streaming Server that supports popular enterprise platforms, such as Linux,

Table 5.2   Comparisons of Server Technologies

Server Features	Real Networks (Helix server)	Microsoft (Windows Media)	Apple (QuickTime)
Unicast	Yes	Yes	Yes
Multicast	Yes	Yes	Yes
Cached content	Yes	Yes	Yes
Launch utility for Web page links	Yes	Yes	No
Proxy support	Yes	Yes	Yes
Access control	Yes	Yes	Yes
Username and password validation	Yes	No	Yes
Media player ID validation	Yes	No	No

Solaris, and Windows NT/2000. It is available for download in source or binary form, and can be ported to other platforms by modifying a few platform-specific source files.

Streaming Media Classification

Streaming media can be classified into two categories: broadcast-type and cached streaming [8].

- *Broadcast-type stream:* This is usually live event streaming that is broadcast to wide audiences, sometimes worldwide.
- *Cached stream:* This is a prerecorded event that is archived and streamed on-demand.

Choosing the streaming type depends on the business requirements and type of event. Live broadcasting requires a more dedicated and faster encoder, since the event is compressed live and streamed on the fly. It has to have the capacity to handle events in real time. Understanding the audience and the nature of the event is important in deciding the classification and cost involved.

Content Creation

Before multimedia data is streamed, the multimedia contents need to be created. Typically, content creation involves three steps. The first step is to acquire broadcast contents from various sources, such as DVDs and videocameras. The second step is the encoding process, which takes the source contents and transforms it into compressed broadcast formats, such as MPEG-1 or MPEG-2. It is important to support different bandwidth levels and many different types of devices that can be used in mobile receptions. Hardware-based encoders for fast compressions usually perform this encoding process. The third step is to load the compressed contents into a media server (streaming server). The media server makes the titles available for streaming service, ensures the delivery of the contents to playback devices, and performs contents and user management functions. Thus, there are two major steps in content creation (the first and second steps can be combined into one):

- *Content production, design, and encoding:* As explained earlier, content production is the process of creating contents. Encoding makes streaming more efficient, since it involves compression of audio and video files. Encoding is performed by codecs, which are the algorithms that handle the compression or decompression of media files. Codecs optimize content and enable audio and video files to be encoded into different formats, so that the data packets can eventually be streamed from the server platform to the client player. While there is no standard compression format, the primary formats include Microsoft Windows Media, Real Network's RealSystem, Apple's QuickTime, and MPEG.

- *Content aggregation:* Once streaming media contents have been created, multimedia streaming has to be managed efficiently. Some companies offer management tools as part of digital media services. Typically, such companies provide tools and services to simplify the process of acquiring, encoding, and indexing content. There is an ecosystem of companies that deliver solutions for the complete value chain of management and deployment of contents and infrastructures.

Some of the standard content creation technologies in use today are proprietary. Some of them are listed here:

- *QuickTime Pro:* This encoding tool, developed by Apple, serves a range of filetypes, from simple slide shows, to complex video and audio encoding, media creation, automation of production, and assembly of different media types.

- *Windows Media Encoder 9 Series:* This production tool, created by Microsoft, can be used for compressing audio and video content into Windows Media format, suitable for multimedia streaming.

- *Helix Mobile Producer Plus:* Developed by Real Networks. Tools are used for encoding media files into Real Media format.

One of the most important adopted standards for multimedia presentations is SMIL. Nokia's cell phones currently support SMIL [9], and major streaming server companies, such as Real Networks and Microsoft, also support it. Let us discuss an authoring tool for the SMIL standard, since multimedia presentation requires coordination among many different media, such as video, text, images, and so forth. Although users can edit an SMIL file, it is desirable to use a graphical user interface, so that authors do not have to remember the syntax of a language. Authoring tools are an integral part of creating presentation contents. SMIL authoring tools can be categorized into general authoring tools and special purpose tools. General authoring tools are designed to be generic tools to create SMIL presentations, such as Digital Renaissance's T.A.G. SMIL Editor [10]. Special purpose tools are designed to work with specific products, such as Sausage Software's SMIL Composer SuperTool [11], which is designed to work with Real Player.

As an example of the SMIL authoring tools, let us discuss one example developed at San Jose State University. Figure 5.2 shows the screenshot of an SMIL authoring tool, called SJSU Authoring Tool (SAT). It is targeted specifically towards

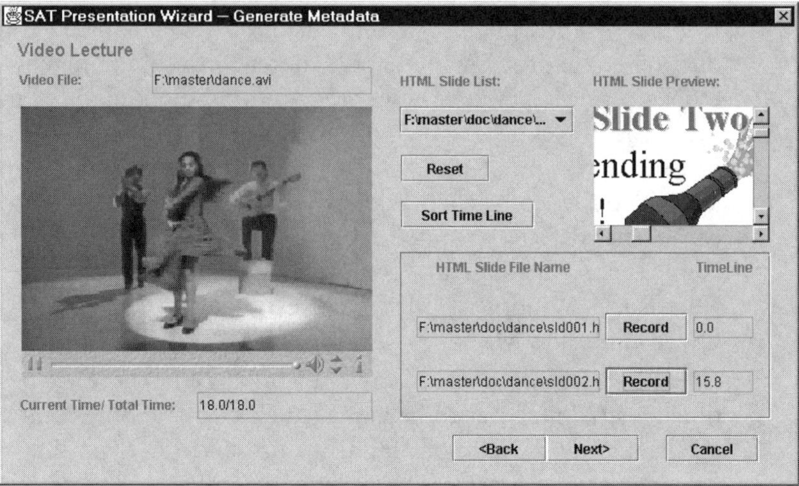

Figure 5.2 Presentation template.

domain users based on SMIL. It provides a high-level point-and-click tool that can create and update the logic of existing SMIL presentations. It can also be used to create and modify existing SMIL presentations. It is designed to provide template-based authoring, which is deemed to be easy to create for domain-specific presentations. For instance, the lecture template can be used to create online course presentations. It simplifies and streamlines the authoring process. It includes a widget to assist in preparing multimedia presentation materials, and common tasks are automated. SAT is a simple and easy-to-use interface, and a platform-independent software, with powerful authoring tools based on SMIL. SAT consists of four major components: presentation wizard, preview manager, metadata manager, and document generator. The presentation wizard is used to automate the common tasks in creating multimedia presentations, and is shown in Figure 5.2. When a presentation is created, preview manager can display each component. The metadata manager will create metainformation for a presentation to represent the timelines of each component. As the last step, the document generator creates SMIL documents. SAT provides a simple but powerful user interface to manipulate the information within the presentation creation process, which includes template selection, multimedia content collection, media timeline design, and automatic SMIL document generation.

Account Management and Program Scheduler

The account management engine maintains user profiles and usage of services. The services that a provider wishes to support must be registered in the account management. Once a user request is recorded and approved by the account management, the program scheduler creates a service plan for the request on appropriate schedule slots. The program scheduler interacts with media servers and the account management to ensure that the media service is delivered as scheduled, and charges are made when it is completed. Many companies are involved in developing services for specific types of streaming media business activities, such as accounting and billing. Table 5.3 shows an example of a comparison of billing.

Security Manager

The security manager is a module that checks for a password and the appropriate security clearance level for a user. If a password is incorrect, the manager will return a message back to the account management indicating that the password is incorrect. Otherwise, it returns the value of a clearance level of the user back to the account management. Consequently, the user's request is either approved or denied on the requested video. It performs other activities related to security verification and management.

Content Management Database

The database provides metadata for all stored videos in media servers. It interacts with the account management, so that a particular video can be located, and a server can be identified. It basically contains all the information on videos, such as lengths, titles, casts, ratings, reviews, and so forth. The database saves the activity log, in order to capture all the activities and VOD information. Typically, media management services are divided into four layers, as shown in Table 5.4 [7].

Applications

The service applications represent a collection of services that are desirable and fundamental for the mobile streaming infrastructure. Examples of these services are the m-commerce transaction platform, and the messaging platform. These services provide a basic infrastructure framework, and can be mixed and matched by the service provider for its third-party developers and target audience. These services are

Table 5.3   Comparison of Billing: Traditional Versus Electronic

Traditional	Electronic
Paper billing	Electronic bill presentment and payment (EBPP)
Biller's PUSH	Subscriber's pull
Self-accomplished	Outsourcing
Convergent billing	Convergent billing

Table 5.4   Media Management Services

Management Layer	Service Provider
Manager (FM)	1. Set up accounts. 2. Manage usernames, passwords, and access rights.
Digital rights manager (DRM)	1. Generate licenses for authorized users. 2. Specify rights, such as the right to download to disk, the number of times it can be played, and the time period in which it can be played.
Statistical manager (SM)	1. Capture statistics. 2. Bandwidth used to transfer contents. 3. Costs.
Broadcast manager (BM)	1. Client logs. 2. Management of IP addresses. 3. Beginning and ending times.

assumed to maintain their own databases and to interact with the content management database. As far as the framework is concerned, the only requirement is that the service follows the framework's predefined APIs.

Another category of service offerings is walled garden. A walled garden application is a confined application, which a third-party application service provider tailors for a focused group of customers. Examples of these applications include horse bidding, interactive games, and so forth. The following are some of the applications that can run as applications.

- *Public relations:* Streaming media is a good way to distribute news about products and services. Applications for streaming media in public relations include press conferences, interviews, product launches, and so forth.
- *Education:* Educational institutions are making learning more accessible through streaming media. Applications for streaming media in education include distance learning programs, lectures, seminars, and online workshops.
- *Live events:* Certain events are popular when streamed live. Live Webcasting allows anyone to view a performance, concert, or ceremony.
- *Gaming:* Games are available as applications. Some games provide interactive play against other players.

5.1.6 Streaming Client

Mobile client is one of the major components of the wireless multimedia architecture, as shown in Figure 5.3. Most of the mobile clients have some limitations in

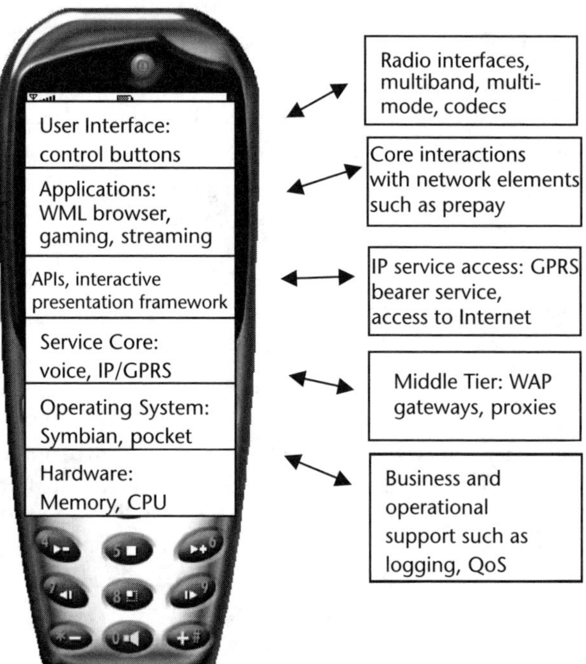

Figure 5.3 Client platform.

terms of CPU power, memory capacity, display speed, and battery life. However, they provide mobility and added functions thorough attached hardware and software. Currently, more open architectures are emerging, in which applications and content can be added in a customized way. The client-operating system can display texts, images, and multimedia contents. It also provides navigation features over the presented contents. Each multimedia element can be hyperlinked to other elements. The player's implementation supports codecs, which decode multimedia data for display on the output devices. After launching the player application separately, users can select the content file or a single stream to navigate through various content files.

An alternative is to click a hyperlink in a standard Web browser that anchors a content file. In either case, the player fetches the file from a Web server via the proxy. The player's content engine interprets the contents of the file and fetches the streams (using the RTSP protocol) and the static content (using HTTP). The operating system launches the content-specific codecs to display the information.

The client platform supports a variety of available services. At the heart of the client platform is the image and video displaying components. It receives streaming videos and other real-time or on-demand contents, and presents the information in display. The mobile streaming service platform must address the requirements and needs for consumers, service providers, and content providers. For consumers, the platform has to (1) mimic a media-rich experience with some interactivity, (2) possess a small footprint tailored for interactive entertainment, and (3) operate in average cell phone settings.

For service providers, the platform has to (1) provide a new scheme to support conditional access for payment and billing over broadband Internet, (2) interact with Web-enabled application services, (3) work off any transport mechanism, and (4) come with a low entry cost.

For content providers, the platform has to: (1) support digital right management (DRM) from a preferred implementation; (2) integrate with major cell phones or mobile devices; (3) provide basic video and e-mail communication; (4) embed with advanced sound control for home entertainment; (5) support walled garden applications, such as gaming; (6) comply with MPEG-4 specifications; and (7) handle audio and video peer-to-peer relay.

The characteristics of these devices include many new categories and form factors. Some are multiapplication devices, some are optimized for specific services (e.g., mobile banking), and many have added functionality and complexity.

Interactive Multimedia Presentation

The interactive multimedia presentation framework (IMPF) in a mobile client allows users to watch movies on a cell phone with customized interactive options. For instance, menus are generated from an extended SMIL file. The ITV framework is general and flexible enough to support many different types of presentations, including interactive VOD presentations, Web-based multimedia presentations, slide shows, and so forth. This approach is similar to GriNS [12]. The SMIL specification 2 is capable of providing user interactions and other capabilities. For example, specifications of buttons and events for buttons are added as *menuitem* and *submenu* tags. In addition, it supports Web-like advertisement banners with a timing control. A DTD file in the framework lists the specifications of buttons, events, and banners. A user can copy the syntax from the DTD file and develop an XML file for interactive VOD presentations.

IMPF is flexible, in that the buttons for user interactivity can be customized dynamically to be consistent with the current programming being displayed. As an illustration, when a movie is played, a set of interactive buttons is displayed along with the movie. They provide more information on its characters or the cast, shopping services for souvenirs, and banner advertisements. However, when a commercial is inserted between the movie segments or before a new movie is played, the existing buttons must be replaced with the new buttons that are related to the current content. In the framework, updating new buttons and events are specified in an extended SMIL file. The SMIL file can be streamed with contents. Thus, the framework generates the subsequent buttons from the file. Since the buttons and events are specified in a plain text file, a service provider can easily insert different interactivity along with program contents.

When the interactive menus are created, a corresponding vector is created from an SMIL file, and later is retrieved to create buttons. Buttons could be transparent or nontransparent. Each of the buttons corresponds to a future event, so that each of the buttons is registered with the listener when they are created. The framework allows any number of banners either to appear or to disappear at a specified time while a video is playing. Since the framework supports multiple channels, each channel is identified with a unique identifier, which is in turn used to identify a particular channel. If a video provides interactive content, then it must have an *Interactive* tag. Otherwise, interactivity is disabled. The framework does not currently support real-time features such as real-time stock quotes.

5.1.7 Solution Providers

The current business model that is fostered by mobile commerce and streaming services is to provide many services in commerce activities. Interactive features provide more choices for a variety of content offerings, along with return path to access additional program-related information. It includes many services, such as real-time information on stock quotes, weather, traffic, and security, and focused advertising based on customer profiles; and walled garden applications, such as shopping, horse bidding, gaming, and so forth.

Many companies deliver solutions for the complete value chain of authoring, encoding, capturing, managing, distributing, streaming, and consuming media. Here is a list of companies that provide different components of mobile commerce and streaming service infrastructures.

- *Video Servers for Enterprise/Education*: Apple, Cisco IP/TV, InfoLibria, InfoValue, Kasenna, Microsoft, Real Networks, 3CX;
- *Video Servers for Carrier*: Concurrent, DIVA, iVast, nCube, Minerva, SeaChange, Streaming21;
- *Middleware*: ACTV, ICTV, iMagicTV, Liberate, EnReach, OpenTV, PowerTV, Prasara, Worldgate;

- *Set Top Box (STB):* EchoStar, Hughes (DirecTV), Matsushita, Motorola (acquired General Instrument), Nokia, Pace, Philips, Sony, Thomson;
- *Personal Video Recording (PVR):* SONICBlue (ReplayTV), TiVo;
- *Ad/Content Service:* AOL TV, Intertainer, Wink.

Many of the wireless multimedia streaming technology developments mimic the user experiences for conventional consumer electronics, such as VCR, channel guide, and so forth. There are also new approaches to create new user experiences, as evidenced by the game console industry and distributed computing paradigm. These new approaches are popularized by a new generation of users who do not have any prior experiences in consumer electronics. The industry will gradually move from unidirectional broadcasting to bidirectional interactive entertainment.

Content providers usually set up a content delivery network (CDN) in order to deliver high-quality programming. This may include suitable delivery of the contents. A CDN service provides increased performance, reliability, and scalability to content providers. Downloading even a single page from a Web site involves traversing the Internet multiple times, passing many router hops. As a result, packet transmissions suffer from varying delivery latencies and inconsistent performance. A CDN service must identify those hot spots and provide content staging or rerouting to provide consistent packet delivery latencies. For example, content portals, such as CNN and MSNBC News, use content staging extensively. Table 5.5 shows a list of mobile streaming vendors [13–17].

5.2   Wireless Network Technologies

Mobile streaming means the delivery of multimedia content over mobile networks, such as GSM, GPRS, and WCDMA. The role of wireless LANs depends equally on market and user requirements. On the market side, wireless LAN technology offers some key datarate advantages for low-mobility market needs; however, there are still some uncertainties about spectrum pricing and availability, security, and technology standards.

Streaming media can be sent over the various wireless networks existing today [18]. Most of them vary from each other by the transmission media and the data rates they support. Table 5.6 shows a set of transmission media along with their data rates suitable for wireless streaming media.

5.2.1   Network Protocols

Wireless streaming technologies use the same protocols that are used to transmit documents in the Internet. All the technologies use user datagram protocol (UDP), transmission control protocol (TCP), or hyper text transfer protocol (HTTP) to transfer data packets between the host server and the client media player. UDP runs on top of Internet protocol (IP) networks, but does not have retransmission and data-rate management services. UDP is fast enough for real-time audio and video delivery. UDP traffic has a high-priority status on the Internet, making it a smooth delivery protocol for media transfer over the public network. HTTP uses TCP/IP

Table 5.5   Mobile Streaming Vendors

Vendor	Business Type	URL	Description
HelloNetwork	Mobile streaming solution provider	http://www.hellonetwork.com	HelloNetwork is the leading developer of the Java-based end-to-end streaming media platform that delivers both live and on-demand video, audio, and data to any Java-enabled device [13].
Oplayo	Mobile streaming technology provider	http://www.oplayo.com	Oplayo is the world-leading provider of streaming video and rich media solutions for wireless devices [14].
Speedera	Global content delivery service provider	http://www.speedera.com	Speedera is a top-tier global provider of Internet content delivery services for static and dynamic content, ranging from streaming media to encrypted data to rich graphics [15].
PacketVideo	Mobile streaming solution provider	http://www.packetvideo.com	PacketVideo is a world leader in the development and deployment of wireless multimedia [16].
Videoedge	Videoedge is a Windows Media service provider	http://www.videoedge.net	Videoedge is a Windows Media service provider, focusing on delivery of streaming video services on behalf of small, medium, and large organizations [17].
Apple	Mobile streaming solution provider	http://www.apple.com	Apple is one of the largest mobile streaming solution providers [8].
Real Networks	Mobile streaming solution provider	http://www.realnetworks.com	Real Networks provides the universal platform for the delivery of any digital media from any point of origin across virtually any network to any person on any Internet-enabled device anywhere in the world [6].

Table 5.6   Data Transmission Rates of Wireless Medium

Transmission Medium	Data Rates (Kbps)
Wireless modem	56
GSM wireless	9.6
GPRS	30
3G WCDMA	384
WLAN 802.11b	11,000

protocol to ensure that all packets are delivered, with retransmitting if necessary. HTTP does not attempt to stream in real time. Streaming by HTTP allows the client to store the data locally and play the content after enough data has arrived. Most firewalls and network configuration schemes will pass HTTP without modification.

Most streaming services use RTP and real-time streaming protocol (RTSP), which provide a method to transmit real-time data such as audio and video. RTSP is intended to control multiple data delivery sessions, and provide a means for choosing delivery mechanisms based on RTP.

RTSP is a control protocol used between client and server, and RTP is the data protocol used to transmit real-time data. Unlike HTTP and FTP, RTP does not download the file to the client, but instead plays it in real time. The main advantage of streaming over RTP is that RTP can be used for live transmissions and multicast. It allows the user to view continuous transmissions without having to locally store more than a few seconds of data. RTP uses UDP protocol as the underlying network protocol. This allows multicasts as well as live streams.

Microsoft uses its own multimedia messaging services (MMS) protocol. MMS is the default method of connecting to Windows Media. Microsoft's streaming server uses the MMS protocol to transfer unicast data. MMS can be transported via UDP or TCP. If the client Windows Media Player cannot establish a good connection using MMS over UDP, then it will resort to MMS over TCP. If that fails, the connection can be made using HTTP over TCP.

5.2.2   QoS and Scalability

QoS refers to the set of metrics to measure the performance and the quality of a network. QoS is often indicated by parameters such as signal-to-noise ratio (S/N), bit error rate (BER), packet latencies, and throughput rate. Other metrics can be defined based on the requirements of an application.

Scalability relates to delivering different bandwidth streams to different QoS networks. This permits an adaptive change in the content bandwidth delivered from a server to a client [19]. This is very useful for delivery over 2.5G and 3G wireless networks, which have highly variable rate characteristics.

5.2.3   Proxy Requirement

The proxy is the control gate between a wireless network gateway and the server systems. The proxy's task includes adapting the streaming bandwidth to the wireless network gateway. When a client requests an interactive multimedia presentation, the proxy initially determines the client's bandwidth and uses a media type that is appropriate for clients' wireless network environment. This requires prior knowledge of the clients' network environment. As the contents are streamed to the client, it may adapt the delivery dynamically to the clients' network, since the wireless network bandwidth may fluctuate, depending on the number of clients serviced at the time [20]. To achieve this dynamic adaptation, the proxy uses feedback information from the player, wireless network, and IP network.

The proxy can be used to configure a different set of preferences for content providers and users. A content provider can specify a minimum bandwidth requirement for each media file type to ensure acceptable quality. This involves ensuring the minimum QoS for each media stream. There is a threshold range for each medium, and the proxy is responsible for ensuring that those requirements are met. The QoS decisions are also made based on the type of customers. For example, premium service customers are guaranteed higher QoS values. When the network parameters go below the QoS metrics, the proxy may interrupt the current stream until it has established the connection with an acceptable QoS. The proxy is also responsible for collecting statistics from network routers for maintenance and overall management.

5.2.4 Unicast Versus Multicast

In a unicast, each client requests its own stream. This results in many one-to-one connections between clients and servers. Upon a client's request, a server sends a stream to a unique customer. The stream is not shared with other users. The main advantage of the unicast connection is the dedicated communication channel between the client and the server.

In a multicast, a single stream is broadcast to many users. A server creates and manages a single stream that is received by multiple users. A new user may join the multicast, or terminate while the multicast is in progress. Thus, a user joins to the stream, not to the streaming server. However, it is difficult to provide individual services similar to the unicast. Users cannot send feedback to the server. The major advantage of the multicast is cost savings in network and server resources, since one stream serves many users at the same time. In order to support the multicast, special routers that support multicasts are often required. Thus, most service providers do not provide this service. Multicast is more popular in a controlled environment, such as an enterprise network, in which the complete knowledge of the network and the uniformity of network equipment is common.

5.2.5 Wireless Multimedia Streaming Standards

Wireless multimedia standards are being defined by a set of standard bodies. Some of them are described here.

- *Third Generation Project Partners (3GPP):* The 3GPP is the new worldwide standard for the creation, delivery, and playback of rich multimedia for the next generation of mobile devices, such as multimedia-enabled cell phones and PDAs. The 3GPP, a group of telecommunications standards bodies, provides uniform delivery of rich multimedia over high-speed wireless networks to mobile devices. The 3GPP is based on MPEG-4, the standard for delivery of video and audio over the Internet, and includes specific tailoring for the unique requirements of mobile devices. The standard creates a base for creators and distributors of 3GPP digital media content by providing a commonly accepted specification [21].

- *Internet Streaming Media Alliance (ISMA):* The ISMA is formed to create specifications that define an interoperable implementation for streaming rich media (video, audio, and associated data) over IP networks. The ISMA is defining an interoperable approach for creating, transporting, and viewing streaming media. The primary goal of ISMA is to publish end-to-end specifications that enable cross-platform and multivendor implementations. The first specification from the ISMA defines an implementation agreement for streaming MPEG-4 video and audio over IP networks.

- *Wireless Multimedia Forum (WMF):* The goal of this forum is the establishment of technology in streaming multimedia over a wireless network. Technical experts from a group of companies have been working toward identifying a set of streaming protocols to use. The first set of identified protocols has been documented in the Recommended Technical Framework Document

(RTFD) Version 1.0, which serves to facilitate and guide other standard bodies' recommendations for wireless streaming [22].

- *Internet Engineering Task Force (IETF):* The IETF is a voluntary standard body that is dedicated to making recommendations for communicating information over IP networks. The IETF has recommended a number of methods based on RTP and RTSP for the delivery of video bit streams with synchronized audio from a server to a terminal.

5.3 Codec

Video files are huge when uncompressed. A broadcast-quality video requires 160 Mbps of network bandwidth. Uncompressed, CD-quality audio requires approximately 2.8 Mbps. The video has to be compressed using a codec. There are two types of codes in video compressions: proprietary compression algorithms created by several manufacturers, such as Microsoft and Real Networks, and standards-based technologies, such as H.263 and the Moving Pictures Experts Group (MPEG).

5.3.1 MPEG

The MPEG is the working group within the International Organization for Standardization (ISO) that has specified compression standards known as MPEG-1, MPEG-2, and MPEG-4. MPEG-4 was finalized in 1998 and became an international standard in 2000.

MPEG-4 builds upon the works of MPEG-1 and MPEG-2. MPEG-4's scope is fairly wide, since it defines not only the compression standards but also the new authoring and delivery paradigm. See Table 5.7.

5.3.2 MPEG-4 Architecture

While audio and video are at the core of the MPEG-4 specification, MPEG-4 can also support 3D objects, text, and other media types. MPEG-4 is designed to support media for low-bit rates for dial-ups, and up to the media for high-bandwidth network. It creates interoperability for video delivered over the Internet and other distribution channels.

Table 5.7 MPEG Comparison Table

MPEG-1	*MPEG-2*	*MPEG-4*
Approved November 1991	Approved November 1994	Approved October 1998
Enabled video CD	Enabled digital TV	Based on QuickTime File Format
Enabled CD-ROM	DVD Quality	Scalable quality
VHS quality	Enabled Digital Versatile Disk (DVD)	Scalable delivery from cell phone to satellite TV
Enabled MP3 audio		

5.3.3 MPEG-4 Visual Profiles

MPEG-4 Visual is a rich standard addressing the compression of natural and synthetic video scenes. The standard can be separated into profiles and levels. For the majority of applications, natural video coding tools are used. The predominant MPEG-4 visual profiles for natural video content is the simple visual profile. It provides error-resilient coding of rectangular objects, which is especially applicable to wireless video communication and transmission. The simple scalable visual profile is suitable for coding of temporal and spatial scalable objects to the simple visual profile. This is appropriate for delivering video over multiple types of communication channels. The core visual profile provides support for coding of arbitrary-shaped and temporally scalable objects to the simple visual profile. The main visual profile adds support for coding of interlaced, semitransparent, and sprite objects to the core visual profile. This profile can be used for broadband entertainment quality video. In addition, the N-Bit Visual Profile adds support for coding video objects having pixel-depths ranging from 4 to 12 bits to the core visual profile. This profile is useful for MPEG-4 applications tied to cameras with extended pixel depths.

5.3.4 MPEG-4 Encoding and Decoding

The MPEG-4 standard describes the main functions and operation of the MPEG-4 encoder and decoder. This topic is very well covered in the standard, so there is no point to reproducing in this chapter. Interested readers can find detailed information in [23].

5.3.5 MPEG-4 File Format

The MP4 file format is a component of MPEG-4 systems. This is the file format used for the storage of encoded audio and video, and is designed for the storage and streaming of MPEG-4 audio and visual information. The MP4 file may be rendered locally, or presented remotely by streaming components. This file format is based on Apple QuickTime.

5.3.6 Audio Compression Standard

Audio compression is a form of data compression designed to reduce the size of audio data files. Audio compression algorithms are typically referred to as audio codecs.

The MPEG group developed advanced audio coding (AAC). MPEG-4 AAC has been specified as the high-quality general audio coder for 3G wireless terminals. The MPEG-4 AAC standard incorporates MPEG-2 AAC, forming the basis of the MPEG-4 audio compression technology for data rates above 32 Kbps per channel. Because of its exceptional performance and quality, it is in the 3GPP specification. It provides audio encoding that compresses much more efficiently than older formats, such as MP3.

5.4 Conclusion

Wireless multimedia technology encompasses a wide range of areas, including encoding and decoding of media formats, protocols for transmitting the media over the network, quality of service for service provisioning, and so forth. Video alone will not provide more advanced interactive viewer experiences. Interactive wireless streaming applications will increasingly involve other media and service programs. Since wireless media streaming is still in its early stages of development, some of the factors, such as presentation style, text size, communication protocol, payment, privacy rules, delivery techniques, and performance metrics, are being refined based on the customer requirements. In the future, the wireless technology must make advancements to minimize the cost and maximize the quality of multimedia streaming. 3G technologies are currently viewed as the best candidate to overcome this problem. With the wide deployment of next generation wireless networks, such as 3G and 4G, that provide higher bandwidth, service providers will be able to offer interactive multimedia services to every cell phone, anytime and anywhere. There are many alternatives to support the necessary interactive wireless multimedia infrastructures. As part of the continuing effort in bringing interactive multimedia services to the cell phones, SMIL provides a common ground towards an open standard. We believe that interoperability and cross-platform support from various solutions of multimedia infrastructure pose a challenge in the near future.

References

[1] http://www.mobile-weblog.com.

[2] "Wireless Streaming Articles," http://www.streamingmedia.com/article.asp?id=5872&c=29.

[3] Xia, J., et al., "Implementation of a Synchronized Multimedia Integration Language Player," *SPIE Internet Imaging '99*, Vol. 3648, January 26–29, 1999, pp. 382–389.

[4] Yankee Group, "A Look at Streaming Media Value Chain," http://www.intel.com/eBusiness/pdf/busstrat/digitalmedia/wp012201.pdf.

[5] Newcomb, S. R., N. A. Kipp, and V. T. Newcomb, "The HyTime: Hypermedia/Time-Based Document Structuring Language," *Communications of the ACM*, Vol. 34, No. 11, November 1991, pp. 67–83.

[6] "Product Documentation Page," http://www.realnetworks.com/resources/documentation/index.html.

[7] "Windows Media 9 Series Homepage," http://www.microsoft.com/windows/windowsmedia/default.aspx.

[8] Apple Inc. Quicktime 6.3, The Digital Media Standard Page, May 20, 2003, http://www.apple.com/quicktime/products/qt/.

[9] Nokia, "MMS Phase 2," http://www.nokia.com/BaseProject/Sites/NOKIA_MAIN_18022/CDA/Categories/Solutions/MobileSoftware/Downloads/_Content/_Static_Files/mms_wp_a4_2706.pdf.

[10] Robinson, D., C. Lester, and N. Hamilton, "Delivering Computer Assisted Learning Across the WWW," *Computer Networks and ISDN Systems*, Vol. 30, No. 1–7, April 1998, pp. 301–307.

[11] Rousseau, F., and A. Duda, "Synchronized Multimedia for the WWW," *Computer Networks and ISDN Systems*, Vol. 30, No. 1–7, April 1998, pp. 417–429.

[12] Bulterman, D., et al., "GriNS: A Graphical Interface for Creating and Playing SMIL Documents," *Computer Networks and ISDN Systems*, Vol. 30, No. 1–7, April 1998, pp. 519–529.

[13] HelloNetworks, Inc., "Mobile Products," http://www.hellonetwork.com/mobile/
        products.jsp.
[14] Oplayo, Inc., "Streaming Service Provider," http://www.oplayo.com.
[15] Speedera, Inc., "Selecting a Streaming Media Content Delivery Service Provider,"
        http://www.speedera.com/pdfs/SelectingaStreamingProvider.pdf.
[16] PacketVideo Corporation, "PacketVideo Multimedia Technology Overview," 2001,
        http://www.packetvideo.com/pdf/pv_whitepaper.pdf.
[17] VideoEdge, Inc., "Streaming Service Provider," http://www.videoedge.net.
[18] Short, M., and F. Harrison, "A Wireless Architecture for a Multimedia World,"
        http://www.gsmworld.com/news/media_2002/short.pdf.
[19] Wee, S., and J. Apostolopoulos, "Research and Design of a Mobile Streaming Media
        Content Delivery Network," HP Labs, http://www.stanford.edu/class/ee398b/handouts/
        GL02-MSMCDN_swee.pdf.
[20] Elsen, I., et al., "Streaming Technology in 3G Mobile Communication System," Ericsson
        Eurolab, http://www.lnt.de/~hartung/ StreamingTechnologyin3GMobileCommunication
        Systems.pdf.
[21] Third Generation Partnership Project, White Paper, http://www.3gpp.org.
[22] Wireless Multimedia Forum, White Paper, http://www.wmmforum.org.
[23] Hantro.com, "MPEG-4 Codec Overview," http://www.hantro.com/pdf/overview.pdf.

Selected Bibliography

Blakowski, G., and R. Steinmetz, "A Media Synchronization Survey: Reference Model, Specification,
    and Case Studies," *IEEE Journal on Selected Areas in Communications*, Vol. 14, No. 1, January
    1996, pp. 5–35.

DeGregorio, S., M. Budagavi, and J. Chaoui, "Bringing Streaming Videos to Wireless Handheld
    Devices," OMAP Software Application Manager Texas Instruments Wireless Terminals Business
    Unit, http://focus.ti.com/pdfs/vf/wireless/streamingvideo.pdf.

Dorsten, D., and J. N. Ader, "Rich Media Rising: Interactive Television, Web Delivery Take Off,"
    Thomas Weisel Partners, November 13, 2000.

Effelsberg, W., and T. Meyer-Boudnik, "MHEG Explained," *IEEE Multimedia*, Vol. 2, No. 1, Spring
    1995, pp. 26–38.

Helin, J., S. Plaza Ltd. Media Lab, "Developing Mobile Multimedia Services to Enable Streaming to
    Handheld Devices," http://www.medialab.sonera.fi/workspace/Wireless_ Multimedia.pdf.

Intel Corp., Intel Java Media Player, http://developer.intel.com/ial/jmedia.

Javasoft, Java Media Framework API, http://www.javasoft.com/products/java-media/ jmf/index.html.

Ma, W., et al., "Video-Based Hypermedia for Education-on-Demand," *IEEE Multimedia*, Vol. 5, No.
    1, January–March 1998, pp. 72–83.

Mahajan, A., P. Mundur, and A. Joshi, "Adaptive Multimedia System Architecture for Improving
    QoS in Wireless Networks,"
    http://www.csee.umbc.edu/~pmundur/courses/CMSC691C/pcm-aug29.pdf.

Microsoft Corp., "Microsoft XML Parser," http://www.microsoft.com/xml/.

Puri, A., and A. Eleftheriadis, "MPEG-4: An Object-Based Multimedia Coding Standard Supporting
    Mobile Applications," *ACM/Baltzer Journal of Mobile Networks and Applications*, Vol. 3, No. 1,
    June 1998, pp. 5–32.

Real Networks, "RealNetworks G2 Player," http://www.real.com/

Schnepf, J. A., J. A. Konstan, and D. H. C. Du, "Doing FLIPS: FLexible Interactive Presentation
    Synchronization," *IEEE Journal on Selected Areas in Communications*, Vol. 14, No. 1, January
    1996, pp. 114–125.

Schnepf, J. A., et al., "Building a Framework for Flexible Interactive Presentation," *Proceedings of the
    1996 Pacific Workshop on Distributed Multimedia Systems*, Hong Kong, June 1996, pp. 296–305.

Seckin, G., "Challenges of Wireless Media Streaming," http://www.ssgrr.it/en/ssgrr2001/
    papers/Gamze%20Seckin.pdf.

Streamingmedia.com Research Center, research reports from http://www.streamingmedia.com/
    article.asp.

VitalStream Inc, "Streaming Service Provider," http://www.vitalstream.com/streaming/ features.html.

World Wide Web Consortium, "Extensible Markup Language (XML) 1.0 Specification," W3C Recommendation REC-xml-19980210, February 10 1998, http://www.w3.org/TR/REC-xml.

World Wide Web Consortium, "Synchronized Multimedia Integration Language (SMIL) 1.0 Specification," W3C Recommendation REC-smil-19980615, June 15 1998, http://www.w3.org/TR/REC-smil/.

Yu, J., and Yuanyuan Xiang, "Hypermedia Presentation and Authoring System," *Proceedings of the 6th International World Wide Web Conference*, April 1997, pp. 153–164, http://gatekeeper.dec.com/pub/DEC/SRC/technical-notes/abstracts/src-tn-1997-003.html.

Zeng, W., and J. Wen "3G Wireless Multimedia: Technologies and Practical Issues," PacketVideo Corporation, http://www.ee.princeton.edu/~wzeng/icip02_3G_preprint.pdf.

Wireless Networks

Wireless networks are telecommunication or equipment networks that use radio as the carrier. Wireless networks provide the communication infrastructure for application and system software. There are three categories of wireless networks: wireless local area network (wireless LAN, or WLAN); wireless wide area network (wireless WAN, or WWAN); and wireless personal area network (wireless PAN, or WPAN). WLAN technologies originate from computer networking, which gives network connections to all computers in the surrounding area. Chapter 6 presents the basic concept of WLAN networks, discusses WLAN infrastructures, protocols, standards, and future development.

The WWAN covers a larger area than a WLAN. WWANs were developed to allow communication through mobile phones. Although WWANs include a variety of networking technologies, Chapter 7 focuses on cellular-based wireless technologies, including cell characteristics, handoff mechanisms, and cellular mobile networks from 1G to 3G, and beyond. The WPAN is a relatively new and promising solution for short-range wireless communication. WPANs can be used to replace cables between computers and peripherals, to establish communities and share resources between devices, and to establish a variety of services. Chapter 8 provides an overview of WPAN networks, and discusses Bluetooth network protocols, operations, and application profiles. The chapter also covers other WPAN technologies (e.g., IrDA, Zigbee, and NFC), and device coordination technologies (e.g., Jini and UpnP).

The WLAN, WWAN, and WPAN technologies are all undergoing rapid development. In the future, each wireless technology is likely to work with other complementary technologies to revolutionize the way people live and work.

CHAPTER 6

Wireless Local Area Networks

A wireless local area network (wireless LAN) is a flexible communication system that extends or replaces wired LANs in a small coverage area, such as in a building or campus (i.e., hot spots). In wireless LAN, users transmit and receive data using radio waves, without wires for last-hop network connectivity; therefore, users enjoy both network connectivity and mobility at the same time.

In the past few years, wireless LANs have gained increasing popularity, and has been recognized as a general connectivity alternative for a range of business customers [1]. Hotels and retail shops, such as Starbucks coffee shops in the United States, have used wireless LANs to offer Internet access to its customers. Network managers have used wireless LANs to install networked computers. Warehouse workers have used wireless LANs to exchange information with a central database to update their inventory. Doctors and nurses have used wireless LANs to deliver patient information to their mobile devices. In the future, wireless LANs are likely to work with other complementary wireless technologies to revolutionize the way people live and work.

This chapter is organized as follows. In Section 6.1, we describe protocol layers implemented in various wireless LAN products. In Section 6.2, we discuss the 802.11 series of wireless LAN standards, such as 802.11a, 802.11b, and 802.11g, and their underlying technologies. After understanding the basics of wireless LAN, we present the deployment of wireless LAN infrastructures in Section 6.3. We discuss the future development trends of wireless LANs in Section 6.4. In Section 6.5, we summarize the key points in this chapter.

6.1 Wireless LAN Protocols

Like traditional Ethernet products, wireless LAN products implement the two lowest layers of the open systems interconnection (OSI) model [2–4]: data link layer, and physical layer.

- *Data link component:* The major task of this layer is to move a *frame* from one end of the physical link to the other end. There are several issues to be addressed in this layer. First, it may need to provide reliable service to the network layer. In other words, it needs to transform the raw transmission link that is inherently unreliable in the physical layer into a link to the network layer that appears free of transmission errors. Second, it needs to prevent a fast transmitter from overrunning a slow receiver. Third, it needs to decide which node can access the link, given that most of the transmission links are broadcast media, with each node overhearing the frames sent by others. Both Ethernet cables and wireless radio waves are such types of links. A special sublayer of the data link layer, commonly called the medium access control (MAC) sublayer, deals with this issue.

- *Physical component:* The major task of this layer is to move the individual *bits* within a link-layer frame from one node to the next. The protocols in this layer depend on the types of transmission media of the link. For twisted-pair copper wire, it is concerned with the voltage that represents a 1 bit and the voltage that represents a 0 bit. For wireless links, it is more concerned with the choice between licensed or unlicensed spectrum bands, how to divide the spectra to be used by individual mobile hosts, and how to represent the "bits" to be transferred (i.e., modem implementation).

In the following, we review how data links and physical layers are implemented in wireless LANs, and how they differ from the implementations of their wired counterparts.

6.2 Wireless LAN Standards

In the world of 802 standards, the physical layer corresponds very well to that in the OSI stack, but the data link layer is split into two sublayers. The first sublayer, logical link control (LLC) layer (802.2), is shared by many 802 variants, such as 802.3 (Ethernet CSMA/CD), 802.5 (token ring), and 802.11 (wireless). This sublayer hides the differences among various 802 standards and makes them appear the same to the upper layers. The second sublayer (i.e., the MAC layer) is specified by various 802 standards, such as 802.3, 802.5, and 802.11. The IEEE 802.11 standard includes a series of wireless LAN protocols covering both MAC layers and physical layers. Figure 6.1 gives an overview of the 802.11 protocol stack and its relationship to other 802 standards.

The base 802.11 standard includes the 802.11 MAC, and three possible physical layers: an infrared layer, a frequency hopping spread spectrum (FHSS) layer, and a

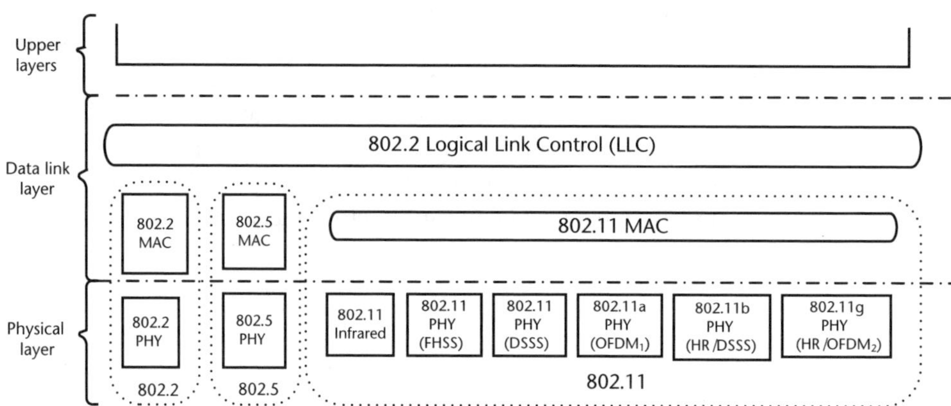

Figure 6.1 Overview of the IEEE 802.11 protocol stack.

direct sequence spread spectrum (DSSS) layer. In 1999, two new techniques were introduced to achieve higher bandwidth: 802.11a, which is based on orthogonal frequency division multiplexing (OFDM); and 802.11b, which is based on a high-rate direct-sequence layer (HR/DSSS). In 2001, 802.11g was introduced. Although 802.11g is based on the 802.11a OFDM modulation standard, it uses the same frequency band as the 802.11b standard. Currently, 802.11g is the dominant standard, while the 802.11a standard is emerging, and the 802.11b standard tends to be used only for low data transfer purposes and for legacy equipment compatibility.

6.2.1 The 802.11 Physical Layer

The 802.11 physical layer specifies six transmission techniques to send frames in wireless networks. Each technique has its own set of advantages and disadvantages. These six transmission schemes come from four broad types of technologies.

Infrared Communication

Infrared communication is used by most consumer devices, such as remote controls, and other data communication devices, such as Infrared Data Association (IrDA) devices on pocket PCs and laptops. There are two transmission modes in infrared communication. One is line-of-sight, where the sender and the receiver have to be visible to each other, and the second is diffused transmission, where the sender and the receiver do not have to be in a clear line of sight. Infrared devices normally oper-ate in indoor environments, since their energy does not pass through doors and external windows. This significantly limits the range of a wireless LAN using infra-red technology, confining it to a single room. On the other hand, it provides good isolation between the wireless LANs running in adjacent rooms, making interference and eavesdropping less likely.

Narrowband Technology

In such systems, the available bandwidth is divided into a number of smaller chan-nels, each of which is allocated to a pair of senders and receivers. Time division multiplexing access (TDMA) and frequency division multiplexing access (FDMA) are two common techniques. TDMA divides the bandwidth into multiple time-slots, and gives each user a unique time-slot for data transmission. The time-slots are assigned to users in a round-robin fashion. Whenever a user has a packet to send, it transmits the packet during the assigned time-slots. Typically, the size of a time-slot is chosen to allow the complete transmission of a single packet. While TDMA shares the channel in time, FDMA divides the spectrum of the channel into narrower frequency bands and assigns one frequency band to each communication pair. The channel bandwidth used in most FDMA systems is typically low, since it only needs to support a single transmission. Because of separate frequency bands, the receiver filters out the signals not on the assigned frequency band, thereby achieving both privacy and noninterference.

Spread Spectrum

This wideband technology was originally developed in the military for mission-critical communication systems. It produces louder and clearer signals, but also consumes more channel bandwidth than narrowband technology. If a receiver is not tuned to the correct frequency, then a spread spectrum signal appears as background noise. Unlike narrowband, spread spectrum signals transmit in the same frequency band, using specially generated code sequences to emulate "channels" in the narrowband techniques. Both the sender and receiver must share the knowledge of the code sequence to allow communication. Spread spectrum techniques have good resistance to interference. For years, the FCC has required all wireless communication devices operating in the unlicensed industrial, scientific, and medical (ISM) bands in the United States to use spread spectrum, minimizing interference among these uncoordinated devices. Wireless LAN systems have used two types of spread spectrum techniques: direct sequence, and frequency hopping.

In direct sequence spread spectrum, a communication pair i ($i = 1, 2, ..., n$) shares a unique m-bit code sequence, $\vec{c}_i$, which is called chip sequence. To transmit a data bit of 1, the sender sends the original chip sequence $\vec{c}_i$. To transmit a data bit of 0, the sender sends the complement of $\vec{c}_i$. To simplify further discussion, let us assume that data bits take the values of 1 or -1. In this case, the transmitted sequence $\vec{S}_{i,j}$ can be represented as the product of data bit $d_{i,j}$ and chip sequence $\vec{c}_i$

$$\vec{S}_{i,j} = d_{i,j} \cdot \vec{c}_i$$

As an example, suppose $m = 6$, and the sender would like to transmit two bits, $d_{i,1} = 1$ and $d_{i,2} = -1$. If the chip sequence $\vec{c}_i = (1, -1, 1, -1, 1, -1)$, then the sender will transmit $\vec{S}_{i,1} = (1, -1, 1, -1, 1, -1)$ for $d_{i,1}$, and $\vec{S}_{i,2} = (-1, 1, -1, 1, -1, 1)$ for $d_{i,2}$.

The receiver obtains the original data bits by taking the inner product of the sending sequence and chip sequence.

$$R_{i,j} = \vec{S}_{i,j} \cdot \vec{c}_i = \frac{1}{m} \sum_{k=1}^{m} S_{i,j,k} \cdot c_{i,k}$$

In the above example, we can easily verify that $R_{i,1} = d_{i,1} = 1$ and $R_{i,2} = d_{i,2} = -1$. For the receiver to correctly decode transmitted signals, the inner product of the same chip sequence has to be equal to one; that is, $\vec{c}_i \cdot \vec{c}_i = 1$. Then we have $R_{i,j} = \vec{S}_{i,j} \cdot \vec{c}_i$. $= d_{i,j} \cdot \vec{c}_i \cdot \vec{c}_i = d_{i,j}$.

The above discussions only apply to a single sender and a single receiver. In practice, it is often the case that multiple senders would simultaneously transmit packets. Therefore, the receiver would receive the sum of the signals from multiple senders. In order for the receiver to correctly decode signals, the chipping signals from different senders must be orthogonal; in other words, they must satisfy the following condition: $\vec{c}_i \cdot \vec{c}_j = 0$, $\forall i \neq j$.

Let us study an example involving two senders and one receiver. The chip sequence shared by sender 1 and the receiver is $\vec{c}_1 = (1, -1, 1, -1, 1, -1)$, and that shared by sender 2 and the receiver is $\vec{c}_2 = (1, 1, -1, -1, 1, 1)$. We can see that $\vec{c}_1$ and

$\vec{c}_2$ satisfy the above-mentioned conditions: $\vec{c}_1 \cdot \vec{c}_1 = 1, \vec{c}_2 \cdot \vec{c}_2 = 1, \vec{c}_1 \cdot \vec{c}_2 = 0$. If the two senders are transmitting a data bit to the receiver at the same time, and $d_{1,1} = 1$, then $d_{2,1} = -1$. The following steps illustrate the sending and receiving processes.

- *Step 1:* Sender 1 transmits sequence $(1, -1, 1, -1, 1, -1)$ to the receiver, and sender 2 transmits sequence $(-1, -1, 1, 1, -1, -1)$ to the receiver.
- *Step 2:* The receiver receives the sum of the signals from sender 1 and sender 2, $\vec{R} = (0, -2, 2, 0, 0, -2)$.
- *Step 3:* To decode the signal from sender 1, the receiver calculates $\vec{R} \cdot \vec{c}_1 = 1$.
- *Step 4:* To decode the signal from sender 2, the receiver calculates $\vec{R} \cdot c_2 = -1$.

Using direct sequence spread spectrum increases the amount of information by m times compared with narrowband techniques. This can be done only if the bandwidth available is increased by a factor of m. If a station sends at bit rate 100 Kbps, and m is equal to 10, then the station needs a bandwidth of 100 kHz using narrowband techniques, while it needs a bandwidth of 1 MHz using spread spectrum, assuming 1 bit per hertz. Therefore, spread spectrum is often considered as a technique that trades off bandwidth efficiency for resistance to interference, integrity, and security.

The second spread spectrum system, *frequency hopping spread spectrum*, hops between available frequency bands according to a specified pattern. Figure 6.2 shows an example of a hopping sequence. The sender and the receiver operate in synchronization, tuning to the same center frequency of a band. In time slot 1, the sender and receiver use frequency band (10,20) kHz to communicate; in time slot 2, they change to frequency band (40,50) kHz to communicate. This applies to later time-slots. Although only a narrow frequency band of 10 kHz bandwidth is needed for a single transmission, the whole system requires much wider bandwidth to allow frequency hopping in different time-slots. The algorithm to generate the hopping sequence may be either random or predetermined, but it must be known on both ends of the communication. This technique was invented in World War II to improve resistance to jamming signals, because the sender and the receiver can change their transmission frequencies in a pattern that is not known to jamming devices.

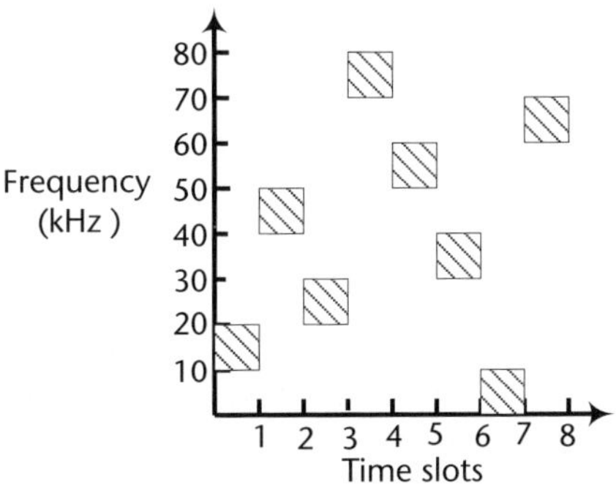

Figure 6.2 An example of a hopping sequence.

OFDM

OFDM can be thought of as a hybrid of narrowband and spread spectrum technologies. Like FDMA, it divides the available bandwidth into many narrow bands for shared access by users. Like spread spectrum, it assigns multiple bands for each user, allowing transmission over multiple frequencies at the same time.

Compared to FDMA and TDMA, OFDM uses the spectrum much more efficiently by more closely spacing the channel bands. This is achieved by making all channel bands orthogonal to each other, preventing interference from closely spaced bands.

In FDMA, approximately 50% of the total spectrum is wasted by allocating extra bandwidth among channels in order to prevent interference. For example, the bandwidth of each voice communication channel ranges from 10 to 30 kHz, although the minimum bandwidth for speech is only 3 kHz. The extra bandwidth is allocated to minimize the interference from adjacent channels, which is filtered out. TDMA partly overcomes this problem by allowing multiple users to share the same wideband channel using different time-slots. However, there are inefficiencies in TDMA that come from switching between users. Extra time must be allocated in the beginning of the time-slots to account for propagation delays and synchronization errors. As a result, this also limits the number of users who can effectively share the channel. OFDM utilizes the spectrum more efficiently than does FDMA and TDMA, by orthogonally dividing the channels and placing them close together. There is no extra overhead in bandwidth allocation as in FDMA, since orthogonal channels are immune to interference. Obviously, there is no overhead at the beginning of time-slots as in TDMA, since it is not based on time multiplexing.

After reviewing the technology background, let us briefly review the physical layers in 802.11.

The *802.11 infrared option* uses diffused, instead of line-of-sight, transmission between 1 and 2 Mbps. At 1 Mbps, the sender encodes a group of 4 bits into a 16-bit codeword using gray code. The codeword contains fifteen zeros and a single one. At 2 Mbps, the sender encodes 2 bits into a 4-bit codeword, which also contains only a single one. Due to the low bandwidth, this option is not widely used.

The *802.11 DSSS option* is also restricted to 1 or 2 Mbps. This scheme uses chip sequences of 11 bits and phase shift modulation at 1 megabauds, transmitting at 1 bit per baud at 1 Mbps, and 2 bits per baud at 2 Mbps. Once again, this option has been superseded by a better DSSS option, due to its low bandwidth.

The *802.11a OFDM option* is the first high-speed wireless LAN standard, delivering up to 54 Mbps in the wider 5-GHz ISM band. Transmission occurs in 52 frequency bands, 48 of which are used for data, and 4 are used for synchronization. As previously discussed, this scheme has a good spectrum efficiency and good immunity to multipath fading. 802.11a products have been readily available in the marketplace, but high prices have prevented their widespread acceptance.

The *802.11b HR-DSSS option* operates in the 2.4-GHz ISM band, supporting data rates of 1, 2, 5.5, and 11 Mbps. In practice, the operating speed of 802.11b is always nearly 11 Mbps. Although the peak transmission speed of 802.11b is slower than 802.11a, its range is up to seven times larger. 802.11b products were the first to come to the marketplace, and they remain a popular option in many situations.

The *802.11g OFDM option* is an enhanced version of 802.11b. It uses the OFDM method of 802.11a, but operates in the same 2.4-GHz band of 802.11b. It supports transmission speeds up to 54 Mbps. 802.11g products already exist in the marketplace. Since it operates in the same frequency band as 802.11b, it can easily coexist with 802.11b products. For this reason, upgrading from 802.11b to 802.11g is easier and cheaper than upgrading from 802.11b to 802.11a.

6.2.2 The 802.11 MAC Sublayer

The most important part of a wireless data link protocol is controlling the access to the common transmission media of radio links. As previously discussed, a sublayer called MAC is responsible for this task. Here, we will focus on describing the protocol operations of the MAC sublayer.

Why Do We Need a Different Wireless MAC Protocol?

Ethernet LAN employs carrier sense multiple access with collision detection (CSMA/CD) in its MAC sublayer. When a station wants to transmit, it uses the following rules.

- A station uses carrier sense to determine whether or not the transmission channel is free, and it only transmits when it senses that the channel is free.
- A station stops transmitting a frame when it detects that another station is also transmitting. This is called collision detection.
- A station retransmits a frame after waiting for a random time interval. This time interval is doubled for each failed transmission attempt.

At first glance, it may seem that the wireless MAC protocol can exactly follow its wired counterpart. In other words, it can use carrier sense to determine whether or not a channel is free, and stop transmission when it detects frame transmissions by other stations. Unfortunately, this is not true, due to some unique challenges arising from radio propagation and power management in wireless networks.

First, there is a hidden-terminal problem in wireless networks. Consider the scenario in Figure 6.3, in which the transmission and reception ranges of each mobile host are depicted by the oval boxes around them. In this case, S1 can send and receive signals to and from R, but S2 cannot. R can send and receive signals to and from both S1 and S2. Similarly, S2 can exchange data with R, but S1 cannot. Suppose S1 starts to transmit a frame to R at time T_0. At time T_1, when S1 is still transmitting, S2 wants to transmit a frame to R. S2 will immediately transmit, because it senses the channel to be free, since it is out of S1's transmission range. Eventually, the frames from S1 and S2 will collide at R. Both S1 and S2 will not be aware of this collision, so both of them will continue their transmission. From this example, we

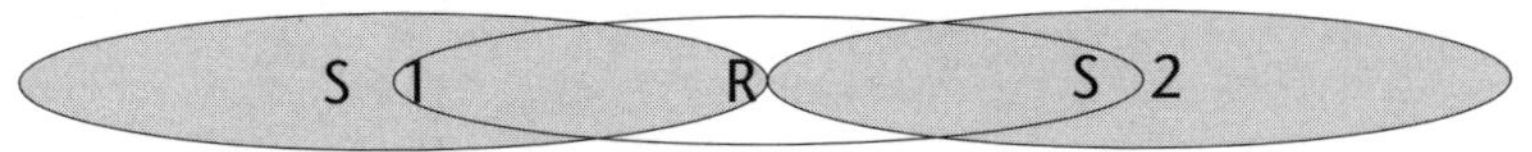

Figure 6.3 The hidden-terminal problem.

can see that both carrier sense and collision detection will not work properly because of the limited transmission and reception range of mobile hosts.

In this simplified example, the range is the same for S1, S2, and R. But in practice, this may not be true, which further complicates the problem.

Second, there is a related exposed-terminal problem, as shown in Figure 6.4. Suppose that S1 is transmitting a frame to R1 at time T_0, and this transmission reaches S2 in its range. Therefore, S2 will not attempt to transmit any frames while S1 is transmitting. This is too conservative for S2. For example, S2 can still transmit to R2, because S2's transmission to R2 will not interfere with R1's ability to receive S1's frame. Similarly, S1's transmission to R1 will not interfere with R2's ability to receive S2's transmission. This example illustrates that the carrier sense capability of a mobile host is not accurate because of the exposed-terminal problem.

Third, multipath fading is another major problem in wireless networks. At a wireless receiver, the total radio wave is a simple superposition of the incoming radio waves from all directions. Since radio waves may experience different delays in their path, the net sum may weaken instead of strengthen the signal. For example, when two incoming waves are approximately the opposite to each other, the received signal is very weak, approaching zero, due to constructive and destructive interference. This problem is very common in wireless networks, since most wireless transmitters and receivers use omnidirectional antennas, which causes the waves to be spread out in every direction and reflected by nearby surfaces. This results in delayed copies of the same transmission from different paths detected by a receiver. This problem can often be alleviated or resolved by changing the position of the receiver. This is the principle behind receive diversity, implemented by having two or more antennas spaced apart, all of which can be automatically switched into the radio receive path.

Fourth, most wireless devices operate on limited power, so power management is another important issue. The ability to detect collision requires the ability to both transmit and receive signals at the same time. The purpose of simultaneous receiving signals is to decide whether transmissions from other stations are interfering with one's own transmission. This can be very costly in power management, which is another reason why collision detection is not considered a good design choice in wireless MAC protocols.

802.11 Wireless MAC Protocol

Given the difficulties in correctly detecting collisions in wireless networks, the IEEE 802.11 [5] defines two modes of operations: point coordination function (PCF), and distributed coordination function (DCF). PCF uses the base station to control transmissions in its coverage area, while DCF is a fully distributed mechanism for transmissions. All standard-compliant implementations must support DCF, while support for PCF is optional.

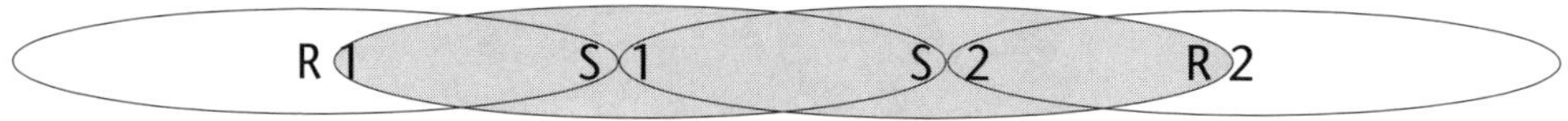

Figure 6.4   The exposed-terminal problem.

In the PCF mode, the transmission order is completely controlled by the base station, assuring that no collisions will occur. The basic mechanism is for the base station to poll each station in the area by broadcasting a beacon frame, which contains certain system parameters, such as clock synchronization and hopping sequences. The PCF mode is relatively simple, so we will focus on the DCF mode.

The DCF mode adopts a multiple access protocol that tries to avoid collisions, as opposed to detecting and recovering from collisions. The protocol operations of IEEE 802.11 are called carrier sense multiple access with collision avoidance (CSMA/CA). In this section, we will review the exchange of control and data frames in CSMA/CA, and how wireless stations recover from channel contentions.

In Figure 6.5, we illustrate the basic protocol operations when transmitting a frame. A successful frame transmission involves the following steps. First, the sender initiates the transmission process by sending a short control frame, known as a request to send (RTS), to the receiver. Second, the receiver responds with another short control frame, known as a clear to send (CTS), if it has successfully received the RTS. Third, the sender transmits its data frame, only after correctly receiving the CTS. The data frame is usually longer than both RTS and CTS frames. Finally, the receiver sends an acknowledgment frame (ACK) to the sender. This ACK frame must be received by the sender before the frame transmission is complete. To understand why IEEE 802.11 adopts this exchange sequence, let us take a closer look into the protocol.

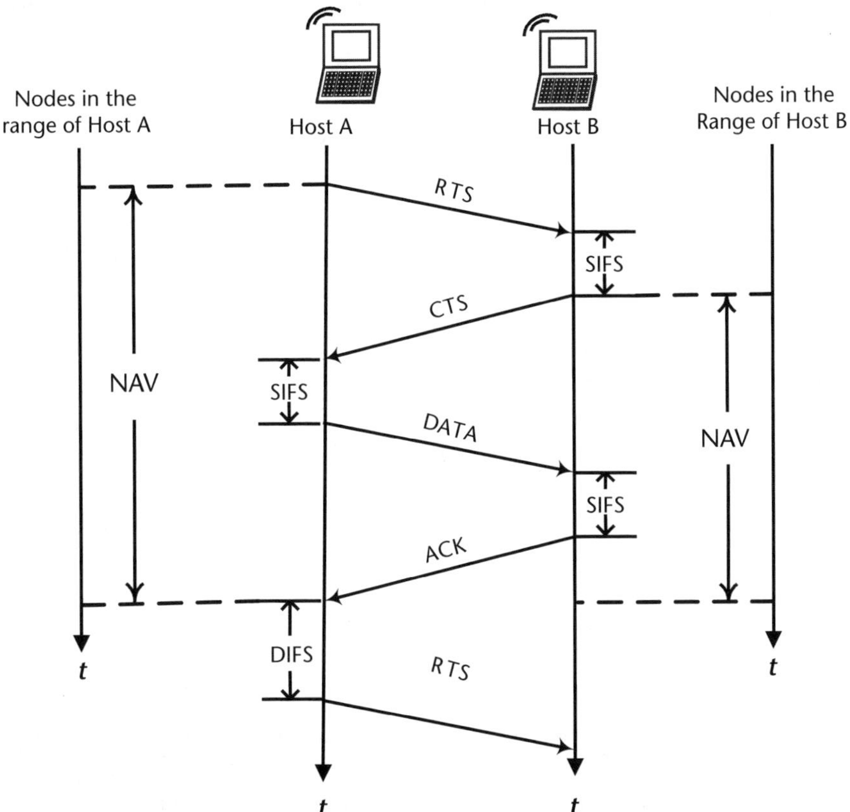

Figure 6.5 Basic frame exchange in IEEE 802.11.

- *RTS and CTS:* The RTS frame serves two purposes. First, it can be thought of as a frame to request permission from the receiver. Second, it helps to avoid collision, because it informs all the nearby stations that a data transmission is about to begin, and that the nearby stations must refrain from transmitting other frames. Similarly, the CTS frame grants transmission permission to the sender, and informs nearby stations that they must refrain from transmitting new frames. In other words, these two control frames clear out the areas in the range of both sender and receiver, avoiding the hidden-terminal problem. There is another reason to adopt RTS and CTS. These two frames are usually very short compared to other data frames, so a collision involving RTS or CTS frame will only last for the duration of the whole RTS or CTS frame. In contrast, a collision involving data frames lasts for a much longer duration, resulting in a large waste of channel usage. In summary, RTS and CTS frames help to "detect" collisions early, which saves channel resources.

- *Network Allocation Vector (NAV):* This is a special field in both RTS and CTS frames, specifying how long the stations in the vicinity of the sender and the receiver should keep quiet in order not to interfere with the current data transmission. The value of the NAV carried in the RTS frame is larger than that in the CTS frame. This is because the stations close to the sender have to wait for both CTS and data frames, while the stations close to the receiver only have to wait for data frames. Both the sender and receiver set the value of the NAV to the time for which they expect to use the medium, which includes any frames necessary to complete the current operation. Other stations hearing the RTS or CTS count down from the NAV to 0. When the NAV is nonzero, it indicates that the medium is busy. When the NAV reaches 0, it indicates that the medium is idle. This functionality is called *virtual* carrier sensing, since the hint of the channel being busy or idle is not derived by really sensing the channel.

- *ACK:* Reliable data transfers are often achieved by acknowledgments and retransmissions. The ACK frame in the exchange sequence helps serve this purpose. The sender starts an ACK timer when it begins to transmit its data frame. If the ACK timer expires before the ACK frame returns, then the entire sequence of RTS-CTS-DATA-ACK will be rerun. It is well known that a similar reliable delivery service has been provided by certain higher layer protocols (e.g., TCP) in the OSI seven-layer. One may wonder why another reliable delivery protocol is implemented in the link layer. There are multiple factors in introducing reliable delivery protocols into the link layer of wireless networks. First, recovery in the transport layer incurs a long delay. Recall that transport protocols are only implemented on end hosts. If the end-to-end transmission involves n links, and transmission error occurs in one of the links, then the error will not be detected until the frame gets across the entire path of n links. The timeout value in the transport layer is usually long, which is chosen to accommodate long-range data transmissions. For example, TCP's timeout value is at least 0.5 second. These large timeout values introduce a long delay in recovering from transmission errors. Second, the probability of data corruption in wireless links is much higher than in wired links. It is much more efficient to perform recovery on the link where the error occurs, since its timeout value can be chosen to fit the short transmission range. On the other hand, transmitting ACK frames can be an unnecessary overhead in wired links, because these links are of low error rates. Third, providing reliable delivery reduces protocol interdependence. It is not wise to assume that reliable delivery will always be available from higher layer protocols. When reliable delivery service is provided in the link layer for high-error-rate links, stations running the OSI seven-layer model can easily interact with stations running other protocol stacks, which may not support reliable transfer.

- *Short InterFrame Spacing (SIFS) and DCF InterFrame Spacing (DIFS)*: In Figure 6.4, we see that there are two interframe time intervals: SIFS and DIFS. Both define the minimum amount of idle time that a station must wait before transmitting a certain kind of frame. They are introduced to prioritize data transmissions. SIFS is used for the highest priority transmissions, allowing the parties in the RTS-CTS-DATA-ACK sequence to proceed first. DIFS is used whenever a station attempts to acquire the channel to send a new frame. As the name indicates, the SIFS interval is shorter than the DIFS interval. This ensures that, once an RTS-CTS-DATA-ACK sequence starts, it finishes before any other station can use the channel to start another RTS-CTS-DATA-ACK sequence to transmit a new frame. This is because both the sender and the receiver can use the transmission medium to send the next frame in the sequence before other stations that must wait for a minimum idle time of DIFS. Given this constraint, the RTS-CTS-DATA-ACK exchange sequence can be thought of as an atomic operation in the 802.11 protocol.

There is yet another topic to complete the picture. As we briefly mentioned above, there may be cases in which a frame is not successfully transmitted, and its RTS-CTS-DATA-ACK sequence has not been completed. For example, two stations may both sense the channel to be idle, and simultaneously transmit their RTS frames. The DATA or ACK frame may be corrupted during transmission. In these cases, the senders are responsible for scheduling retransmissions by starting a new RTS-CTS-DATA-ACK sequence. If there are multiple senders with unsuccessful frames, and they restart to access the channel at the same time, then their retransmitted frames will again collide. To avoid such collisions, the 802.11 MAC protocol defines a contention window, and adopts a backoff algorithm similar to Ethernet. The contention window is the amount of time that a station must wait before attempting to sense the channel. This window is divided into slots. Slot length is dependent on the medium, and higher speed physical layers use shorter time-slots.

As in Ethernet, a station randomly picks a slot number from a list of possible contention window numbers, and waits for that slot before attempting to access the medium. When several stations are attempting to transmit, the station that picks the earliest slot (i.e., the station with the lowest random number) wins. The list of possible contention windows becomes larger each time a transmission fails. Suppose that a station attempts the first time to retransmit a frame. Its contention window K will be chosen with equal probability from the set $\{0, CW_{min}\}$. If $K = 0$, then it immediately begins to sense the medium, and waits for an idle period of DIFS. If $K = 1$, then it waits for one slot to sense the medium. If the transmission fails again, the contention window K will be chosen from a contention window $\{0, 1, 2, 3, ..., 2CW_{min} + 1\}$ that is doubled in size. After another failure, K will be chosen from a contention window $\{0, 1, 2, 3, ..., 2(2CW_{min} + 1) + 1\}$ that is again doubled in size. Thus, the size of the contention window from which K is chosen grows exponentially with the number of collisions, until it reaches the maximum number, CW_{max}. The physical layer limits the maximum size. Table 6.1 lists CW_{min} and CW_{max} for different physical layers.

This exponential backoff scheme is a mechanism to schedule retransmissions based on the number of colliding stations in the network. If there are only a small number of colliding stations, then it is good to choose K from a small set of values in order to effectively utilize the channel. On the other hand, if there are a large number of colliding stations, then it is good to choose from a large and diverse set of values to avoid further collisions. In the beginning, the station has no idea of the number of stations that are colliding, so it starts from a small value. As more transmissions fail, it realizes that there are many stations colliding in the network, so it adapts to the situation by exponentially increasing its range of contention window.

6.3 Wireless LAN Infrastructures

In practice, 802.11 networks can be deployed in two ways. The basic building block of 802.11 networks is the *basic service set* (BSS), which consists of a number of mobile stations communicating with each other. The second deployment covers a larger area, connecting multiple overlapping BSSs in an *extended service area*.

6.3.1 Basic Service Areas

A station in a BSS communicates with other mobile stations in the same BSS that cover a basic service area. There are two types of BSS: ad hoc BSS, and infrastructure BSS.

- *Ad hoc BSS:* In this setup, a station is directly communicating with other hosts, without central control or coordination points. Since the communication is direct, the stations must be within the transmission range of each other. Typically, an ad hoc BSS is immediately formed and lives for a short period of time. For example, it might be formed when people with wireless devices meet in a conference room and exchange information in the absence of either a wired network or a central access point. In recent years, ad hoc networks have gener-

Table 6.1 Contention Window Boundaries

Physical Layer	CW_{min}	CW_{max}
FH	15	1,023
DS	31	1,023
IR	63	1,023

ated much interest, because more portable devices are appearing on market. In 802.11 terminology, an ad hoc BSS is also referred to as an independent BSS.

- *Infrastructure BSS:* This setup consists of a number of mobile stations and a base station, known as an access point (AP) in 802.11 terminology. All communications in an infrastructure BSS must go through the access point. If a mobile station needs to communicate with another mobile station, it first transfers the frame to the access point, which then transfers the frame to the destination station. With all communications being relayed through an access point, the mobile hosts do not have to be within transmission range of each other. Thus, the range of its access points defines the basic service area of an infrastructure BSS. The 802.11 standard does not specify a limit on the number of mobile stations that can be served by an access point. However, the relatively low throughput of wireless networks is likely to limit the number of stations accessing a single access point.

6.3.2   Extended Service Areas

BSSs can cover small areas, such as offices and homes, but they cannot provide network coverage to larger areas. In 802.11 networks, extended service areas are created by linking multiple BSSs.

Stations from different BSSs within the same extended service set (ESS) may communicate with each other, even though they may move between BSSs. Figure 6.6 illustrates an ESS as the union of BSS1, BSS2, BSS3, and BSS4. As long as stations remain in the extended service set, they can directly communicate. For example, Station 1 initially resides in BSS1, and it is communicating with Station 2 in BSS4. Next, when Station 1 moves to BSS3, it still maintains its connection with Station 2. In order for stations in an ESS to communicate with each other, the ESSs need to have two kinds of support. First, APs within the same ESS must be connected through a backbone network. Second, all the APs within the same extended service set must function as a bridge, to offer a single link layer connection to different BSSs. The backbone network, together with the bridging engine, comprises a *distribution system*, which provides mobility for stations connected through different access points in the same extended service set.

The distribution system achieves mobility through the management of associations between mobile hosts and access points. When a mobile host joins a BSS, it initiates the association process by registering with its access point. The access point

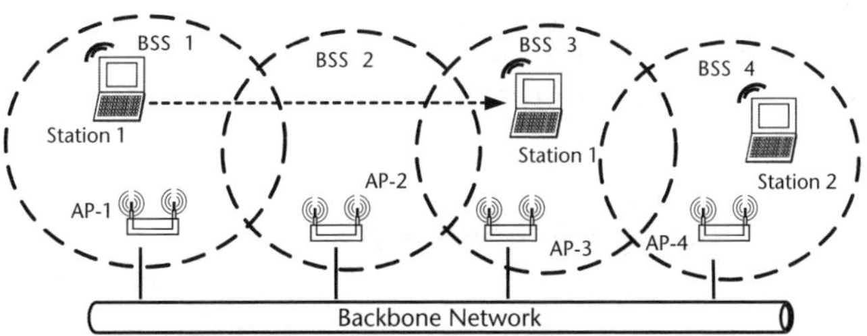

Figure 6.6   An example of an extended service set.

may choose to grant or deny access, based on the contents of an association request. A mobile station can be associated with only one access point at a time. In order for Station 1 in Figure 6.6 to communicate seamlessly with stations in other BSSs, each access point must know all the stations associated with other access points in the extended service area. For example, when Station 1 in BSS1 sends a frame to Station 2 in BSS4, it first sends its frame to AP1. By consulting its association table, AP1 learns that the destination is associated with AP4, so it forwards the frame through the backbone network to AP4. Similarly, AP4 refers to its association table, and determines that the destination is in its own BSS, so it transmits the frame directly to Station 4.

To propagate association information, access points must implement an interaccess point protocol. However, there is no existing standard for this, and vendors are using proprietary protocols to send association information.

6.3.3 Mobility

With the proliferation of portable devices, such as mobile phones, PDAs, and laptops, people are constantly on the move. Deploying wireless networks is an important step towards supporting that mobility. Since mobility has taken on many meanings in different settings, it helps to first give a precise definition of mobility.

In many cases, people interpret mobility as the capability to access network resources whenever and wherever they move. This ubiquitous network access takes two forms. In the first case, when a mobile user moves across networks, he closes his existing network connections; shuts down his computer; takes his portable device to the new network (wired or wireless); connects to the network; acquires an IP address through dynamic host configuration protocol (DHCP) [6, 7]; and begins to access the network. In the second case, the user only moves without the need to restart everything, because his network connections remain active during the move. It is the second form that offers true mobility. The first form merely removes physical constraints on providing network accesses.

Let us consider three scenarios to determine the extent of true mobility in 802.11 networks. First, when a user moves around in the same BSS, his network connections stay alive; therefore, he enjoys true mobility. Second, when a user moves between BSSs in the same ESS, the 802.11 networks also provide true mobility in the MAC layer. When a user moves to a new BSS, he will reassociate with the AP in the new BSS, and the new AP will update the other APs with its new association information. Therefore, other users can still find this user using the same MAC and IP addresses. As long as these two addresses are the same during the transition, a user's network connections can stay alive. Third, when a user moves across ESSs, 802.11 networks no longer provide mobility in the MAC layer. The user needs to use mobile IP [8] to have seamless transition between ESSs.

6.4 Future for Wireless LAN Networks

Wireless LAN is only one of the technologies that free people from wired connectivity. The future of wireless LAN largely depends on its interoperation with other relevant technologies, such as Bluetooth, ultrawideband, 3G cellular networks, and 802.16 broadband wireless networks. In this section, we review such technologies and discuss their relationships to wireless LANs.

6.4.1   Bluetooth

The Bluetooth project was initiated by a small special interest group (SIG) of five companies: Ericsson, IBM, Intel, Nokia, and Toshiba [9]. Today, the Bluetooth SIG has evolved into a network of nearly 2,000 companies. The objective of the Bluetooth project is to connect mobile phones, PDAs, and other wireless devices. In 1999, the Bluetooth SIG published a 1,500-page specification. Soon after that, the IEEE standards group began to work on the specification of IEEE 802.15 wireless personal area networks (PAN) [10], largely based on the initial Bluetooth document.

A Bluetooth device operates at low power, with a low data rate, within a small range. By default, a Bluetooth device consumes only 1 mW for communication. In packet-switched mode, Bluetooth's data rate is between 57.6 and 721 Kbps. In circuit-switched mode, its data rate is fixed at 64 Kbps. Compared to the 11 Mbps of 802.11b networks and the 54 Mbps of the 802.11g networks, Bluetooth's data rate is indeed very low. The basic unit of a Bluetooth system is a *piconet*, which consists of a master node and up to seven active slave nodes within a radius of 10m. Multiple piconets can interconnect to form a *scatternet*, linking up to 80 Bluetooth devices.

Bluetooth is not designed to compete with wireless LANs. It is considered as an enabling technology for PANs. The goal of a wireless PAN system is to enable devices around a human being to communicate information to each other without human intervention. In a wireless PAN, a Bluetooth-enabled PDA or mobile phone is able to, for example, turn on a TV, adjust heating settings, lock and unlock doors, or get an updated shopping list from a refrigerator. Although the original purpose of the Bluetooth project had little overlap with 802.11 wireless LANs, it still created competition for wireless LANs, especially for ad hoc data communications.

Bluetooth also operates in the 2.4-GHz unlicensed ISM bands, which are also used by most 802.11 devices. The electrical interference created an illusion that these two technologies were competing with each other, which is not the case.

6.4.2   Ultrawideband

A new technology, ultrawideband (UWB), which was initially used in the military, is now shifting towards public and commercial use for wireless devices.

Unlike Bluetooth and 802.11 wireless LAN systems that operate within a relatively narrow bandwidth, UWB operates across a wide range of the frequency spectrum by transmitting a series of narrow- and low-power pulses [11]. The wide spectrum used by UWB has caused military, aviation, fire, police, and rescue officials to voice concerns that interference from wireless UWB devices could potentially disrupt public services and military operations. However, the UWB industry argues that the combination of broader spectrum, lower power, and pulsed data means that UWB causes less interference than does conventional narrowband radio. The future for UWB depends on the FCC regulations on the use of UWB devices.

UWB has the potential to dominate the markets of wireless PANs, wireless LANs, and wireless wide area networks (WANs). Operating in short-range and low-power state, UWB can offer from 400 to 500 Mbps in bandwidth. It can potentially replace Bluetooth, which only offers up to 2 Mbps in the wireless PAN market. High-power and medium-range UWB can also be a competitor for wireless LANs. UWB is also a good candidate to provide cheap, fast, last-mile access in wireless WANs. Because of the power limitations imposed by the FCC, UWB is not in an immediate position to replace wireless LANs and wireless metropolitan area networks (MANs).

Currently, UWB is most likely to revolutionize home networking of multimedia devices, where high bandwidth is crucial. UWB could wirelessly connect virtually every multimedia device in a home network. For example, UWB can wirelessly connect digital cameras and camcorders to stream videos to TVs, PCs, and DVD players. It can also wirelessly connect computer peripherals, to a computer.

6.4.3   IEEE 802.16 Broadband Wireless

The IEEE 802.16 standard, the "Air Interface for Fixed Broadband Wireless Access Systems," is designed to provide last-mile broadband access in the MAN using wireless technology. It is intended to wirelessly connect residential and commercial customers to the Internet, replacing wired solutions, such as traditional cable, DSL, and T1.. Currently, various limitations prevent cable, DSL, or T1 from reaching many potential broadband customers. DSL can only reach approximately 3 miles from the office switch; the cost of deploying cables to underserved rural areas is very expensive; and the time to provision a T1 line for a business customer is approximately 3 months, if the service is not already available in the building. To fill existing gaps in broadband coverage, the IEEE 802.16 wireless technology provides a fast, cost-effective, and flexible way to provide broadband services.

In January 2003, the IEEE approved the 802.16a standard, an extension of the IEEE 802.16 standard published in April 2002. The most common 802.16a setup consists of a base station mounted on a high point, which communicates with subscribers at homes or business offices. It has a range up to 31 miles. With data rates up to 75 Mbps, an 802.16a base station can support up to 60 connections with T1-like performance, and hundreds of connections with DSL-like performance.

Broadband wireless providers, such as Airspan Networks, are starting to utilize 802.16a-compliant products, but interoperability is still a big issue. The sales of 802.16 products are predicted to surpass \$1.5 billion by 2008, according to the Allied Business Intelligence research firm.

It should be clear that the IEEE 802.16 standard is not a competitive, but rather a complementary, standard to IEEE 802.11 wireless LANs and IEEE 802.15 wireless PANs. Each standard specifies the best technology to address the needs of a different market, and helps to drive the market development of the other standards. For example, the proliferation of wireless LANs is driving the demand for wireless broadband connectivity to the Internet that is provided by 802.16 devices. On the other hand, the customers of IEEE 802.16 will likely choose IEEE 802.11 devices to build a completely wireless access solution.

6.4.4   3G Cellular Networks

While traditional networking is moving towards wireless connectivity, the telecommunications industry is also pushing for mobile telephony. In 1992, the International Telecommunication Union (ITU) issued a blueprint, International Mobile Telecommunications (IMT-2000), to characterize future trends of telephony. At that time, ITU had envisioned that data traffic will eventually exceed voice traffic, and that multimedia information would dominate the bandwidth. Reflecting this vision, IMT-2000 proposes to support four types of basic services: (1) high-quality voice transmission; (2) text and voice messaging; (3) multimedia information (e.g., video phone, gaming, and music); and (4) Internet access (e.g., Web access and e-mail).

There have been two major technologies that support IMT-2000. Ericsson proposed the first one. Ericsson's system operates in 5-GHz range using direct sequence spread spectrum. It was designed to be compatible with GSM networks, which are popular in Europe. Qualcomm proposed the second one. It also operates in 5-GHz range using direct sequence spread spectrum. However, it is based on IS-95, which is widely deployed in the United States. After numerous lawsuits, the two companies finally agreed to a single 3G standard. After resolving the differences, major mobile carriers in the United States, such as Sprint, AT&T, and Verizon, began to deploy 3G infrastructures and devices.

Since both wireless LAN and 3G networks offer data networking, they are competing technologies. Some experts feel that 3G will be a big success, dramatically changing the way people live and work [12, 13]. Others argue that 802.11 wireless LAN infrastructure has already been widely established, while the 3G companies were still resolving the differences. It is very difficult for 3G networks to replace existing wireless LANs, so its future is not very optimistic [4, 14].

6.4.5   Discussions

Among the technologies we have discussed, Bluetooth and ultrawideband are perceived to be competing technologies for wireless PANs; while wireless PANs, wireless LANs, and wireless MANs are complementary solutions. The advent of 3G cellular networks has created competition for wireless data networking provided by PANs, LANs, and MANs, since all of them aim to provide wireless data and voice communication to end users. The wide deployment base of the 802.11 LAN has created a certain edge in the competition so far, but the battle has just begun.

6.5   Summary

People have long dreamed of a wireless world. The 802.11 wireless LAN has been an initial success in realizing this dream. The 802.11 wireless LAN standard basically specifies two protocol layers, the physical layer and the MAC sublayer of data link layer, in the seven-layer OSI model. The physical layer has included several options, including infrared, frequency hopping spread spectrum, direct sequence spread spectrum, high-rate direct sequence spread spectrum, and orthogonal frequency division multiplexing in both 2.4- and 5-GHz ranges. The MAC layer also has multiple functional modes, including DCF, which is a contention protocol based on carrier sense multiple access with collision avoidance; and PCF, which is a contention-free protocol with centralized control.

In recent years, although the 802.11 wireless LAN faced significant challenges in security, interference, and quality of service, it has thrived to gain a large installation base, in which both basic service areas and extended service areas coexist to provide wireless access to end users. In the future, it is likely to be complemented by other wireless technologies, to fully realize the dream of providing wireless information access anytime and anywhere.

References

[1]   "What Is a Wireless LAN?" http://www.wlana.org/learn/educate1.htm.

[2]   Tanenbaum, A. S., *Computer Networks,* 4th ed., Upper Saddle River, NJ: Prentice Hall, 2003.

[3]   Peterson, L. L., and B. S. Davie, *Computer Networks: A Systems Approach,* 3rd ed., San Francisco, CA: Morgan Kaufmann, 2004.

[4]   Kurose, J. F., and K. W. Ross, *Computer Networking: A Top-Down Approach Featuring the Internet,* Reading, MA: Addison-Wesley, 2003.

[5]   The IEEE 802.11 Standards, http://standards.ieee.org/getieee802/802.11.html.

[6]   Droms, R., and T. Lemon, *The DHCP Handbook,* Indianapolis, IN: Macmillan Technical Publishing, 1999.

[7]   Internet Software Consortium, "Internet Domain Survey," January 2002, http://www.isc.org/ds/.

[8]   Perkins, C., (ed.), "IP Mobility Support for IPv4," RFC3220, January 2002, http://www.rfc-editor.org/rfc/rfc3220.txt.

[9]   Bluetooth Special Interest Group, https://www.bluetooth.org/.

[10] IEEE 802.15 Working Group for WPAN, http://grouper.ieee.org/groups/802/15/.

[11] Taylor, J. D., *Ultra-Wideband Radar Technology,* Boca Raton, FL: CRC Press, 2000.

[12] Lu, W., *Broadband Wireless Mobile: 3G and Beyond,* New York: John Wiley & Sons, 2002.

[13] Vriendt, D., et al., "Mobile Network Evolution: A Revolution on the Move," *IEEE Communications Magazine,* Vol. 40, April 2002, pp. 104–111.

[14] Garber, L., "Will 3G Really Be the Next Big Wireless Technology?" *Computer,* Vol. 35, January 2002, pp. 26–32.

Selected Bibliography

802.1x, "Port Based Network Access Control," Supplement to ISO/IEC 15802-3:1998 (IEEE Std. 1D-1998), http://standards.ieee.org/getieee802/download/802.1X-2001.pdf.

AirSnort, version 4b, http://sourceforge.net/projects/airsnort.

Bharghavan, V., et al., "MACAW: A Medium Access Protocol for Wireless LANs," *Proc. ACM SIGCOMM,* 1994, pp. 212-225.

Blunk, L., and J. Vollbrecht, "PPP Extensible Authentication Protocol (EAP)," RFC2284, March 1998, http://www.faqs.org/rfcs/rfc2284.html.

Fluhrer, S., I. Mantin, and A. Shamir, "Weaknesses in the Key Scheduling Algorithm of RC4," *Lecture Notes in Computer Science,* Vol. 2259, 2001, pp. 1–24.

Hachman, M., "Wi-Fi to Face Interoperability Challenges," *eWeek magazine,* December 4, 2003.

Heiskala, J., and J. Terry, *OFDM Wireless LANs: A Theoretical and Practical Use,* Indianapolis, IN: SAMS, 2002.

Lawrey, E., "The Suitability of OFDM as a Modulation Technique for Wireless Telecommunications with a CDMA Comparison," B.E. thesis, James Cook University, 1997.

CHAPTER 7

Wireless Wide Area Networks

Wireless wide area networks (WWANs) were developed to allow communication through mobile phones. Since the establishment of the first mobile phone system, WWANs have rapidly grown. The major trends in WWAN evolution have been from analog to digital technology, from voice to data communications, and from circuit to packet switching. (Packet switching, however, already implies data communications.) Currently, data traffic dominates voice traffic on the fixed network. The same trend is expected with WWANs in the near future.

The WWAN covers a larger area than a wireless local area network (WLAN). Although WWANs include networks other than cellular mobile networks (e.g., satellites), this chapter focuses on cellular-based wireless technologies. The WWAN offers a wide variety of wireless access and personal mobility services, with the goal of enabling communications at any time, at any place, and in any form. Many cellular systems have been developed to meet this goal.

This chapter is organized as follows. In Section 7.1, we describe cellular network architecture and mobility, along with some background on communications. Practical and existing cellular systems are then presented. Section 7.2 presents the cellular network evolution up to 2.5G systems, including D-AMPS/TDMA, GSM, and GPRS. Section 7.3 discusses 3G and future WWAN networks, including CDMA2000, WCDMA, and 4G systems. A list of the standardization organizations is presented, and then Section 7.4 summarizes the key points in this chapter.

7.1 Wireless WAN Overview

7.1.1 Common Wireless WAN Architecture

Cellular mobile systems are the most important components of WWANs. The cellular concept ensures efficient utilization of the available spectrum. The area covered by a cellular network is divided into cells. A cell is the geographical area that obtains wireless coverage from a single base station (BS). A base station consists of a computer and transmitter/receiver connected to an antenna, to which all the mobile stations (MSs) in the cell transmit. Hexagonal shapes give the most efficient area coverage in terms of number of cells. For the convenience of modeling, cells are usually considered to be hexagonal [1, 2].

Cellular networks are divided into two major parts: the radio access network, and the core network. Radio access network defines communication from the MSs to the BSs in the network through radio frequency (RF) channels. The core network manages the communications among BSs, mobile switching centers (MSCs), and database servers. Most old cellular mobile networks attempted to jointly develop the access network and the core network as a system with unified infrastructure. The trend in modern cellular systems is to keep the core network functions independent from the technologies utilized at the radio access network.

Components of Core Network

The common architecture of a cellular system is shown in Figure 7.1. BSs are connected to an MSC, the nerve center of the system. In a larger operation, multiple MSCs could form a hierarchical system. The MSC acts as an interface between the cellular radio system and the public switched telephone networks (PSTNs). The MSC performs overall supervision and control of the mobile communications, which include location update, call delivery, and user identification.

Databases are incorporated into cellular mobile systems to store information. For the purpose of location management, each subscriber is permanently associated with a home location register (HLR) in a cellular network. The HLR contains the user profile, which includes the service subscribed, and location information. The visitor location register (VLR) maintains the information of visiting roaming users in the cell. The authentication center (AuC) validates the MSs by verifying their identity with the equipment identity register (EIR). Recent improvements in location management include the use of a distributed database and replication of user profiles to enable faster access to user information [2].

Cell Capability Enhancement

The cellular concept ensures efficient utilization of the available spectrum. Multiple cellular cells are clustered together for large areas. Clustering ensures frequency reuse, which is achieved by allowing cells that are separated by a minimum distance

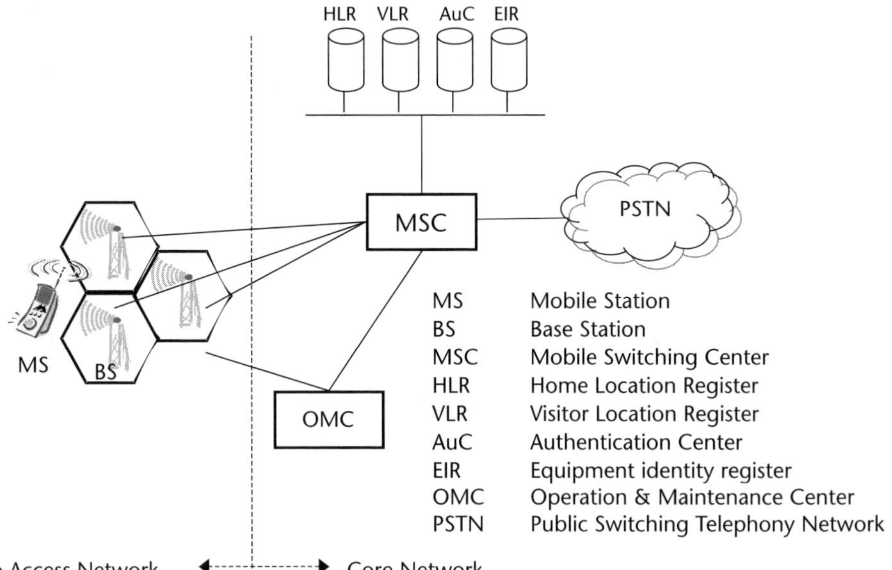

Figure 7.1 A common cellular architecture.

(i.e., reuse distance) to use the same frequency. On the other hand, the coverage area for a single cell depends on the network protocol, signal power, and any obstructions. Smaller cells mean that less power is needed, which leads to smaller and cheaper BSs and MSs. However, a major disadvantage to implementing a system with smaller cells is the large number of cells required for infrastructure expansion. To improve the capacity of a single cell, three techniques are adopted: cell splitting, sectorization, and power control.

First, cell splitting is effective in dealing with traffic from the hotspot areas. An overloaded regular macrocell is covered by smaller microcells to permit more frequency reuse. A typical microcell is approximately 100m in diameter. Picocells, with a diameter of only a few meters, are used to cover microcells for very heavy traffic (e.g., conference/convention centers, transport hubs, and so forth). Second, modification of antennas could also improve the capacity of the cell. Instead of using omnidirectional antennas, sectorized antennas would beam signals only in a particular sector within a cell. This greatly reduces the link interference, and allows shorter reuse distance.

As an MS moves farther away from the BS, the signal strength becomes weaker. Without power control, the strength of signals is determined by the distance between the BS and MS. Strong signals can drown out weak signals of some MSs on an adjacent frequency. This is called the near-far problem of a cellular network. One way to solve this problem is for each MS to transmit to the BS at the inverse of the power level it receives from the BS. In other words, an MS receiving a weak signal from the BS will use more power than one getting a strong signal. Thus, the BS will receive a fairly constant, equal power from each MS.

7.1.2   Mobility Management

Mobility management is one of the most important tasks in a cellular network. Many aspects relate to mobility management, such as location, resource management, traffic, call admission, and quality of service (QoS). The mobility in the WWAN can be classified into the personal mobility and terminal mobility. Some cellular systems provide a subscriber identity module (SIM), which can identify a particular subscriber. This type of identification implies personal mobility support. Terminal mobility means that consistent services are provided to frequently moved MSs [2].

To support mobility management, many protocols have been defined for cellular networks. Here, we introduce the fundamental operations of mobility management for most cellular networks. At any instant, each MS is logically under the control of a single BS of a specific cell. When an MS moves from one cell to another, its original BS notices the fading signal. The BS then transfers ownership to the cell receiving the strongest signal. This process is called a handoff (also called handover, or automatic link transfer). The handoff allows for the uninterrupted service, which is a critical requirement of all mobile communications. The handoff process consists of three phases: (1) handoff detection, (2) channel assignment, and (3) radio link transfer.

Handoff detection is based on wireless link measurement. The received signal strength, word error, and other quality indicators determine the quality of a channel. Path loss, shadow fading, multipath fading, and so forth, affect these quality measurements. Three strategies have been proposed for handoff detection: mobile-controlled handoff (MCHO), network-controlled handoff (NCHO), and mobile-assisted handoff (MAHO). In MCHO, the MS continuously monitors the signals of the surrounding BSs, and initiates the handoff process when certain handoff criteria are met. MCHO is popular for low-tier systems. In NCHO, the surrounding BSs measure the signal from the MS, and the network initiates the handoff process when certain handoff criteria are met.

MAHO is the most popular handoff detection strategy. The MS measures the signal from the surrounding BSs. The network (i.e., MSC) makes the handoff decision based on reports from the MS. MAHO is employed in the GSM and IS-95 CDMA. The procedure involved in a MAHO between BSs is outlined in Figure 7.2. During the handoff, the BSs are responsible mainly for radio relaying, and the MSC is responsible for channel assignment.

The second step in handoff is the channel allocation. Several channel allocation strategies have been developed to reduce forced terminations at the cost of increasing the number of lost or blocked calls. One of the popular strategies is called channel subtracting. The subtracting scheme creates a new channel on a blocked BS for a handoff access attempt by subtracting an existing call. An occupied full-rate channel is temporarily divided into two channels, each at one-half of the original rate. One of the temporary channels serves the existing call, and the other serves the handoff request.

The final step in handoff is the radio link transfer. The radio link transfer may be either a soft or a hard link transfer. In soft handoff, the MS is acquired by the new BS before disconnecting from the old BS, preventing a loss of continuity. The disadvantage is that the MS needs to have the ability to tune to two frequencies at the same time. In hard handoff, the old BS drops the MS before the new BS acquires it. If the new BS is unable to acquire it, the connection is abruptly disconnected. In general, the handoff quality is measured by delay, duration of interruption, success rate, and

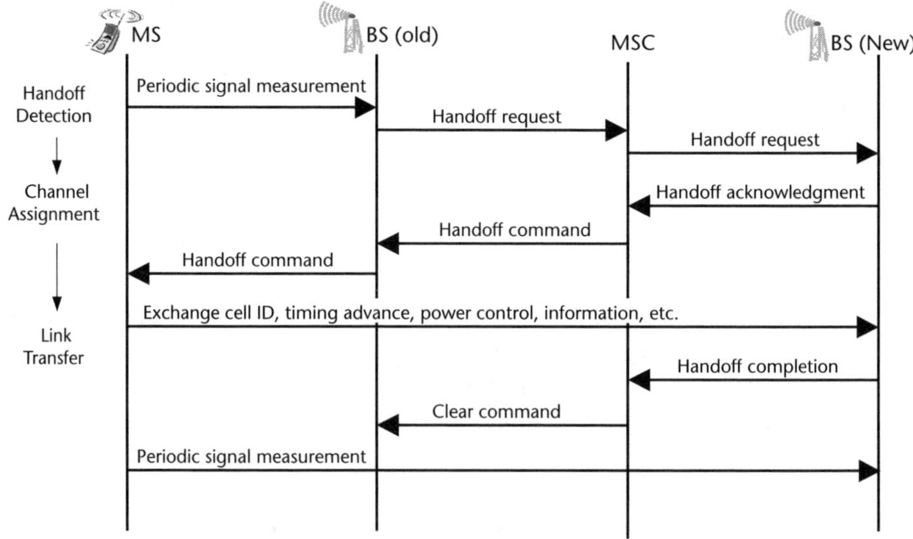

Figure 7.2   The MAHO handoff procedure.

unnecessary handoff rate. Neither the first generation (1G) nor the second generation (2G) cellular systems can achieve soft handoff.

The above-mentioned handoff process works for the inter-BS handoff. The operations of inter-MSC handoff are the same. Furthermore, when MS moves (i.e., roams) to a cell that is owned by a different operator, even if the owner is in another country, it shares a similar process. Two basic operations handling roaming are registration and location tracking. The registration (also called location update) occurs when an MS informs the systems of its current location. Then, the system could keep tracking the location of the MS. The VLR temporarily stores subscription information for the visiting subscribers. All services provided to the original operator, such as the short message service (SMS) and the calling number presentation, are also expected to be provided during roaming.

7.1.3 Communication Background

Before presenting real-world cellular systems, this section introduces two important physical and network layer technologies of the wireless communication and systems: multiple accessing and switching technologies. Multiple access schemes are responsible for sharing the limited bandwidth in a communication system. Similar to the 802.11 wireless LAN (introduced in Chapter 6), the radio multiple access techniques for wireless WANs also include frequency division multiple access (FDMA), time division multiple access (TDMA), code division multiple access (CDMA), and orthogonal frequency division multiple access (OFDMA). The switching technologies determine how data is transmitted between two parties. Data transfer switching can be either circuit switching or packet switching. The first generation cellular network started with circuit switching only for voice transmission. As the system evolved from analog to digital, and as the data traffic dominated the traffic, packet switching became much popular (e.g., the GPRS generation).

FDMA

Each signal on the channel is assigned a unique frequency for communication. The caller and the receiver tune to the same frequency to communicate. Frequency slots are divided into uplinks (i.e., transmissions from MSs to BSs), and downlinks (i.e., transmissions from BSs to MSs). Unused transmission time in frequency bands go idle. No other MSs can use a channel that is already assigned until the transmission is complete. FDMA encounters interference, inefficiency, and transmission interruption problems. Only the 1G analog cellular systems are purely FDMA-based.

TDMA

TDMA channels are divided into separate time slots. TDMA takes multiple calls and assigns each call to a different time-slot on the same radio frequency, and then accesses each channel in a round-robin fashion. The receiver interprets the appropriate time-slot (channel) to receive the information. Each MS gets a fixed length slot (i.e., packet transmission time) in each round, while unused slots go idle. TDMA allows for variation in the number of signals sent along the line, and constantly adjusts the time intervals to maximize bandwidth. Many of the current 2G cellular systems adopt TDMA, since it provides efficient use of spectrum with minimal interference. However, TDMA is inefficient for light-loaded systems.

CDMA

CDMA is a form of direct sequence spread spectrum (DSSS) technology, allowing many users to occupy the same time and frequency allocations in a given spectrum. Multiple simultaneous transmissions are separated using coding techniques. Each user is assigned a chip sequence. The sender and receiver synchronize by having the receiver lock into the chip sequence of the sender. All the other unsynchronized transmissions are treated as random noise. An example of DSSS is illustrated in Section 6.2.1. Compared to FDMA and TDMA, CDMA consistently provides better spectrum efficiency and lower levels of interference for voice and data communications. CDMA is used in several 2G cellular networks, and is the basis for nearly all third generation (3G) networks. The main disadvantage is the near-far problem, requiring fast and efficient power control.

OFDMA

OFDMA is a multiple access scheme for orthogonal frequency division multiplexing (OFDM) systems. As introduced in Section 6.2.1, OFDM can be thought of as a hybrid of narrowband and spread spectrum technologies. OFDM schemes tend to be used in FDMA systems (e.g., digital video broadcasting), or systems where other collision avoidance are used (e.g., 802.11a and 802.11g). OFDM is not yet widely used in wireless WANs. However, OFDMA is a major candidate for the fourth generation (4G) systems.

Circuit Switching

A reliable connection between the two communicating parties is established in circuit-switched network. This connection is maintained and cannot be used by any other parties for the duration of the connection. Circuit-switched networks provide an ideal environment for constant bit rate delay-sensitive applications, such as stream voice communication. However, circuit switching fails to provide efficient bandwidth utilization in the network. It is very inefficient for bursty data communication. The nature of the circuit-switched network implies that the user has to pay based on the time of the connection. Circuit switching is widely used in most 1G and 2G cellular systems.

Packet Switching

Packet-switched networks do not require dedicated connections. Data in packet-switched networks is divided into small packets. The destination address itself is a field of the packet. Data packets do not have to follow any specific path to get to their destination. Each packet is routed independently through the network to its destination. Thus, multiple users can share the data path. Packet switching allows packets to take the optimal path, which improves bandwidth efficiency. Eventually, data packets are reassembled into their original format at the destination. Packet switching relieves users of having to establish a network connection each time they need to transfer data. More importantly, users are charged only for the data they transmit, rather than for connection times, which can result in significant savings. Packet switching is the basis of the always-on connectivity provided by 3G cellular networks.

7.2   Cellular Mobile Networks—Up to 2.5G

Cellular mobile networks have gone through three distinct generations, with different technologies: analog voice, digital voice, and digital voice and data. In this section, cellular mobile systems before 3G are introduced. 3G and subsequent systems are introduced in Section 7.3.

1G analog networks were only used for voice communication, and were deployed commercially in many countries. In the early 1980s, the FCC mandated the Advanced (or Analog, or American) Mobile Phone System (AMPS) for the United States. The system was also used in England, where it was called the Total Access Communications System (TACS); in Japan, where it was called MCS-L1; and in Scandinavia, where it was called Nordic Mobile Telephony (NMT).

2G networks introduced digital capabilities to cellular systems in the early 1990s. This resulted in higher-quality voice, as well as basic data services. There are four major 2G systems currently in use: D-AMPS, PDC, GSM, and cdmaOne.

Digital AMPS (D-AMPS) directly inherited many fundamental properties of AMPS, allowing for backward compatibility. D-AMPS, also called TDMA networks, is the leading 2G technology in America, and is the only national network in many South American countries. Personal digital cellular (PDC) is a modified D-AMPS system in Japan. PDC offers a 9.6 Kbps data transfer rate over circuit-switched networks. In the 1990s, Japan's largest telecom company, NTT DoCoMo, added a packet-switched addition to PDC that provides packet data capabilities. It launched the i-mode service, introduced in Section 4.4, which provides wireless Internet service. The evolution of PDC goes directly to 3G networks using WCDMA. There is no intermediate transition to 2.5G networks.

The Global System for Mobile Communications (GSM) is the most extensively used 2G system in the world. European countries developed GSM. CDMA is both a multiplexing technique and a system. Qualcomm introduced the CDMA-based 2G system, called cdmaOne, in 1995. The cdmaOne system was based on two versions: the EIA/TIA IS-95 standard, namely IS-95A and IS-95B. Hutchison (Hong Kong) first deployed IS-95A in September 1996. IS-95B provided faster data rate, and was first deployed in September 1999 in Korea. The Sprint PCS in the United States also provides cdmaOne service.

While waiting for the slow progress of 3G, some operators are taking a cautious step in the direction of 3G by going to 2.5G. The most widely used 2.5G system is the General Packet Radio System (GPRS) (see Section 7.2.3). The evolution of wireless WAN cellular systems is illustrated in Figure 7.3. 3G and subsequent systems are introduced in Section 7.3.

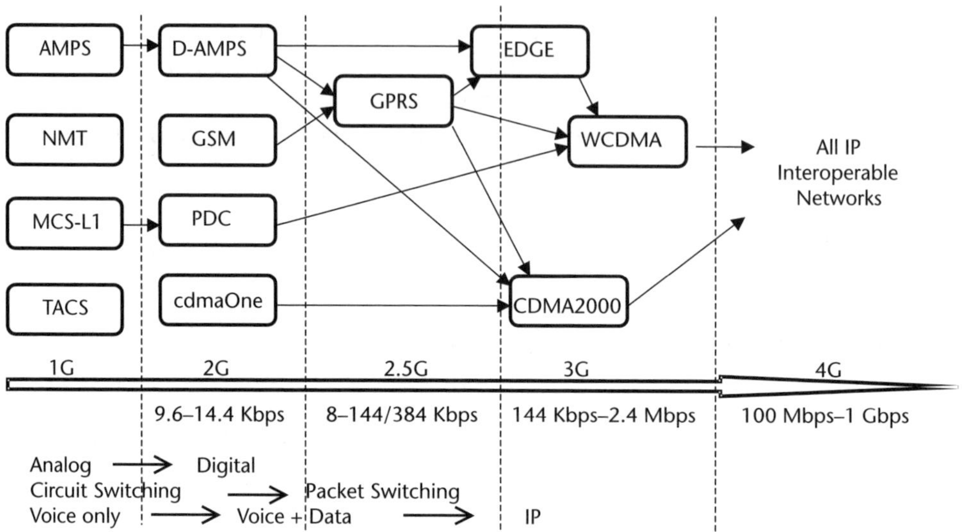

Figure 7.3   Evolution of WWAN networks.

7.2.1   1G Networks: AMPS

1G networks were all analog for voice communication, and suffered from high levels of interference, poor handoffs, and almost no security. The devices also had to be quite large to incorporate the radio receivers necessary to capture the analog signal.

The AMPS system is an FDMA-based system in which channels are separated by frequencies. The AMPS system uses 832 full-duplex channels, including 832 transmission channels from 824 to 849 MHz, and 832 simplex receiving channels from 869 to 894 MHz. Since the same frequencies cannot be reused in nearby cells, the actual number of voice channels available per cell is approximately 45. The channels are divided into four categories:

- *Control channels:* for system management (from base to mobile);
- *Paging channels:* for mobile users alerting (from base to mobile);
- *Access channels:* for call setup and channel assignment (in both directions);
- *Data channels:* for voice, fax, or data (in both directions).

Each AMPS phone implements the 21 control channels hardwired in programmable read-only memory (PROM). The bandwidth of each simplex channel is 30 kHz, which is much narrower than the later broadband systems. AMPS cells are typically 10 to 20 km in diameter, which is much larger than the later digital systems. As introduced in the previous section, the MSC of the AMPS system manages handoff completely without help from the mobile devices (i.e., NCHO). Each handoff takes about 300 ms.

The AMPS network was widely deployed across the United States. AMPS and other 1G networks do not play much of a role in the current wireless WAN environment. They are considered mainly for backward compatibility of 2G networks, or for broader coverage in rural areas for voice coverage. Due to the limitations of the

1G networks, European countries were quick to move to a digital-based 2G technology.

7.2.2　2G Networks: D-AMPS, GSM, cdmaOne, and More

The 2G cellular systems introduced digital capabilities to wireless mobile networks in the early 1990s. The advancement of high-speed digital signal processing (DSP) capability was the gateway to the use of 2G technology. Digital signals can transfer more data over the same amount of spectrum with less power consumption. Digital signals resulted in better quality voice with less interference. Filters were used to remove the interference noise in digital signals. Additionally, digital networks provided the means for stronger security, which was missing in the 1G analog networks. The 2G operators also offered some services that were not in 1G, such as voice mail, call waiting, caller ID, three-way calling, and so forth. Most of all, 2G systems started offering some basic data services. However, 2G systems were mostly circuit switch–based. Due to their inefficient performance, expensive charges, and limited data rates, 2G systems are inadequate for data-driven applications.

The four 2G networks (D-AMPS, PDC, GSM, and cdmaOne) are all based on different standards, making them incompatible with each other. Moreover, even the networks with the same network protocol may not be compatible with each other because of frequency regulations. Since PDC is essentially another D-AMPS–based system, only D-AMPS, GSM, and cdmaOne are introduced, which is followed by a brief description of two related technologies: data-only systems, and low-tier systems.

D-AMPS/TDMA

The D-AMPS system is a fully digital system that evolved from AMPS. D-AMPS uses time slices to divide a signal among multiple users (i.e., TDMA-based multiplexing). This system is described in IS-54 and its successor, IS-136. D-AMPS uses the same 30-kHz channels and the same frequencies as AMPS. D-AMPS and AMPS could operate simultaneously in the same cell.

Some operators introduce D-AMPS as a new service. A new frequency band has been made available to handle the expected increased load. The upstream channels are in the 1,850 to 1,910 MHz range, and the corresponding downstream channels are in the 1,930 to 1,990 MHz range. At these frequencies, the wavelength is shorter than AMPS, so that a standard quarterwave antenna is shorter, leading to smaller phones. However, many D-AMPS phones can use both the 850 MHz (i.e., the band used by AMPS) and the 1,900 MHz bands to get a wider range of available channels. There are also dual-mode phones, which allow for roaming between the digital D-AMPS network and the analog AMPS network.

On a D-AMPS phone, the voice signal picked up by the microphone is digitized and compressed using a vocoder circuit. The vocoder compresses human voice to a bandwidth of 8 Kbps or less, which reduces the number of bits sent over the air link.

Unlike AMPS, D-AMPS and other 2G systems adopt the MAHO strategy for handoff detection, and the D-AMPS phone uses idle slots to measure the line quality. If it detects that the signal is weak, then the MSC is notified.

The D-AMPS system does not provide any data capabilities by itself. The packet data services are provided by the cellular digital packet data (CDPD) as an additional layer of the existing AMPS and D-AMPS. This allows carriers to offer both voice and data services with their existing network infrastructure. Data can be transmitted on a channel only if there is no voice call using that channel. Voice transmission can preempt data transmission. When the voice traffic is heavy, a data transmission may spread over several channels. CDPD is the only way to send data over the analog AMPS network.

CDPD has a maximum download speed of 19.2 Kbps and upload speed of 9.2 Kbps. Since multiple users share cells, these speeds are actually lower. As discussed in Chapter 3, this transfer rate may be fast enough for simple Internet applications, such as WAP 1 on the SMS, but it is not adequate for more data-intensive applications. Despite the low data transfer rate, CDPD can run IP applications without modification. Every user connected to a CDPD network has a unique IP address. CDPD assumes that the cellular service providers centrally operate the network. CDPD network coverage is found in most major cities of North America [3].

GSM

The multiple access scheme used in GSM is a combination of FDMA and TDMA systems. Similar to D-AMPS, GSM uses FDMA, with each phone transmitting on one frequency and receiving on a higher frequency (55 MHz and higher). The GSM system has 124 pairs of 200-kHz-wide simplex channels. Each simplex channel supports eight separate connections, using TDMA. Each currently active station is assigned one time-slot on one channel pair. Transmitting and receiving do not happen in the same time-slot.

Each time-slot has a specific structure, and groups of time-slots form multiframes, also with a specific structure. Each time-slot consists of a 148-bit data frame that occupies the channel for $577\ \mu$s, including a $30\text{-}\mu$s guard time after each slot. Taking off control overhead and error correction, 13 Kbps is actually left for voice transfer, which is substantially better than D-AMPS.

GSM operates on three distinct frequencies: 900, 1,800, and 1,900 MHz. The 900 and 1,800 MHz frequencies are used in Europe and Asia. The GSM-based personal communications services (PCS) 1900 system in North America uses 1,900 MHz. DCS 1800 is also based on GSM technology, but is configured around a lower powered mobile terminal and a smaller cell size. DCS 1800 provides a maximum of 374 radio channels, compared to 124 for GSM 900. To accommodate these differences, dual-band and triband handsets are available for roaming, including international roaming, with a single phone set [4, 5].

The data transfer rate of GSM networks is 9.6 Kbps. Similar to the CDPD, this is sufficient for simple text messaging applications and WAP 1 browsing, but inadequate for other types of data applications. High-speed circuit-switched data (HSCSD) is a transitional circuit-based technology, which evolved from GSM and provides a higher data transfer rate. HSCSD allowed for multiple timeslots to be used in a single connection, for data transfer rates up to 57.6 Kbps. HSCSD is rarely considered as the future for GSM networks. Instead, through an upgrade to the packet-based GPRS technology, GSM networks have a clear path to 3G.

One of the unique features of the GSM systems is the SIM, which is a removable smart card that contains the following information: phone number (MSISDN), international mobile subscriber identity (IMSI), status of SIM, service code, authentication key, personal identification number (PIN), and personal unlock code (PUK). A user can easily remove a SIM card from a mobile device and insert it into any other GSM-compatible device to receive the same type of services. Thus, personal mobility is achieved by the use of a SIM card [6].

cdmaOne

Qualcomm developed cdmaOne as a CDMA-based circuit-switched technology. CDMA is very efficient in spectrum utilization. CDMA provides better capacity for voice and data communications, allowing more subscribers to connect at any given time.

The cdmaOne technology has been shown to increase the user capacity of an AMPS network by a factor of up to 10. Compared to GSM, cdmaOne delivers better spectral efficiency and power management, leading to longer battery life.

IS-95A was the first CDMA cellular standard. In addition to voice, IS-95A provided a data rate of 14.4 Kbps. Like other 2G systems, this rate is sufficient for text messaging and simple applications, but insufficient for richer multimedia applications. Subsequently, IS-95B, considered as 2.5G technology, provided higher data transfer rates, up to 115 Kbps [7]. The cdmaOne system operates over two main frequencies: 800 and 1,900 MHz. The 800-MHz system competes with D-AMPS in the United States, while the 1,900-MHz system competes with GSM. Each channel of a cdmaOne system is 1.25 MHz wide.

CDMA has also been selected by the International Telecommunication Union (ITU) as the basis for 3G networks. The CDMA Development Group (CDG) and the Universal Mobile Telecommunication System (UMTS) Forum now handle the evolution and promotion of CDMA. The newer CDMA 3G standards (CDMA2000 and WCDMA) are described in the next section.

2G Data-Only Networks

Beside the voice-data coexisting 2G systems, several packet-switched narrowband wireless data access networks are developed for data-only applications. Mobitex and Data Total Access Communications (DataTAC) are the most famous applications. Ericsson first developed the Mobitex application in the 1980s, and the application has since been deployed in more than 30 networks worldwide. It provides low data transfer rates, up to 8 Kbps, at a low cost. Mobitex operates on three different spectrum bands: 400, 800, and 900 MHz. Any application developed for the Mobitex specification can operate equally well on any of these frequencies. It is mainly used by military, police, firefighter, and ambulance services with Palm and RIM Blackberry devices.

DataTAC offers data transfer rates up to 19.2 Kbps over an 800-MHz frequency. The DataTAC 4000 networks deployed in North America, known as ARDIS, cover more than 90% of the business population. The RIM Blackberry devices support the DataTAC network for wireless connectivity.

Low-Tier Digital Systems

During the period of 2G infrastructure establishment, some cordless systems were also developed. The well-known examples include: CT-2, Digital Enhanced Cordless Telecommunications (DECT) in Europe; and Personal Handyphone System (PHS) in Japan. These systems do not have a network component. A typical system configuration includes a BS and a group of handsets. The BS is attached to some other network, which can be either a fixed or mobile network. The coverage area is often quite limited. Among them, PHS is a more advanced system and provides comprehensive data services [2].

7.2.3 2.5G Networks: General Packet Radio System and Enhanced Data Rates for Global/GSM Evolution

Data applications on 2G networks are typically text-based with limited graphics. Due to the high cost of building a 3G network, 2.5G networks provide an intermediate step during the evolution to 3G. The 2.5G networks can be deployed on existing 2G systems, with higher data rates and lower costs [8]. The following are four important characteristics of 2.5G networks.

- *Better performance:* 2.5G systems accommodate more users on the same network through more efficient modulation algorithms and multiplexing. 2.5G networks can provide transfer rates up to 144 Kbps, which is a significant improvement over 2G systems.
- *Packet-based, always-on capability:* Since most 2G systems are circuit switching–based, data services on a 2G network are charged by the minute. The packet switching–based 2.5G systems provide always-on capabilities that allow users to remain connected to the wireless network for extended periods of time without incurring large usage fees. Packet switching also allows for applications that can push data to the user, rather than have users pull it.
- *Upgrade from 2G Systems:* 2.5G technologies are upgrades to existing cellular networks. Users still have the same voice capabilities as with 2G, but now have high-speed data access with the same network coverage. In most cases, moving from 2G to 2.5G involves a only software upgrade for the wireless carrier, as opposed to building new infrastructure.
- *Foundation of 3G:* From a business point of view, it is expected that as users experience the benefits of 2.5G high-speed data, they will require even more speed and capacity. It is natural to advance to 3G as the next step. From a technical point of view, 2.5G systems provide some of the core 3G infrastructures in the form of packet-switching networks.

This section introduces the two main 2.5G networks: the General Packet Radio System (GPRS), and the Enhanced Data Rates for Global/GSM Evolution (EDGE). Actually, EDGE provides 3G services with GSM/GPRS/D-AMPS infrastructure upgrade. In this book, we classified EDGE as a 2.5G technology, since it has a close relationship to both 2G and 3G systems. Some people also considered CDMA2000 1x as a 2.5G technology from a performance point of view. We introduce CDMA2000 in Section 7.3.1.

GPRS

GPRS introduced packet-switched capabilities to wireless networks. It is a simple, cost-effective upgrade to GSM/D-AMPS networks, which provides increased bit rates and an improved user experience. Users are able to roam between the GPRS and GSM networks. GPRS also provides a foundation for EDGE and WCDMA systems.

GPRS allows an MS to send and receive IP packets in a cell. While GPRS is in operation, some time-slots on some frequencies are reserved for packet traffic. Depending on the ratio of voice to data traffic in the cell, the number and location of the time-slots can be dynamically managed by the BS. The available time-slots are divided into several logical channels, and are used for different purposes. The BS determines the mapping between logical channels and time-slots. One logical channel is for downloading packets from the BS to some MS, with each packet indicating its intended destination. To send an IP packet, an MS requests one or more time-slots by sending a request to the BS. If the request arrives without damage, then the BS announces the frequency and time-slots allocated to the MS for sending the packet. Once the packet has arrived at the BS, it is transferred to the Internet by a wired connection.

GPRS is trying to transform the 2G system to a more data-friendly architecture. As illustrated in Figure 7.4, two types of gateways are added into the GPRS systems: serving GPRS supporting node (SGSN), and gateway GPRS supporting node (GGSN). SGSNs and GGSNs are responsible for billing, relaying, routing, encapsulating, tunneling, address translating, mapping, and managing mobility.

BSs are connected to SGSNs. The SGSNs also work on authentication and admission control, ciphering, and compression. Thus, SGSNs can be considered as access points to the GPRS core network. When moving to another SGSN (i.e., location update), GGSN and HLR are informed about the new routing context. GGSN is a gateway interfacing the GPRS core network to other packet data networks (PDNs). GGSNs must screen messages between two connected networks.

The GPRS protocol is designed to support the bursts of data traffic over the cellular network. For GSM services, every MS call causes a query to HLR, and all ser-

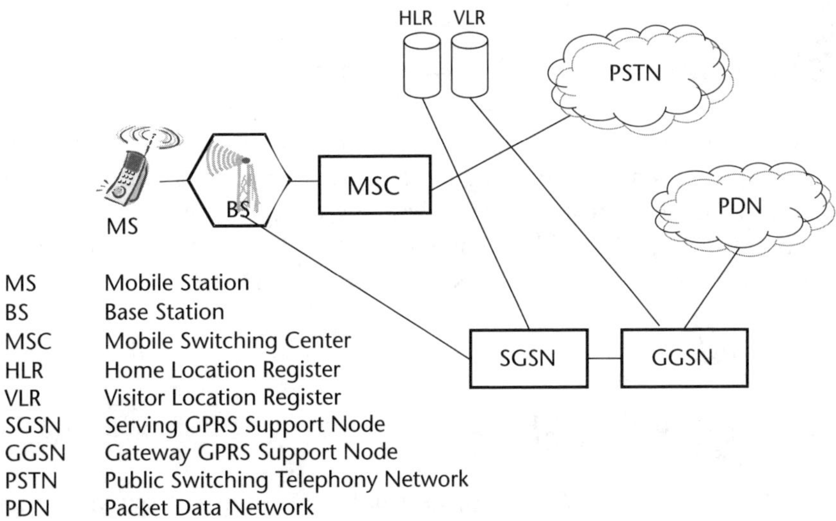

MS	Mobile Station
BS	Base Station
MSC	Mobile Switching Center
HLR	Home Location Register
VLR	Visitor Location Register
SGSN	Serving GPRS Support Node
GGSN	Gateway GPRS Support Node
PSTN	Public Switching Telephony Network
PDN	Packet Data Network

Figure 7.4   The GPRS reference architecture.

vices activated at the initiation of IMSI. Every network element knows where to route the GPRS service packet. Each GPRS packet does not need to access HLR. Users can activate each service separately.

GPRS operates on the same frequencies as the 2G networks that it upgrades: 900, 1,800, and 1,900 MHz. It can provide high-speed data transfer, theoretically from 115 to 144 Kbps, but in practice from 40 to 56 Kbps. This is still fast enough for a broad range of new applications. Often, the receiving device becomes the limiting factor for the total data rates. In order to achieve the theoretical data rate, the device has to utilize all available time-slots for both uploading and downloading data. Most devices do not provide this capability.

Three classes of terminal devices have been defined for GPRS services. In addition to mobile handsets, GPRS terminal devices may also be wireless cards or other modules. Class A devices handle both voice and packet data at the same time. Class B devices can handle both voice and packet data, but not at the same time. This requires that one transceiver be used for either voice or data. Class C handsets can handle either voice or data. Each Class A device requires two transceivers. Class C devices may be a low-end handset, or, more likely, a wireless modem [9].

EDGE

EDGE is a packet switching technology in which D-AMPS and GSM/GPRS converges. EDGE allows 3G services on 2G infrastructure with existing frequencies. D-AMPS or GSM/GPRS infrastructure can be upgraded to EDGE with minimum impact, mainly on error correction. Since EDGE is a narrowband (i.e., 200-kHz channels) technology, wireless operators are able to deploy 3G EDGE services without obtaining a 3G license. This makes EDGE a low-cost but fast solution for providing 3G services.

EDGE is able to increase over-the-air data transfer rates using the same spectrum as used by 2G and 2.5G systems. With a spectrum-efficient modulation scheme (i.e., QPSK), EDGE can achieve a peak data rate of 384 Kbps, with actual download rates ranging between 75 and 150 Kbps [10].

7.3 Cellular Mobile Networks—3G and Beyond

3G networks started with the vision of developing a single global standard for high-speed data and high-quality voice services. In 1999, the ITU approved an industry standard for 3G wireless systems, called the International Mobile Telecommunication-2000 (IMT-2000) [11]. The IMT-2000 has defined three levels of mobility. All 3G networks must support the following minimum requirements:

- *High Mobility:* 144 Kbps for users traveling at speeds greater than 120 km/h (75 mi/h);
- *Full Mobility:* 384 Kbps for users traveling from 10 to 120 km/h in urban areas (from 6 to 75 mi/h);
- *Limited Mobility:* 2 Mbps for users traveling at less than 10 km/h (approximately 6 mi/h).

Many of the applications designed for 3G networks are similar to those developed for 2.5G systems. Both generations of networks support IP applications, providing an easy migration path for application developers. The basic services that the IMT-2000 network is supposed to provide are:

- High-quality voice transmission;
- Messaging (replacing e-mail, fax, SMS, chat, and so forth);
- Multimedia (downloading and playing music, videos, television, and so forth);
- Internet access (Web access, with audio and video pages).

Third generation networks could also provide more advanced services through mobile devices, such as video conferencing, workplace collaboration groups, voice-over-IP (VoIP), and m-commerce (e.g., electronic payments). These services are supposed to have worldwide availability, instantaneous access (i.e., always-on), and QoS guarantees. 3G systems allow users to establish agreements with network operators for certain QoS properties, such as data transfer rate and network latency [12].

The goal of 3G was to have a single standard that would allow for true global roaming. Having a single technology would also make life much simpler for network operators, device manufacturers, and users. Unfortunately, companies and standards bodies could not agree on a single protocol for 3G systems. After long negotiations, it was realized that backward compatibility with 2G networks and frequency differences among countries were too difficult to overcome. Instead of reaching an agreement on a single 3G standard, two 3G systems were created: CDMA2000 and wideband CDMA (WCDMA).

Many manufacturers have already released mobile devices for 3G networks. These devices usually support multimedia content, such as video streaming with large high-resolution screens. These devices also support mobile operating systems that allow for sophisticated client-side applications. The details of mobile operating systems and applications are introduced in Chapter 2.

7.3.1   3G Networks: CDMA2000 and WCDMA

The 3G technologies are mostly CDMA-based. Although basic principles behind both CDMA2000 and WCDMA systems are the same, they are different in chip rate, frame time, spectrum, and time synchronization method.

CDMA2000

CDMA2000, proposed by Qualcomm, is basically an extension of, and backward compatible with, IS-95 cdmaOne. CDMA2000 offers a migration path from both GSM and D-AMPS networks. The CDMA2000 suite of technologies includes: CDMA2000 1x, CDMA2000 1x Evolution Data Optimized (CDMA 1x EV-DO), CDMA2000 1x Evolution Data and Voice (CDMA 1x EV-DV), and CDMA20000 3x.

As defined by the ITU, CDMA2000 1x is officially a 3G technology. The 1x in the CDMA2000 1x signifies that it uses one 1.25-MHz channel. Due to improved modulation, power control, and overall design, CDMA2000 1x can achieve theoretical data transfer rates of 307 Kbps on a single 1.25-MHz channel, but only from 40 to 56 Kbps in practice. Although CDMA2000 1x provides nearly double the capacity of previous cdmaOne systems, it does not meet the minimum speed requirements for 3G networks. Its performance is similar to that of GPRS. Some people considered CDMA2000 1x as a 2.5G technology. CDMA2000 1x also uses packet switching to provide always-on capabilities.

CDMA2000 1x is operating on the same 800- and 1,900-MHz frequency bands as cdmaOne. It is a cost-effective upgrade to existing cdmaOne, paving the way for CDMA2000 1xEV and CDMA2000 3x networks. CDMA2000 1x also provides an intermediate step to WCDMA. SK Telecom of Korea launched the first CDMA2000 1x commercial system in October 2000. Since then, CDMA2000 1x has been deployed in Asia, the United States, and Europe.

CDMA2000 1x EV-DO is a data-optimized version of CDMA2000 that can deliver data rates of up to 2.4 Mbps using a single CDMA 1.25-MHz channel. With the introduction of improved technology for modulation and dynamically assigned data rates, 1x EV-DO is able to achieve downloading rates near the theoretical levels.

CDMA2000 1x EV-DO supports the complete range of available frequencies, including 450, 700, 800, 1,800, and 1,900 MHz, and 2 GHz. It can work on all IP networks, giving wireless operators an opportunity to gain experience with IP-based technologies before moving their voice networks to IP. SK Telecom of Korea also deployed the first commercial 1xEV-DO network in January 2002.

CDMA2000 1x EV-DV networks provide similar wireless data capabilities as do CDMA2000 1x EV-DO networks, but with the addition of integrated voice capabilities. The 1x EV-DV network will provide peak data rates of 3 Mbps, with a typical throughput of 1 Mbps. This technology will enable real-time packet services for two-way conversational communication. CDMA20000 3x networks use three 1.25-MHz channels simultaneously, thereby providing increased data throughput. The CDMA2000 core network is considered IP-friendly, and it is efficient for Internet services [13].

WCDMA/UMTS

WCDMA, proposed by Ericsson, has been designed to work with the GSM suite of networks. The European Union advocated this system, which it called UMTS.

Two types of WCDMA are in use: frequency-division duplex (FDD), and time-division duplex (TDD). TDD is a hybrid of CDMA and TDMA technologies that is better suited for indoor usage. The FDD version is being deployed commercially in several countries. The WCDMA being deployed in Europe (i.e., UMTS) is also called DS-WCDMA, where DS indicating direct sequence.

The channel bandwidth of WCDMA is 5 MHz, which is four times larger than that of CDMA2000 1x (1.25 MHz), and 25 times larger than that of GSM (200 kHz). The wider bandwidth allows for higher data transfer rates. WCDMA supports all three modes of mobility defined in the IMT-2000.

Unlike the upgrade to GPRS and EDGE, WCDMA requires coverage to be built from scratch. Since this requires significant capital investments by carriers,

WCDMA coverage is initially available only in urban centers. Fortunately, WCDMA is compatible with GSM, GPRS, and EDGE networks, so users do have the ability to roam across the various networks. As users move out of WCDMA coverage, they will automatically be handed off to a GSM, GPRS, or EDGE network, and continue communicating at a slower data rate.

In Japan, NTT DoCoMo has deployed a WCDMA-based 3G network called Freedom of Multimedia Access (FOMA), which is slightly different than the WCDMA version used in Europe. FOMA was designed as an upgrade from PDC without an intermediate 2.5G network, and does not require backward compatibility with GSM-based networks. In North America, the 2-GHz frequency band is already in use, so WCDMA either has to be introduced into the existing frequencies, or a new spectrum has to be allotted. This is one of the reasons that a single 3G network protocol cannot be adopted worldwide.

7.3.2   4G Networks and Beyond

It is becoming increasingly difficult to find a common spectrum to enable 3G global roaming. The recommended spectrum allocation in the IMT-2000 could not be implemented in many countries. Deploying 3G systems has been found to be not cost-effective for some countries (e.g., India) and operators (e.g., Nextel in the United States). These countries and operators aim to do directly to 4G wireless WAN technology. Historically, each cellular generation requires approximately 10 years to be completed. Japan will start to reallocate part of the spectrum for 4G in 2005, and prepare for the full deployment in 2010.

The 4G wireless WAN networks are mostly a concept. There are no standards under development. The requirements of 4G approved by ITU-R in June 2003 are: (1) 100 Mbps data rate for high-speed mobile data, and (2) 1 Gbps data rate for a static system. Under such a broadband WAN environment, multimedia and peer-to-peer (P2P) applications are expected to be two types of typical applications on 4G systems.

The 4G wireless-broadband systems can be considered in two ways. From the wireless accessing point of view, 4G is a rival to wireless LANs that offers wider coverage with high mobility. From a wired accessing viewpoint, 4G offers a wireless alternative to the cable and digital subscriber line (DSL) technologies. Current consensus indicates that 4G will be seen as a unification of different wireless networks, including wireless LANs (e.g., IEEE 802.11); public cellular networks (including 2G, 2.5G, and 3G systems); and personal area networks (e.g., Bluetooth). 4G calls for a wide range of mobile devices roaming across different types of networks. Figure 7.5 shows the 4G interconnections [14, 15].

To seamlessly integrate with different networks (especially IP), and to achieve high data rates and high mobility, 4G systems require advanced radio multiple access technologies.

Advanced Radio Access Network

The 4G access technology aims at a 1-Gpbs data rate, with a 20- to 30-MHz channel spectrum. Advanced antenna and multiplexing technologies are expected to be

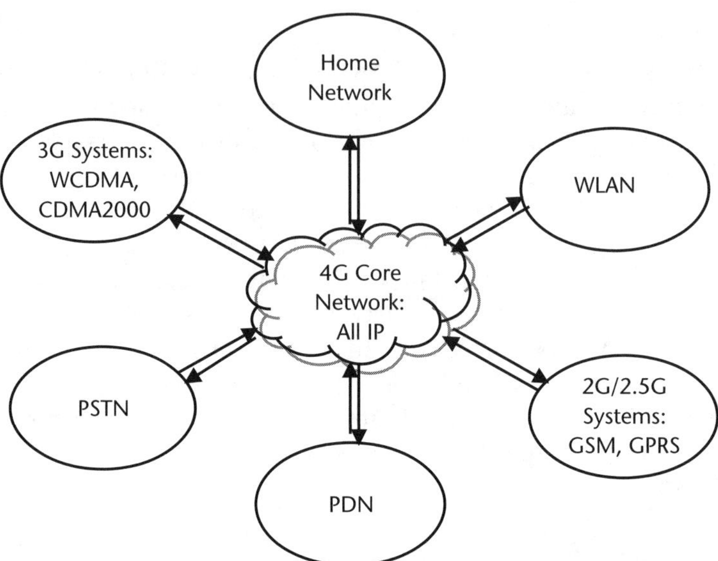

Figure 7.5 4G network interconnections.

adopted. Adaptive/intelligent antennas with multiple-input multiple-output (MIMO) channels are expected to increase the system capacity, coverage, and signal quality. Major candidates for 4G multiple access technology include OFDM, and multicarrier CDMA (MC CDMA, a combination of CDMA and OFDM). A number of 4G air interfaces are ready for beta deployments by leading wireless operators [16, 17].

All IP P2P Core Network

Although many aspects of 4G have yet to be finalized, the 4G core network will support all-IP P2P networking. An all-IP network indicates end-to-end IP protocol support. It makes sense because users are likely to use the same Internet applications as in wired networks. P2P applications are considered to be one of the most significant applications for 4G. P2P systems avoid the single point-of-failure problem of traditional client-server systems. From the application viewpoint, every peer is both a client and a server for other peers. From the core network viewpoint, every device is both a transceiver and a router for other devices in a 4G P2P network.

4G Pervasive Services

Since the 4G network integrates different wired and wireless technologies, 4G operators will provide not only controlled services of their own, but also services from other networks. To seamlessly integrate with different networks, computing devices are expected to share information, and are embedded in virtually all devices. Software architecture and development technologies, such as WAP 2, UAProf, CC/PP, and many XML-related technologies, are also developed to meet this goal. These service architecture and development technologies are introduced in Chapters 3 and 4.

7.3.3   Standardization Organizations

The research and development of wireless WANs would be meaningless without standardization. It is important to have a basic knowledge about those standardizing forums and their responsibilities. These organizations also serve as good resources for information on specific technologies.

Unlike the open mobile alliance (OMA) introduced in Section 3.5.2, which is a single standardization body for wireless applications, services, and architectures, there are many organizations that oversee wireless WAN cellular technologies. On the other hand, cellular operators based on one technology are consolidating with companies that offer complementary technology. Maintaining an accurate list of the cellular carriers is not easy. Web sites of these standardization organizations maintain the most recent information on the carriers in each region for their respective technologies.

ITU

The ITU is an international organization within the United Nations that coordinates global telecommunication networks and services, mainly through advice and recommendations from governments and private industries. The ITU is divided into three main sectors: the radio communication sector (ITU-R), the telecommunication development sector (ITU-D), and the telecommunication standardization sector (ITU-T). The ITU-T replaced the role of the former International Telegraph and Telephone Consultative Committee (CCITT) in 1993 [18].

GSM Association

GSM is rapidly approaching 1 billion users, with 1 in 7 people on the planet already connected. The GSM Association is a global trade association serving the worldwide GSM mobile operator community by promoting, protecting, and enhancing user interests and investments. The GSM Association members are 2G and 3GSM mobile network operators, implementing the GSM technologies (GSM, GPRS, EDGE, and 3GSM). The 3GSM technology is the latest addition to GSM, and is built around a core GSM network with a WCDMA-air interface [19].

CDG

The CDG, founded in December 1993, is an international consortium of companies that have joined together to lead the adoption and evolution of CDMA wireless systems. The CDG is comprised of CDMA service providers and manufacturers. By working together, the members help ensure the interoperability among systems, while expediting the availability of CDMA technology to consumers. CDG technical teams focus on the development of cdmaOne and CDMA2000 advanced features and services [20].

UMTS Forum

UMTS/WCDMA represents an evolution in terms of services and data speeds from 2G mobile networks. UMTS is the natural evolutionary choice for operators of GSM networks. The UMTS Forum is a forum for researchers and developers of the UMTS technology, providing updates of news, activities, and implementations around the world. The UMTS Forum is also an open, international body for promoting UMTS 3G mobile systems and services. The UMTS Forum recognizes the importance of all players in the mobile value chain, including new entrants [21].

Third Generation Partnership Project

The Third Generation Partnership Project (3GPP) is an IMT-2000 initiative of ITU. It is a collaboration agreement aimed at bringing together a number of telecommunication bodies, which are known as "Organizational Partners," including the Association of Radio Industries and Business (ARIB), and the Telecommunication Technology Committee (TTC) of Japan; the Telecommunications Technology Association (TTA) of South Korea; the China Wireless Telecommunication Standard Group (CWTS) of China; and ANSI T1 1, by the Committee of the Telecommunication of American National Standard Institute. 3GPP produces globally applicable technical specifications and reports for 3G mobile systems, based on evolved GSM core networks, including GPRS and EDGE [22].

Third Generation Partnership Project 2

The Third Generation Partnership Project 2 (3GPP2) works on a 3G standard, based on ANSI/TIA/EIA-41 systems. Some specifications prepared by 3GPP2 are related to CDMA2000. 3GPP is an observer to 3GPP2 [23].

Wireless World Research Forum

The Wireless World Research Forum (WWRF) is a global organization founded in August 2001. Members of the forum include manufacturers, network operators/service providers, R&D centers, universities, and small and medium enterprises. The WWRF provides a global platform for discussion of results and exchange of views to initiate global cooperation toward systems beyond 3G [24, 25].

7.4   Summary

The evolution of wireless WANs (i.e., cellular mobile systems) involves many wireless network technologies. It has been more than 20 years since the 1G analog networks were commercially deployed. The emergence of digital technology created the capability to transfer data with cellular networks. The packet switching networks provided the always-on capabilities. Most of all, for any new cellular system to succeed in the market, it must be back-compatible with a previous installed system. This chapter presented the architecture and evolution of cellular mobile networks. Cell characteristics and handoff mechanisms were introduced. The cellular systems from 1G up to 4G, including AMPS, GSM, D-AMPS, cdma2000, GPRS, CDMA2000, and WCDMA, were then presented.

While network operators continue to deploy 3G networks, WLANs are rapidly being deployed. Although WLANs provide a high-speed, low-cost solution for wireless data access, the revenue stream is uncertain. The mobility limitation is another problem for WLANs. 3G operators expect new income and profits from WWANs, but the voice communication remains the main source of revenue. Furthermore, the fast growth of P2P VoIP applications seems to be a major threat to both wired and wireless telecommunication industries in the near future.

References

[1] Tanenbaum, A. S., *Computer Networks*, 4th ed., Upper Saddle River, NJ: Prentice Hall, 2003.

[2] Lin, Y. B., and I. Chlamtac, *Wireless and Mobile Network Architectures*, New York: John Wiley & Sons, 2000.

[3] CDPD Industrial Coordinator, Cellular Digital Packet Data Specification, Release 1.0, Kirkland, WA, 1994.

[4] Redl, S. M., M. K. Weber, and M. W. Oliphant, *An Introduction to GSM*, Norwood, MA: Artech House, 1995.

[5] Rahnema, M., "Overview of the GSM System and Protocol Architecture," *IEEE Communication Magazine*, Vol. 31, No, 4, 1993, pp. 92–100.

[6] GSM 02.17 (ETS 300 509), European Digital Cellular Telecommunications System (Phase 2); Subscribers Identity Modules (SIM) Functional Characteristics.

[7] Lee, J. S., and L. E. Miller, *CDMA System Engineering Handbook*, Norwood, MA: Artech House, 1998.

[8] Siva Ram Murthy, C., and B. S. Manoj, *Ad Hoc Wireless Networks: Architectures and Protocols*, Upper Saddle River, NJ: Prentice Hall, 2004.

[9] Bettstetter, C., H. Vogel, and J. Eberspacher, "GSM Phase 2+, GPRS: Architecture, Protocols, and Air Interface," *IEEE Communications Surveys*, Vol. 2, No. 3, 1999.

[10] Furuskar, A., et al., "EDGE: Enhanced Data Rate for GSM and TDMA/136 Evolution," *IEEE Personal Communication Magazine*, Vol. 6, No. 3, 1999, pp. 56–66.

[11] International Mobile Telecommunication-2000 (IMT-2000), http://www.imt-2000.org.

[12] Mallick, M., *Mobile and Wireless Design Essentials*, New York: John Wiley & Sons, 2003.

[13] Garg, V. K., *IS-95 CDMA and cdma2000: Cellular/PCS Systems Implementation*, Upper Saddle River, NJ: Prentice Hall, 2000

[14] Lu, W., *Broadband Wireless Mobile: 3G and Beyond*, New York: John Wiley & Sons, 2002.

[15] Jamalipour, A., and S. Tekinary, "Fourth Generation Wireless Networks and Interconnecting Standard," *IEEE Personal Communication Magazine*, Special Issue, Vol. 8, No. 5, 2001.

[16] Glisic, S., *Advanced Wireless Communications: 4G Technologies*, New York: John Wiley & Sons, 2004.

[17] Tsoulos, G., "Adaptive Antennas and Mimo Systems for Wireless Communications: Part I," *IEEE Communications Magazine*, Vol. 42, No. 10, October 2004, p. 26.

[18] International Telecommunication Union (ITU), http://www.itu.int/home/.

[19] GSM Association, http://www.gsmworld.com.

[20] CDMA Development Group, http://www.cdg.org.

[21] Universal Mobile Telecommunication System (UMTS) Forum, http://www.umts-forum.org.

[22] Third Generation Partnership Project (3GPP), http://www.3gpp.org.
[23] Third Generation Partnership Project 2 (3GPP2), http://www.3gpp2.org.
[24] Wireless World Research Forum (WWRF), http://www.wireless-world-research.org.
[25] Roberto, J., "WWRF Visions and Research Challenges for Future Wireless World," *IEEE Communications Magazine*, Vol. 42, No. 9, September 2004, pp. 54–55.

CHAPTER 8

Wireless Personal Area Networks

Compared to the local area network (LAN) and the wide area network (WAN), the personal area network (PAN) is a relatively new networking solution. A PAN could be thought of as the interconnection of devices within the short range of an individual person, and it typically uses wireless technologies. Wireless PANs (WPANs) can be used to replace cables between computers and peripherals, to establish communities and share resources between devices, and to establish a variety of services [1, 2].

The WPAN technology is undergoing rapid development. There are many devices currently on the market that have incorporated WPAN capabilities, such as mobile phones, notebooks, and headsets. It is also common to allow a wired device without built-in WPAN technologies to join a WPAN by just adding an adapter. A WPAN is automatically set up as soon as two enabled devices come within the range of each other. It even provides a way to establish ad hoc networks among mobile devices. The IEEE 802 committee initiated the 802.15 working group to standardize protocols and interfaces for WPAN [3].

Besides supporting short-distance wireless connectivity, WPAN devices are expected to be cheap, low power consuming, and interoperable. Device manufacturers are willing to incorporate WPAN into a broad range of products, since the price of WPANs are cheaper than WLANs. Low power consumption is critical for any mobile device, and interoperability is also imperative. Since there are many WPAN technologies, devices are not only capable of communicating with other devices using the same technology, but also with devices using different technologies. Thus, the creation of application-level middleware and corresponding protocols for device coordination is critical.

The rest of this chapter is structured as follows. In Section 8.1, we provide an overview of application scenarios and standards of current WPAN technologies. In Section 8.2, we focus on the most popular WPAN technology–Bluetooth. The protocol suite, operations, and application profiles are illustrated in detail. A relatively new technology, near-field communication (NFC) for very short-distance applications is also introduced. In Section 8.3, we discuss two of the most widely adopted device coordination and service discovery schemes: Jini and Universal Plug-and-Play (UPnP). The home service gateway is also introduced. Finally, Section 8.4 summarizes the WPAN.

8.1 Wireless PAN Overview

A WPAN is a wireless network used for communication among devices within a short range of distance. Wireless LANs (WLANs) have a greater data rate and range than most WPANs. A WLAN is more suitable for cable replacement than WPAN. However, WPANs usually support a simple setup and administration, which implies the immediate deployment of a dynamic network. Furthermore, the WPAN technology is not limited to cable replacement, but also applies to more advanced applications. The current WPAN usage scenarios include the following:

* Synchronizing objects and files (e.g., e-mail, calendar, and contacts) between a desktop and a portable device;
* Exchanging information (e.g., name card) directly between two mobile devices;
* Connecting a peripheral (e.g., keyboard, mouse, printer) wirelessly to a desktop/notebook PC;
* Connecting a WPAN-enabled headset to a mobile phone or portable media player;
* Using WPAN features to connect a device to the Internet;
* Using WPAN-enabled devices to form a dynamic network.

A key issue in WPAN technology is device coordination (also known as service discovery, plugging in, or wireless access protocol). In the ideal scenario, any two WPAN-enabled devices that come into close proximity can communicate as if connected by a cable. Within the physical limited range, every device in a WPAN should be able to plug into any other device in the same WPAN, or further discover the available services. To avoid unauthorized access, each device should be able to lock out selected devices. Device coordination issues and schemes are introduced in Section 8.3 [4].

The number of mobile devices is growing significantly. It is expected that there will be over 2 billion mobile subscribers in the world by July 2006. As a result, more peripherals and sensors (e.g., health monitoring sensors, contactless IC card) are being integrated with mobile devices. WPAN researches show a strong interest in new application ideas and unresolved technical problems. Ongoing research topics include application scenarios, scalability issues, infrastructure issues, quality of service, and security issues. The mobile ad hoc networks (MANET), a popular research topic, can be viewed as a complex network using WPAN technology. The initial research focused on developing routing protocols, and was conducted by the Internet Engineering Task Force (IETF) MANET working group. There are also quite a few research activities on scalability issues [5].

8.1.1 IEEE 802.15

Today, there are many WPAN technologies available. Each technology has its own strengths and weaknesses, making it suitable for specific applications. The IEEE 802.15 WPAN working group was established in 1999 as a part of the Local and Metropolitan Area Networks Standards Committee, which only focuses on lower-layer protocols. The higher-layer protocols are specified by industry working groups, such as Bluetooth special interest groups (SIGs) [6] and the ZigBee Alliance [7]. Beyond the scope of 802.15, some WPAN technologies have their own standardization organizations, such as the Infrared Data Association (IrDA) for infrared communication [8, 9], and the newly formed NFC Forum for very short-distance near-field communication (in Section 8.1.2) [10].

The 802.15 specifications broadly cover surrounding WPAN technology, including Bluetooth, Bluetooth coexisting with WLAN, high data rates, low data rates, and low-power consumption solutions. Four task groups (TG1, TG2, TG3, and TG4) have been established. Each task group works on specific components of the 802.15 specification [3], as described in the following sections.

WPAN/Bluetooth Task Group (TG1)

TG1 created the WPAN 802.15.1 standard based on the Bluetooth specification. The IEEE licensed the technology only to PHY and MAC layers from the Bluetooth SIG. The Bluetooth SIG itself pushed ahead of the 802.15.1 standard, and developed specifications with higher data rates and a wider range. Whether or not newer Bluetooth versions will also be ratified under the 802.15 standard is unclear. The details of Bluetooth technology are discussed in Section 8.2.

Coexistence Mechanisms Task Group (TG2)

TG2 is developing the recommended practices to facilitate the coexistence of WPAN (IEEE 802.15.1/Bluetooth) and WLAN (IEEE 802.11b/g). Both technologies use the 2.4-GHz industrial, scientific, and medical (ISM) unlicensed spectrum, in which interference seems unavoidable. However, if devices are far enough apart, or if adaptive frequency hopping is implemented, then the interface problem will be minimized. Bluetooth devices are usually low-power devices, so the effects that they may have on an 802.11 network are not far-reaching, except in localized scenarios (e.g., Bluetooth and WLAN in mobile phones). Furthermore, Bluetooth hops (1,600 hops/s) much faster than 802.11 does. Therefore, it is far more likely that a Bluetooth device will ruin 802.11 transmissions than the other way around [3].

High Rate WPAN Task Group (TG3)

TG3 specifies 802.15.3 for high data rates (up to 55 Mbps), and low-power, low-cost WPAN technologies that are capable of handling multimedia content with a high quality of service (QoS). As introduced in Section 6.5.2, the ultrawideband (UWB) is being considered by the IEEE as the 802.15.3a (WPAN Alternate Higher Rate) standard for applications with data rates of up to 440 Mbps. At the time of this book's writing, there are two separate UWB implementations that may appear as rivals in the marketplace, and this leads to potential interoperability issues in the future [11].

Low Rate, Long Battery Life Task Group (TG4)

TG4 is chartered to establish a low data rate (200 Kbps maximum) solution, with a long battery life (months or even years) and low complexity. It is intended to operate in an unlicensed international frequency band, and is targeted toward sensors, interactive toys, smart badges, home automation, and remote controls. The ZigBee standard specified the upper layer of the 802.15.2, and has been endorsed by the IEEE as the official 802.15.4 standard. ZigBee is aimed at low-cost, low-power ad hoc applications.

In May 2005, the Bluetooth SIG announced its intent to work with UWB manufacturers to develop the next generation Bluetooth technology, using UWB technology and delivering UWB speeds. This will enable Bluetooth technology to be used in delivering the high-speed network data exchange rates required for wireless VoIP, music, and video applications.

8.1.2　Near-Field Communication

Near-field communication (NFC) is a short-range radio frequency (RF) technology for very short-range half-duplex wireless connectivity without user configuration. In order to make two devices communicate, the users bring them either close together or in contact. NFC evolved from a combination of contactless identification (RF identification) and networking technologies. It is compatible with both Philips' MIFARE and Sony's Felica contactless smart card platforms. Philips, Nokia, and Sony founded the NFC Forum to standardize the NFC technology. At the time of this book's writing, Nokia has announced a Nokia NFC shell with four tags, which is a functional cover developed for the Nokia 3220 phone. Users get easy access to services by simply touching service shortcut tags in the mobile phone [10].

Applications

One major application of NFC technology is its use in configuring other faster technologies, such as Bluetooth, WiFi, and so forth. NFC takes the responsibility of corresponding and aligning with all of the initial communications. Once configured, NFC will switch off, and communications will take place in the other technologies. Whether or not this application of the technology becomes widely adopted will depend on other technologies.

More than just setting up a wireless connection, NFC also provides a secure storage of confidential personal data, such as credit card numbers, coupons, membership data, or digital rights. NFC allows users to intuitively interact with the electronic environment. Touch-based interactions offer mobile device users an intuitive and easy way to connect, collect, and share information. NFC-based applications at least include the following:

- *Touch and Go:* This is for applications such as access control or ticketing, in which the user only needs to bring the device storing the ticket or access code close to the reader. This is also for simple data-capture applications, such as picking up an Internet URL from a smart NFC–enabled label on a poster.
- *Touch and Confirm:* This is for applications such as mobile payment, in which the user has to confirm the interaction by entering a password or accepting the transaction [12].
- *Touch and Connect:* This is for the application link of two NFC-enabled devices to enable peer-to-peer transfer of data, such as downloading music, exchanging images, or synchronizing contact information.

Protocol and Operation

NFC is operated at the unregulated speed of 13.56 MHz. The specified transfer rates are 106, 212, and 424 Kbps. Higher rates are expected in the near future.

NFC-enabled devices can operate in active or passive modes. Mobile devices operating primarily in passive mode can achieve significant power savings. Active NFC devices can supply all the power needed for communication with passive devices in exactly the same way contactless smart cards are powered, which ensures that the data remains accessible even when the mobile device is switched off.

The NFC system consists of a tag and a reader. The reader, when activated, emits a short-range radio signal that powers up a microchip on the tag, and allows the reading of a small amount of data that is stored on the tag. The effective distance of an NFC solution, which is typically only a few centimeters, depends on the design of the tag and the reader. NFC connection will be terminated on command or if the distance is greater than, for instance, 20 cm.

8.2   Bluetooth

Bluetooth got its name from the King of Denmark, who united Denmark, Norway, and Sweden in the tenth century. Bluetooth was likewise intended to unify different devices and accessories. Ericsson, together with IBM, Intel, Nokia, and Toshiba, formed the Bluetooth SIG to develop a wireless standard for Bluetooth technology. The Bluetooth SIG has over 2,000 members at the time of this book's writing [6].

The Bluetooth Version 1.0 had many problems. Manufacturers had great difficulties in making their products interoperable. Later, Version 1.1 fixed the problems, and made some modifications to improve the performance of Bluetooth devices. Version 1.2 supported a higher transmission speed, and was backwards-compatible with Version 1.1. In addition to performance enhancements, Bluetooth Version 1.3 now supports nonhopping narrowband channel(s) and broadcasts/multicasts.

The Bluetooth SIG issues two types of documentation: specifications and profiles. The specification describes the device, infrastructure (see Section 8.2.1), protocol layers (see Section 8.2.2), and operations (see Section 8.2.3). It also defines Bluetooth security at the link level (see Section 12.2). Not all Bluetooth devices, however, have the same functionality implemented. The functionality of Bluetooth is defined by the Bluetooth profiles (see Section 8.2.4) [13].

8.2.1   Introduction

Bluetooth was originally designed for short-range wireless connections between two devices. Unlike infrared, a clear line of sight is not required between Bluetooth devices. Bluetooth-enabled devices apply to wide and diverse application areas, such as gaming, medical practice, GPS, cars, and so forth. The number of new applications and products has grown rapidly. Not long ago, most computing devices used an add-on adapter (also called a dongle) to support Bluetooth communication. Presently, notebooks and smart phones with built-in Bluetooth support are quite common. In the following discussions, we look into Bluetooth device classifications, addressing, and network architecture.

Devices

The number of Bluetooth devices on the market is constantly increasing. Low-cost Bluetooth chipsets are developed to meet this increased market demand. Bluetooth hardware devices vary by their connection capabilities and the required power class.

Some devices support point-to-point communication, and others support multipoint communication. A point-to-point device (e.g., headset) can only communicate with a single Bluetooth device at a time. A multipoint Bluetooth device is able to communicate with up to seven devices at the same time within a defined range. Bluetooth devices also differ by the power class. The following are the three classes of Bluetooth devices, according to power class:

- Class 1 device, with a power rating of 100 mW and a range of 100m;
- Class 2 device, with a power rating of 2.5 mW and a range of 20m;
- Class 3 device, with a power rating of 1 mW and a range of 10m.

Class 2 or 3 devices are usually small and powered by batteries (e.g., mobile phones and headsets). Class 1 Bluetooth devices are mostly plugged into an ac power supply, or are built into the hardware of another unit.

Addressing

There are four types of device addresses in Bluetooth:

- *Bluetooth Device Address (BD_ADDR):* Each Bluetooth device is allocated a unique IEEE 802 standard 48-bit device address.
- *Active Member Address (AM_ADDR):* AM_ADDR is also called the medium access control (MAC) address of a Bluetooth device. This 3-bit address is only valid as long as the slave is active on the channel. The all-zero AM_ADDR is for broadcast messages.
- *Parked Member Address (PM_ADDR):* This is an 8-bit member address for parked slaves. The PM_ADDR is only valid when the slave is parked.
- *AR_ADDR (Access Request Address):* This is used by the parked slave to determine the slave-to-master half-slot in the access window, to which it is allowed to send access request messages. It is only valid as long as the slave is parked, and it is not necessarily unique.

Bluetooth Network Architecture (Piconet and Scatternet)

Piconets and scatternets determine the Bluetooth network architecture. Bluetooth devices connect in an ad hoc fashion to form a piconet. Up to eight devices can be networked into a piconet, and up to 255 devices can be connected while in a parked mode. Several piconets may form a larger scatternet, with each piconet maintaining independence. An example of Bluetooth architecture is shown in Figure 8.1.

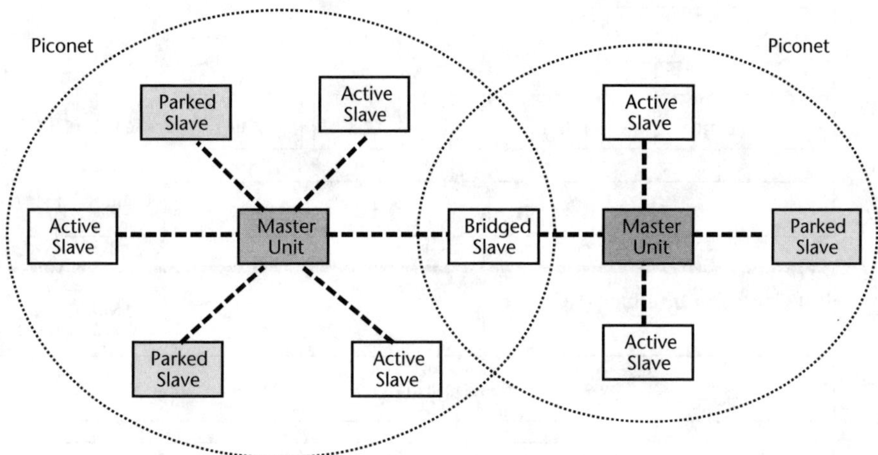

Figure 8.1   Bluetooth scatternet and piconets.

The master unit of the piconet is the device that initiates the connection. The device that accepts the connection automatically becomes the slave. The master determines which device is able to communicate. The master unit synchronizes the slaves with its clock and its hopping sequence. In certain conditions, a role switch between the master and slave is allowed. All communication is between the master and the slave; direct slave-to-slave communication is not possible. Master and slave roles are not predefined. The slaves are usually fairly dumb. The master/slave design lowers the price of a Bluetooth network.

If a Bluetooth device wants to join the piconet after the master has acquired seven slaves, the master will not invite new members to join until at least one of the slaves leaves or goes into an inactive state. Alternatively, if one of the slaves is multipoint-capable, then the new device can create a piconet with that slave. Thus, these two piconets form a scatternet. A scatternet will also be created if the master of the existing piconet becomes a slave to the newcomer. There may be a maximum of 10 fully loaded piconets in a scatternet.

8.2.2   Protocol Layers

Bluetooth standard documents specify the Bluetooth layered architecture. As shown in Figure 8.2, most of these layers are called "protocols," and some are programming interfaces. The Bluetooth core protocols include radio, baseband, link manager protocol (LMP), logical link control and adaptation protocol (L2CAP), and service discovery protocol (SDP). All other protocols, which are called adopted protocols, are not originated by the Bluetooth SIG, but have been incorporated into the Bluetooth. Each Bluetooth layer in ascending order is now illustrated.

The Bluetooth Radio Layer

The bottom layer is the Bluetooth radio layer, which corresponds fairly well to the physical layer in the OSI and 802 models. The radio layer moves the bits between the master and the slave.

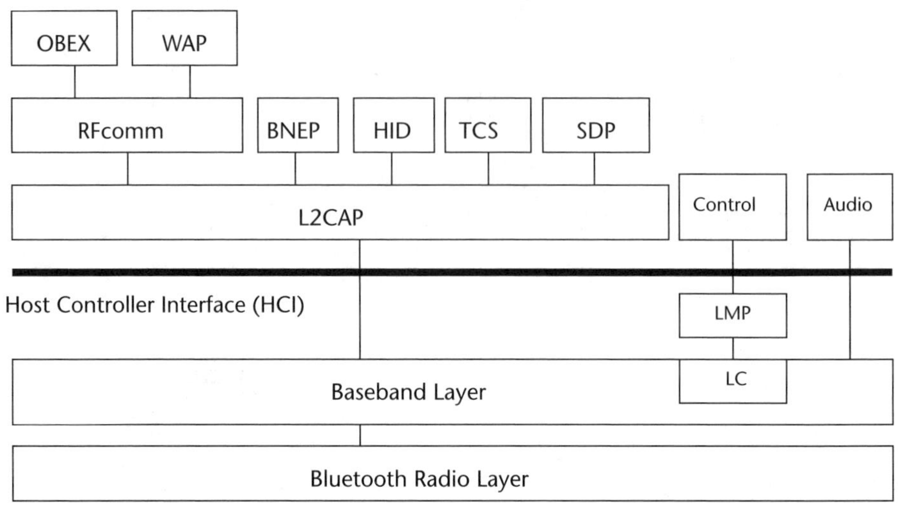

HCI: Host Controller Interface
LC: Link Controller
L2CAP: Logical Link Control and Adaptation Protocol
HID: Human Interface Device Protocol
BNEP: Bluetooth Network Encapsulation Protocol
OBEX: Object Exchange Protocol (OBEX)

LMP: Link Manager Protocol
BNEP: Bluetooth Network Encapsulation Protocol
TCS: Telephony Control Protocol Specification
SDP: Service Discovery Protocol
RFcomm: Radio frequency communications
WAP: Wireless Application Protocol

Figure 8.2   The Bluetooth layered architecture.

Bluetooth devices use the 2.4-GHz unlicensed spectrum. The air band is divided into 79 channels of 1 MHz each, with a dwell time of 625 μs. All the nodes in a piconet hop simultaneously, with the master dictating the hop sequence. The standard range is 10 cm to 10m, and can be extended to at least 100m by increasing the transmission power. Implementations with Bluetooth Versions 1.1 and 1.2 can reach 723.1 Kbps. Bluetooth Version 2.0 implementations feature an enhanced data rate (EDR+), and can reach 3 Mbps. Technically, Version 2.0 devices require a higher power consumption, but the faster rate reduces the transmission times. Thus, Version 2.0 can effectively reduce power consumption under equal traffic loads.

The Baseband Layer

The baseband layer roughly corresponds to a MAC layer. It deals with data from the upper L2CAP layer on the sending side, which is then delivered to the L2CAP layer on the receiving side. The communication between the master and the slave is based on the point-to-point time division multiplexing (TDD) scheme.

The baseband layer manages how the master should control time-slots, and how these slots are grouped into frames. In the simplest form, the master always transmits starting from the even slots, and the slaves transmit starting from the odd slots. The master gets one-half of the slots, and the slaves share the other half. The frequency hopping requires a settling time per hop, to allow the radio circuits to become stable. For a single-slot frame, 366 of the 625 bits remain after settling. Of these, 126 are for the access code and the header, leaving 240 bits for data. When multiple slots are grouped together, only one settling period is needed, and a slightly shorter settling period is used. Thus, longer frames are much more efficient than single-slot frames [6, 13].

The Bluetooth baseband combines both circuit and packet switching. Each frame is transmitted over a logical channel, called a link, between the master and the slave. The baseband provides two types of links: the synchronous connection-oriented (SCO) link, and the asynchronous connectionless link (ACL). The SCO is used primarily for voice/speech, while the ACLs are primarily used for packet-switched data at irregular intervals.

The bandwidth is controlled by the master, which determines how much of the total each slave can use. Because of the need for smoothness in voice transmission, SCO packets are generally delivered via reserved intervals. The master can support up to three simultaneous SCO links, and frames sent over SCO links are never retransmitted. Instead, error correction can be used to provide high reliability. Each SCO link can transmit one 64-Kbps PCM audio channel.

The ACL link is a point-to-multipoint link between the master and all the slaves participating in the piconet. The slaves cannot transmit data until the master has polled them, and the master can broadcast messages to the slave units via the ACL. If the slots are not reserved for the SCO links, then the master can establish an ACL on a per-slot basis to any slave. The ACL traffic is delivered on a best-efforts basis. Frames can be lost and may have to be retransmitted. ACLs support both symmetric and asymmetric transmissions. A slave may have only one ACL to its master.

The Bluetooth baseband layer also defines three error correction techniques: the one-third rate forward error correction code (FEC), the two-thirds rate FEC, and the automatic repeat request (ARQ). The FEC methods are designed to reduce the number of retransmissions. However, the overhead significantly slows down transmissions, so it is generally used for packet headers. The ARQ scheme requires that the header error and cyclic redundancy check (CRC) be passed. Then an acknowledgement is sent. If there is a header error or CRC error, the data is resent.

Finally, the baseband security is provided in three ways: pseudorandom frequency band hops, authentication, and encryption. Frequency band hops make it difficult for anyone to eavesdrop. Authentication, encryption, and other security mechanisms are introduced in Section 12.2.

LMP

The LMP is responsible for link setup and ongoing link management. The basic implementation of the Bluetooth system consists of the radio chip (with controller) and the link manager (LM) software. The LM handles the logical channels setup, link configuration, power management, authentication, and quality of service.

The hardware underlying the LM is the link controller (LC). The ACL and SCO links can be implemented with first-in, first-out (FIFO) queues for flow control and synchronization. The LM fills the queues, and the LC automatically empties the queues. Together, the LM and LC perform the following tasks: sending and receiving data, paging and receiving inquiries, setting up connections, authenticating, negotiating and setting up link types, determining the frame type of each packet, and placing a device in sniff or hold mode. The connection operations and state transitions are introduced in Section 8.2.3.

L2CAP

The L2CAP is a core layer through which all data must pass. It is analogous to the 802 LLC sublayer, but is technically different from it. Notice that only data will pass through the L2CAP layer; audio links have direct access to the host controller interface (HCI).

The L2CAP layer mainly performs three tasks: packet segmentation, frame reassembling, and protocol multiplexing. It accepts packets of up to 64 KB from the upper layers and breaks them into frames for transmission. At the receiving end, the frames are assembled into packets again. The L2CAP also handles the multiplexing and demultiplexing of multiple packet sources, and it can accept data from more than one upper protocol at the same time. When a packet has been reassembled, the L2CAP layer determines which upper-layer protocol (e.g., RFcomm, SDP, or telephony) will hand it out. L2CAP also handles the quality of service. The maximum payload size allowed is negotiated at setup time to prevent overloading a small-packet device.

HCI

The HCI is a layer of software interface that passes all data from the host (e.g., a PC) to the controller (i.e., attached Bluetooth device). If the Bluetooth device is attached to the USB port of a PC, then the HCI should understand the USB messages and send that information to the upper layers of the stack. Voice and data both pass though the HCI.

As shown in Figure 8.2, the next layer up is the middleware layer, which contains a mix of different protocols. The 802 LLC was inserted here by the IEEE for compatibility with its other 802 networks. The service discovery, RFcomm, and telephony protocols are native.

Service Discovery Protocol

The service discovery protocol (SDP) is used to locate available services within the network. It is one of the core protocols of Bluetooth. The details of the generalized WPAN service discovery are introduced in Section 8.2.4.

Radio Frequency Communications

Radio frequency communications (RFcomm), also called the wireless serial port or the cable replacement protocol, emulates the standard serial port on PCs for connecting the peripherals. It has been designed to allow legacy devices to easily use it Bluetooth. For example, a Bluetooth-enabled smart phone would use the RFcomm layer to synchronize its data with a Bluetooth-enabled PC as if they were physically connected.

Telephony Control Protocol Specification

The telephony control protocol specification (TCS)/TCS-Binary is aimed at real-time voice-oriented applications. TCS manages call setup and termination.

Control signals are sent to devices. For example, a Bluetooth phone will send signals to the base station, indicating that the user has requested an action, such as to hang up, to use call waiting, to place a three-way call, and so forth.

Bluetooth Network Encapsulation Protocol

The Bluetooth network encapsulation protocol (BNEP) allows other networking protocols to be transmitted over Bluetooth. It encapsulates TCP/IP packets in L2CAP packets before handing off the data to the L2CAP layer in the stack. Thus, BNEP is a good choice for implementing TCP/IP networking in a Bluetooth device.

Wireless Application Protocol

The wireless application protocol (WAP) is an adopted protocol for Bluetooth. The Bluetooth SIG has incorporated the existing WAP protocol into the Bluetooth protocol to fit Bluetooth's needs. The details of WAP are illustrated in Chapter 3.

Object Exchange Protocol

The object exchange protocol (OBEX) is an adopted communication protocol initially defined by IrDA. It is useful in transferring objects (e.g., files) between Bluetooth devices. The OBEX does not require that the TCP/IP be present in the stack, but the manufacturer is free to implement OBEX over the TCP/IP.

Human Interface Device Protocol

The human interface device (HID) protocol was originally defined in the USB specification, and it indicates the rules and guidelines for transmitting signals and data for human interface devices, such as keyboards, mice, remote controls, and video game controllers.

The top layer is where the applications and profiles are located. They make use of the protocols in lower layers to get their work done. Each application has its own dedicated subset of the protocols. Specific devices, such as a headset, usually contain only those protocols needed by that application. Thus, a Bluetooth vendor does not need to implement all the Bluetooth protocol layers. For example, a Bluetooth cordless phone may only have HCI, SDP, L2CAP, and TCS implemented into its stack. The functions specified in other layers would not be required.

8.2.3 States and Operations

Bluetooth devices operate in one of two states: standby, and connection.

Standby State

The standby state is the default low-power state in the Bluetooth unit. Only the native clock is running, and there is no interaction with any other device. Devices not connected in a piconet are in standby mode. In this mode, they listen to messages every 1.28 seconds over 32 hop frequencies (less than 32 in Japan, Spain, and France). To avoid overloading the 2.4-GHz spectrum and to limit power consumption, a standby Bluetooth device can be in one of the three following modes.

- *Generally Discoverable Mode:* The Bluetooth device in this mode is allowed to be detected by other Bluetooth devices within its proximity.
- *Limited Discoverable Mode:* Only well-defined devices will be able to detect a device in this mode. This mode will be used when a user has many Bluetooth devices, and wants them to automatically discover each other.
- *Nondiscoverable Mode:* The device in this mode is invisible to other devices, so it cannot be detected.

Device Discovery and Connection Setup

There are seven modes that are used to make connections in the piconet. The master and the slaves are in one of the seven modes during the connection setup process. These modes are: inquiry, inquiry scan, inquiry response, page, page scan, slave response, and master response. As shown in Figure 8.3, a connection between two devices is established via the following sequences:

- The master (in inquire mode) uses the inquiry access code (IAC) to discover other Bluetooth devices in the range.
- If any nearby Bluetooth device is listening for inquiries (inquiry scan mode), then the device responds to the master by sending its address and clock information (in inquire response mode) to the master. After sending the information, the slave (in page scan) may start listening to page messages from the master.
- After discovering the inrange Bluetooth devices, the master may page these devices (in page mode) for a connection setup.
- The slave in page scan mode will respond (in slave response mode) with its device access code (DAC).
- After receiving the response from the slave, the master may respond by transmitting a packet containing the master's real time clock, BD_ADDR, and the device information. Once the slave has received this packet, both the master and the slave enter the connection state.

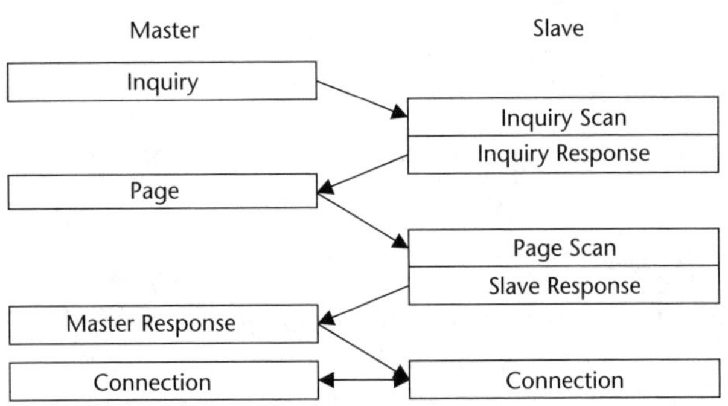

Figure 8.3   Bluetooth device discovery and connection setup.

Connection State

After connection setup, the master polls the slaves to verify the timing and channel frequency hopping. For power conservation, a Bluetooth device in a connection state is in any one of four modes: active, hold, sniff, and park.

- *Active Mode:* The master and the slave are kept synchronized with each other. They both listen and exchange packets.
- *Hold Mode:* At the request of either the master or the slave, a device could get into a hold mode. A device in the hold mode does not temporarily support ACL packets, and it goes into low-power sleep mode to make the channel available for paging, scanning, and so forth. The device may still participate in SCO exchanges while in the hold mode.
- *Sniff Mode:* The sniff mode is applicable only to slave units. In this mode, the slave listens at a reduced interval. The slave can go to sleep in the free slots, thus saving power. This is usually a programmable setting.
- *Park Mode:* The park mode is a more reduced level of activity than the hold mode. The device gives up its Active Member Address (AM_ADDR) and gets a Parking Member Address (PM_ADDR). While in the park mode, the slaves only listen to keep their synchronization with the master, and check for broadcast messages.

One of the challenges in Bluetooth is to move between these modes, especially from park to active modes, and vice versa.

8.2.4   Application Profiles

Most network standards specify channels and communicating entities, but not individual applications. Bluetooth specifies profiles for each type of supported applications. Bluetooth profiles are not meant to be forcibly used, but are rather intended to provide standards for implementers upon which to build. It is generally the responsibility of the manufacturer to implement the profiles necessary for their target customers. These profiles help ensure that Bluetooth products are built on a single foundation. Thus, true interoperability can be achieved by using these profiles.

All Bluetooth devices have a minimum set of functionality that allows them to locate another Bluetooth device. To utilize services provided by that device, certain additional profiles are required. Currently, there are more than 20 profiles, covering a wide diversity of uses. These profiles are heavily dependent upon each other. The building block relationship of Bluetooth profiles is shown in Figure 8.4. Every profile depends upon the generic access profile (GAP). A Bluetooth-enabled headset must support the headset profile, the serial port profile, and the GAP. Some important Bluetooth profiles and their applications are now discussed.

GAP

The GAP is not really an application, but is rather the root in the Bluetooth profile. All Bluetooth devices must implement the GAP. It defines the generic procedures for

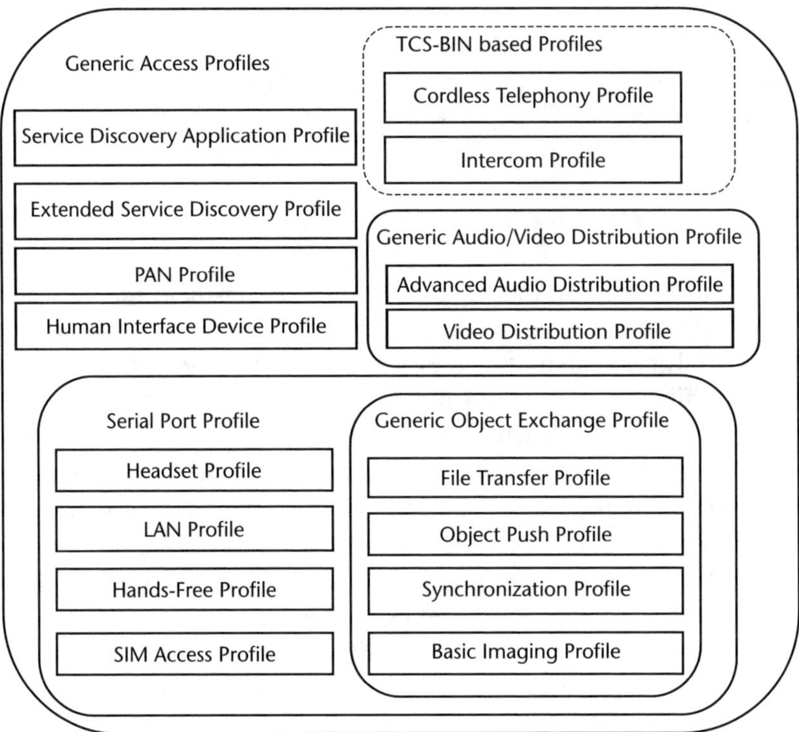

Figure 8.4 Bluetooth profiles.

discovering Bluetooth devices, and provides the basic mechanism to establish and maintain secure links between the master and the slaves.

The following are the four major groups of profiles resting on the GAP:

- The serial port profile (SPP) group, which uses RFcomm serial port emulation;
- The generic object exchange profile (GOEP) group, which uses OBEX object exchange;
- The telephony control protocol (TCS) specification group, which uses TCS binary;
- General audio/video distribution profiles (GAVDP).

Service Discovery Application Profile

The service discovery application profile (SDAP) rests on the GAP. Devices use SDAP to discover what services other devices have to offer in their neighborhood. The SDAP itself is also relatively generic. All Bluetooth devices are expected to implement the SDAP. Some profiles, such as the LAN and PAN profiles, implement a newer version of the SDAP that is called the extended service discovery profile (ESDP).

Serial Port Profile

The serial port profile (SPP) emulates a serial line that most of the remaining profiles use. It is a profile that interacts directly with the RFcomm layer in the Bluetooth protocol stack. This profile is used to create a virtual COM port on a Bluetooth-enabled device. With a serial port profile, the application treats Bluetooth links as virtual COM ports. It also supports the GOEP group, and other Bluetooth profiles. The SPP defines a gateway that provides access to a service, and a terminal that uses that service. The SPP is the building block for many other profiles, such as the headset profile and the LAN access profile.

Generic Object Exchange Profile

The generic object exchange profile (GOEP) is a generic profile that defines a client/server-based object movement. The client is the unit that initiates an operation. A Bluetooth slave can either be a client or a server. The GOEP sets the rules for using the OBEX protocol over Bluetooth links. Similar to the SPP, the GOEP is also a building block for many profiles, such as the file transfer profile, the object push profile, the synchronization profile, and the basic imaging profile.

File Transfer Profile

The file transfer profile is a robust profile for transferring objects. It allows Bluetooth devices to use OBEX's facilities to create, delete, and move objects from one Bluetooth-enabled device to another.

Object Push Profile

The object push profile provides the functionality with which a device can push and pull a predefined standard data object, such as a virtual business card (vCard) or a virtual calendar (vCal).

Synchronization Profile

The synchronization profile provides a standard way to synchronize personal information manager (PIM) items, such as contacts, calendars, e-mail, notes, and tasks. Synchronization can be automatically triggered at any particular time of the day, even without the user being aware of it (e.g., when the devices come within range of one another). This hidden (or unconscious) nature is considered to be unique for WPAN technologies.

Basic Imaging Profile

The basic imaging profile allows a Bluetooth-enabled imaging device (e.g., camera) to engage in remote control, image transfers, and downloading.

Headset Profile

The headset profile is primarily designed for connecting Bluetooth-enabled headset terminals to Bluetooth-enabled wireless phone gateways. It relies on the SCO for audio. The signaling system uses AT commands, which is the format used by modems. This is the part that relies on the Serial Port Profile.

Local Area Network Access Profile

A Bluetooth-enabled device (e.g., notebook PC, PDA, or smart phone) will use the local area network (LAN) access profile to connect to a network access point, which itself is connected to a LAN. The LAN access profile has a gateway that provides the Bluetooth with a device link to a fixed LAN. This profile is a direct competitor of 802.11.

Hands-Free Profile

Automobiles use the hands-free profile (HFP) to allow the driver to place and receive calls from a Bluetooth-enabled phone. The mono, PCM audio is carried on the SCO link.

SIM Access Profile

The SIM access profile (SAP) allows devices (e.g., car phones with built-in GSM transceivers) to connect to a SIM card in a Bluetooth-enabled phone, so that the device itself does not require a separate SIM card. The SIM access server has direct access to the SIM card, and acts as the SIM card reader. The SIM access client can access and control the SIM card via the Bluetooth link.

Telephony Control Protocol Specification Profile

The telephony control protocol specification (TCS) defines ways to send audio calls between Bluetooth devices. It further extends the functionality of Bluetooth-enabled mobile phones. However, one weakness of the TCS is that no handover facilities are defined. This means that a call will be dropped when the handset is out of the range of the gateway.

Cordless Telephony Profile

The cordless telephony profile enables a gateway to be connected to a fixed landline phone network to relay calls to a mobile phone handset. A mobile phone user can thus use a landline phone through a Bluetooth-enabled handset. Outgoing calls can be placed without paying cellular charges. Eventually, cordless and mobile phones may merge. With the VoIP enhancement, landline phones may disappear in the future. It is central to the Bluetooth SIG's 3-in-1 phone scenario.

Intercom Profile

The intercom profile, also called the walkie-talkie profile, provides a basic point-to-point audio service, allowing direct voice calls between Bluetooth handsets in a short range (e.g., at home or in the office).

- *General Audio/Video Distribution Profile (GAVDP):* The GAVDP provides the basis for A2DP and VDP.

- *Advanced Audio Distribution Profile (A2DP):* The A2DP is designed to transfer an audio stream (e.g., music from an MP3 player) to a headset or car radio. This profile includes support for subband codec (SBC) and MP3.
- *Video Distribution Profile (VDP):* The VDP profile allows the transport of a video stream. This profile supports the H.263 baseline and MPEG-4.

HID

The human interface device (HID) profile is resting on the HID protocol. This profile defines the case scenarios for using Bluetooth-enabled human interface devices (e.g., keyboards and mice). A device that conforms to the HID profile should run for three months using three AAA batteries.

Personal Area Networking Profile

The personal area networking profile is similar to the LAN access profile, except that it also has support for devices to form ad hoc networks among themselves. This profile allows the use of the BNEP for transport over a Bluetooth link.

Bluetooth working groups are currently defining new profiles. In time, many more profiles will be defined. Some of the proposed profiles include the WAP over BT, the local positioning profile (LPP), unrestricted digital information (UDI), and the video conferencing profile (VCP). New profiles expand the usefulness and interoperability of Bluetooth devices. Furthermore, a Bluetooth profile can be implemented on any platform.

On the other hand, the Bluetooth profile mechanism is controversial. Each Bluetooth working group focuses only on its specific problem and generates its own profile. It is not necessary to provide different protocol stacks for each profile. Two profiles, one for file transfer and one for streaming real-time communication, would be sufficient.

Java API for Bluetooth

Although Bluetooth profiles can be implemented in any language, Java is the preferred choice, due to its platform-dependence nature. Programmers are able to write applications without any knowledge of the underlying Bluetooth hardware or protocol stack.

The Java community process (JCP) is the formal procedure to get a simple concept incorporated into the Java standard. A Java specification request (JSR) is within the JCP. The JSR-82 is the formal JCP name for the Java APIs for Bluetooth [14].

The Bluetooth control center (BCC) is a special component that is required to exist in a JSR-82–compliant implementation. A BCC should:

- Include base security settings of the device;
- Provide a list of Bluetooth devices that are known or trusted;
- Provide a mechanism to pair two devices that are trying to connect for the first time;
- Provide a mechanism to grant authorization of connection requests.

The official Java Bluetooth specification provides no guidelines about how the BCC should be implemented. One vendor could implement the BCC as a set of Java classes, and another vendor could implement it as a native application in the Bluetooth host.

8.3 Device Coordination (Wireless Access Protocol)

With the rapid progress in wireless technologies, many new devices (e.g., information appliances) and services are developed. Mobile devices are capable of using the wireless network to share resources and services with one another, whether in the home or office, and whether or not they are attached to the Internet. Since devices can be very specialized and can follow different protocols, device coordination and the service discovery framework become critical for future wireless applications. A standardized discovery and coordination mechanism will bring convenience and flexibility to users, developers, and system managers. Therefore, the balance between a standardized device coordination scheme and various proprietary protocols on devices is an important issue.

A device coordination mechanism should provide the following capabilities: presence announcement, device/service discovery, device/service description and query, and self-configuration. To achieve interoperability, the coordination frameworks are usually based on a common foundation (e.g., the Internet); support common API interface (e.g., Java) for applications; and are extendable. Various industry leaders developed different coordination frameworks. Here, we focus on Jini, pioneered by Sun Microsystems, and Universal Plug-and-Play (UPnP), pioneered by Microsoft. As can be seen later in this section, Jini provides the standard API programming interface, while UPnP is built upon HTTP and the standard XML representation. Other popular schemes, such as HAVi (on top of IEEE 1394 and pioneered by the consumer electronic industry) [15], are beyond the scope of this study.

8.3.1 Jini

Jini, developed by Sun Microsystems, is a layer of open software architecture that provides a flexible environment for device coordination and service discovery. Taking advantage of Java's machine-independent property (as introduced in Section 4.3), Jini services can run on any device with Java Virtual Machine (JVM) installed. A Jini federation is a group of services available on a home or office network. Jini provides simple mechanisms that enable users to join and leave a federation without any prepared planning or human intervention. The members of the federation are assumed to agree on basic notions of trust, administration, identification, and policy. Joining and leaving a Jini federation is easy, natural, and often spontaneously occurs.

A major advantage of the Jini architecture is that it allows programs to use services in a network without knowing the protocol used by the service [16, 17].

A Jini system can be seen as a network extension of three main components: services, infrastructure, and programming model. These three components are designed to work together, and are constructed using each other. Jini services appear as objects written in the Java programming language. A Jini service is defined by its API, which is declared as a Java interface. The service can translate an API request into its internal format using RMI, simple object access protocol (SOAP) [18], or some other private protocol.

The Jini infrastructure refers to the set of components that enables the creation of a federated Jini system, while the services are the entities within the federation. Jini services use the infrastructure to make calls to each other, to discover each other, and to announce their presence to other services and users. The programming model is a set of interfaces that enables the construction of reliable services, including those which are part of the infrastructure and those which join the federation.

Infrastructure and Operations

The Jini infrastructure defines the minimal Jini technology core. It enables and uses the programming model. The infrastructure includes discovery/join protocols and lookup services, and a distributed security system. The discovery/join protocols are for presence announcements, which define the way a service becomes part of a Jini system. Lookup occurs when a client or user needs to locate and invoke a service.

The lookup service, which serves as a repository of services, reflects the current members of the federation. Entries in the lookup service are Java objects. The distributed security model identifies entities and gives them the rights to perform actions. The Jini infrastructure and operation sequences are shown in Figure 8.5.

1. Discovery/join is the process of adding a service to a Jini system. The service provider first locates a lookup service by multicasting a request on the local network, so that any lookup services may identify themselves. Discovery is shown as Step 1 in Figure 8.5.

2. A service object (in Java) for the service is then loaded into the lookup service, as shown in Step 2 of Figure 8.5. This service object, also called a proxy, contains a Java interface for the service, including the methods that users and applications will invoke to execute the service and any other descriptive attributes.

3. The service is now ready to be looked up and used. A client locates an appropriate service by its interface written in the Java and the descriptive

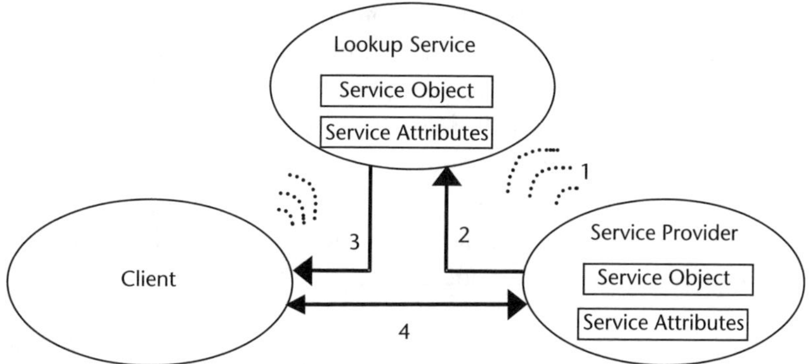

Figure 8.5   Jini service discovery/join and lookup.

attributes. As shown in Step 3 of Figure 8.5, a copy of the service object is moved to the client.

4. The final stage is to invoke the service, as shown in Step 4 of Figure 8.5. No matter what services are downloaded into the local address space, or implemented by software objects or in the hardware of a different machine, all services will appear as Java objects to the client. As long as the correct functions are provided, one implementation could be replaced by another without the knowledge of the client.

The service object's methods may implement a private protocol between itself and the original service provider. Different implementations of the same service interface can use completely different interaction protocols. This implies that Jini can accommodate devices with different proprietary protocols.

Based on the Java language, Jini uses the remote method invocation (RMI) protocols to move service objects, including their behavior, around the network. The service provider always directly supplies the service object. The service object movement provides flexible communication patterns between the service provider and clients, and ensures synchronization between the proxy service object held by the lookup service and the client. If all the manufacturers of a certain device agree on standard RMI interfaces, then one piece of the service object that follows that standard would be enough for manufacturers. Thus, the driver will be significantly reduced. However, manufacturers may not like this idea, due to concerns about Java's performance or political reasons.

A client could use peer lookup if no lookup service is found. The client can send the lookup request to service providers and ask for registration. Service providers will then attempt to register the client as if it were a lookup service. The client can select the required services from the registration requests, and drop the other services.

Programming Model

Both infrastructure and services consist of a set of interfaces. These interfaces extend the standard Java to the Jini programming model. The three core interfaces are the following.

- *Leasing Interface:* The leasing interface defines a renewable duration-based model to allocate and free resources. Devices should periodically renew their leases for the lookup services to maintain their registration information.

- *Event and Notification Interface:* The event and notification interface is an extension of the JavaBean event model that enables event-based communication between Jini services.

- *Transaction Interface:* The transaction interface enables entities to cooperate in such a way that either all of the changes made to the group occur atomically, or none of them occur at all.

The Jini programming model rests on the ability to move codes. Entries in the lookup service are leased, allowing the lookup service to accurately reflect the set of currently available services. When services join or leave a lookup service, events are signaled, and objects that have registered interest in such events get notifications when new services become available or when old services cease to be active.

The Jini system can be easily extended. Non-Java devices can participate in a Jini federation through the gateway, bridge, and proxy protocol translations. It is also possible to extend the Jini lookup to larger interfederation systems [19]. However, Jini does not specify the important coordination feature of self-configuration. When an IP device is plugged into a network, an IP address and a subnet mask should be assigned, which may optionally be a gateway and DNS. The lookup operation then can proceed. The self-configuration or auto-configuration is better supported by a TCP/IP-based UPnP technology.

8.3.2 Universal Plug-and-Play

Universal Plug-and-Play (UPnP) is a set of computer network protocols developed primarily by Microsoft. The UPnP forum is an industry initiative body of UPnP. The goals of UPnP, similar to those of Jini, are to allow devices to seamlessly connect (with wire or wirelessly) to the network, and to simplify the network implementation in the home/office environment. Users can use a new device without configuring the settings and installing new drivers [20, 21].

UPnP is built upon standard Internet protocols (e.g., IP, TCP, UDP, and HTTP), and standard data presentation format (i.e., XML). UPnP provides developers with a common set of interfaces for accessing services on a home/office network. Thus, UPnP middleware is independent of the developing language and physical network media. UPnP also accommodates networks that run non-IP protocols, which make different types of legacy devices discoverable. In the following discussions, we will look into UPnP network components.

Network Components

The key logical components in a UPnP middleware environment are control points, controlled devices, and services.

- *Control Points:* Control points (also called user control points) can manipulate network devices. When a new control point (e.g., a PC) is added to a network, the device may ask the network to find a UPnP-enabled device. The control point is always the initiator of the communications session. It responds with its URL and device description.

- *Controlled Devices:* Controlled devices are directed by control points. Controlled devices include home appliances, security systems, networking gateways, peripherals, automated light controllers, and so forth.

- *Services:* Services, such as time services, are provided by devices, which are provided by alarm clocks. In UPnP, the services are described in XML format. The control points can set or get services information from controlled devices.

Even though the implementation of UPnP requires a very small amount of system resources, not all devices have sufficient hardware resources for full TCP/IP support. Examples include power-line–controlled equipment, thermostats, and inexpensive toys. For non-UPnP compliant legacy devices, a software bridge is required to communicate with native UPnP devices. These devices are also called bridged devices. Proxies sitting on the TCP/IP network can also perform registration and discovery for such devices.

Addressing

UPnP supports both IPv4 and IPv6 addressing. The dynamic host configuration protocol (DHCP) and domain name service (DNS) servers are optional. For a DHCP-enabled IPv4 client, the device may obtain a name from the DNS server during the DHCP transaction. If no DHCP server is available, then the device must use Auto IP to get an address from a reserved IP address range, which is known as the LINKLOCAL net (169.254.0.0/16). Note that this is not a routable IP address, and therefore, packets will not cross gateways.

UPnP also supports the network address translation (NAT) traversal, which can automatically solve many NAT-unfriendly problems. The applications assign the dynamic port mappings to the NAT gateway, and delete the mappings when the connections are complete. No matter which addressing scheme is used, a device can automatically leave a network without leaving behind any unwanted state information [1, 2].

UPnP Operations

When a new device is connected to a UPnP network, the presence announcement and service discovery processes are initiated. The details of the device and its services are described in a description file. The control point and device will then exchange control messages and event notifications. The controlled device will provide a presentation Web page as an entry to control it. Once the discovery process is finished, proprietary protocols can communicate with the devices.

Service Discovery

When a device is added to a network, it will announce its presence over the network. The control point can also discover the services provided by devices. The UPnP service discovery is based on the simple service discovery protocol (SSDP). The announce/query process sends and receives data in the HTTP format. The traditional HTTP is built upon the TCP protocol. SSDP uses HTTP over UDP. A device joining the network can send out a multicast announcement message that is embedded in HTML. This message must also contain a URI that identifies the resource, and a URL that contains the description file in XML format. The latter essentially uses a style sheet tailored to various types of devices. A query for device discovery can also be multicast or directed toward a directory service (also called lookup service or proxy), if one is present.

Description

After a device is discovered, the control point must then retrieve the description of the device. The UPnP description for a device is expressed in XML. The description includes vendor-specific information and a list of embedded devices or services. It also includes URLs for control, event notification, and presentation. For each service, the description includes a list of the commands and variables of the service, or parameters for each action.

Control

Control points through a control message manipulate devices. After a control point has retrieved a description of the device, the control point then sends a control message to the control URL for the service. The control messages are expressed in XML using SOAP [22, 23]. In response to the control message, the service returns any action-specific values. The effects of the action are modeled by changes in the variables of the service.

Event Notification

The next step in UPnP networking is event notification, which is also called eventing. When the service changes, the service publishes updates by sending event messages. Event messages contain the names of one or more state variables, along with the current value of these variables. These messages are also expressed in XML, and are formatted using general event notification architecture (GENA).

Presentation

Each device can provide its control interface by use of a URL, allowing users to go to the device's presentation Web page to control this device.

8.3.3   Home Service Gateway

One important application domain for wireless PANs and the device coordination scheme is home networking. The main applications of a home network include resource sharing (e.g., content, peripherals, and Internet connection); home control (e.g., temperature, lighting, security/health monitoring), which can either be local or remote; and entertainment (e.g., multiuser gaming, streaming media redirecting). Home networks connect a variety of computing devices and information appliances to a high-speed Internet service through a home/residential service gateway [24, 25].

Home networks are usually built upon multiple technologies. In-home lower-layer enabling technologies include wired (e.g., Ethernet, IEEE 1394, USB, HomePNA, and power-line), wireless LAN (see Chapter 6), and wireless PAN technologies. A combination of WLANs and WPANs appears to be one of the most effective options to enable a home network. The device coordination and service discovery protocols (see Sections 8.3.1 and 8.3.2) are considered as upper-layer middleware protocols of home networks.

Similar to the WAP proxy gateway (see Section 3.2.4), the home gateway does the bridging/routing, and protocol and address translation between the external broadband network and the in-home networks. The home gateway must also perform media translation, authentication/filtering, and various system management tasks. The physical home gateway can be created from intelligent computing devices, such as a home PC, set-top box, cable/DSL modem, or a WAP gateway device. From the service provider's point of view, a home gateway provides a focal point to deliver services to client devices on the home network. Thus, it also called a service gateway.

The open services gateway initiative (OSGi) is a consortium for the development of open standards and specifications for the services gateway. OSGi specifications are now used from mobile phones to the new version of the open-source Eclipse IDE. Devices that implement OSGi specifications will enable service providers and others to deliver various services to home networks. In addition to various in-home devices, networks, and the home service gateway, the OSGi model identified four additional components of a complete end-to-end operation, which are the following:

- *Service Provider*: The service provider furnishes a range of services to its users. The service is delivered and enabled by downloading a software application to the home service gateway.
- *Service Aggregator*: The service aggregator is a new type of service provider that offers a set of services (e.g., various utility meter readings) bundled together.
- *Gateway Operator*: The gateway operator manages and maintains the status of the home gateway and its services. A gateway operator is responsible for starting, stopping, updating, and removing services.
- *WAN and Carrier/ISP*: The wide area network provides the necessary communications between the service gateway, the gateway operator, the service aggregator, and the service provider. This communication platform is provided and managed by a telecommunications carrier or by an Internet service provider (ISP).

8.4 Summary

WPAN is a new and promising domain of wireless networking for short-range communication. WPAN enables communication between devices with relatively slow data rates, as compared to communication using WLAN technologies. Many WPAN technologies have been developed in recent years. The future of WPAN heavily depends on the compatibility and interoperability of these technologies. Device coordination and service discovery mechanisms are required to seamlessly connect a variety of devices and network technologies. In this chapter, we described two open architectures: Jini and UPnP. Both frameworks support flexible operations, provide dynamic architecture, and have strong support by the industry.

Bluetooth is the most widespread WPAN technology. The Bluetooth protocol supports both voice and data. A master Bluetooth device can communicate with up to seven slave devices. Bluetooth specifies a complete set of protocol suites and application profiles. The Bluetooth profiles describe the implementations of user models, and are used to decrease the potential interoperability problems between products from different manufacturers.

WPAN technologies are rapidly evolving. This chapter introduced IEEE 802.15 and NFC. Each technology has its own strengths and weaknesses, which makes each of them suitable for specific applications. Therefore, more inexpensive, low power–consuming, and interoperable WPAN devices are expected in the future.

References

[1] Tanenbaum, A. S., *Computer Networks*, 4th ed., Upper Saddle River, NJ: Prentice Hall, 2003.

[2] Kurose, J. F., and K. W. Ross, *Computer Networking: A Top-Down Approach Featuring the Internet*, Reading, MA: Addison-Wesley, 2003.

[3] IEEE 802.15 Working Group for WPAN, http://grouper.ieee.org/groups/802/15/.

[4] Richard III, G. G., *Service and Device Discovery: Protocols and Programming*, New York: McGraw-Hill, 2002.

[5] Siva Ram Murthy, C., and B. S. Manoj, *Ad Hoc Wireless Networks: Architectures and Protocols*, Upper Saddle River, NJ: Prentice Hall, 2004.

[6] Bluetooth Special Interest Group, https://www.bluetooth.org/.

[7] ZigBee Alliance Web site, http://www.zigbee.org/.

[8] Knutson, C. D., and J. M. Brown, *IrDA Principles and Protocols: The IrDA Library, Vol. I*, Salem, UT: MCL Press, 2004.

[9] Infrared Data Association Web site, http://www.irda.org/.

[10] NFC Forum Web site, http://www.nfc-forum.org/.

[11] Taylor, J. D., *Ultra-Wideband Radar Technology*, Boca Raton, FL: CRC Press, 2000.

[12] Near Field Communication, http://www.semiconductors.philips.com/markets/identification/products/nfc/.

[13] Morrow, R., *Bluetooth: Operation and Use*, New York:McGraw-Hill, 2002.

[14] Hopkins, B., and R. Antony, *Bluetooth for Java*, Berkeley, CA: Apress, 2003.

[15] Home Audio Video Interoperability Web site, http://www.havi.org.

[16] Jini Network Technology Web site, http://java.sun.com/developer/products/jini/.

[17] Li, S., et al., *Professional Jini (Programmer to Programmer)*, Hoboken, NJ: Wrox Press, 2000.

[18] Newcomer, E., *Understanding Web Services: XML, WSDL, SOAP, and UDDI*, Reading, MA: Addison-Wesley, 2002.

[19] Tseng, W. T., and H. Mei, "Inter-Cluster Service Lookup Based on Jini," *Proc. of the 17th IEEE Int. Conference on Advanced Information Networking and Applications*, March 2003, pp. 84–89.

[20] UPnP Forum Web site, http://www.upnp.org/.

[21] Jeronimo, M., and J. Weast, *UPnP Design by Example: A Software Developer's Guide to Universal Plug and Play*, Santa Clara, CA: Intel Press, 2003.

[22] Mueller, J., *Special Edition Using SOAP*, Indianapolis, IN: Que, 2001.

[23] Simple Object Access Protocol, http://www.w3.org/TR/soap/.

[24] The OSGi Alliance, *OSGi Service Platform, Release 3*, Washington, D.C.: IOS Press, 2003.

[25] OSGi Web site, http://www.osgi.org/.

Engineering Wireless-Based Application Systems

Similar to the development of conventional application systems, the engineering of wireless-based software systems and understanding and application of well-defined engineering processes and well-known development methodologies. Here, we use a derived development process based on the conventional process models, such as Waterfall and Prototyping models. It consists of five phases.

- *Understanding wireless application domain:* In this phase, engineers need to understand and learn different wireless networks, diverse mobile technologies, and operation platforms.
- *System requirements engineering:* The objective of this phase is to communicate with customers and mobile users, in order to collect and document system requirements, including function and nonfunction requirements, required standards, and technologies.
- *System analysis and modeling:* During this phase, engineers conduct system analysis and modeling, based on the collected requirements for wireless-based applications.
- *System design and construction:* In this phase, engineers first work on different system design tasks in architecture design, mobile user interface, mobile database, wireless security, and application server. Then, they construct a wireless application system based on the given system design.
- *System validation and deployment:* After system construction, the product must be validated and deployed against the given system requirements to assure its quality.

This part includes five chapters to address different topics in the first four phases. Chapter 9 focuses on system requirements analysis and modeling. Chapter 10 describes system infrastructures and application architecture design. Chapters 11 and 12 discuss wireless security concepts, issues, and solutions. Chapter 13 addresses mobile client design issues and solutions.

CHAPTER 9

System Requirements Engineering, Analysis, and Modeling

This chapter is dedicated to requirements analysis and modeling of wireless-based software application systems. It helps readers understand how to perform effective system requirements analysis and modeling for wireless-based software application systems. This chapter is structured as follows. Section 9.1 discusses the understanding of wireless application domains. Section 9.2 focuses on system requirements engineering for wireless applications. Section 9.3 covers different system analysis and modeling methods for wireless-based application systems. Finally, Section 9.4 summarizes the chapter.

9.1 Understanding the Wireless Application Domain

Understanding wireless application systems includes the following subjects:

- Wireless networks and connectivity;
- Mobile devices and platforms;
- Mobile technologies;
- Mobile storage technologies;
- Wireless security.

1. *Understanding wireless networks and connectivity:* It is necessary for engineers to have the basic understanding and concepts about wireless networks, including network infrastructures and components, connectivity standards, features, and limitations. The chapters of Part III have covered the basics of different types of wireless networks and connectivity standards. For a given wireless network, engineers need to understand the network coverage area, network connectivity standards, throughput range, bandwidth limit, and typical applications. For the standard protocols, they need to pay attention to its frequencies, data transfer rate, and network coverage.

2. *Understanding mobile devices and platforms:* Mobile device vendors provide many new device models to meet the increasing needs of mobile users. Since a single mobile device may support different operation platforms, types of wireless network connectivity, specific mobile technologies, and frameworks, it is essential to understand mobile devices

and supporting platforms. The chapters in Part II provide the necessary background and knowledge about mobile devices. For each selected mobile device model, engineers must pay attention to its basic features:

- Physical features, such as size, weight, processor speed, memory, screen size, battery life, and power consumption;
- Mobile operating system support, such as Windows CE and J2ME;
- Network connectivity, such as GSM and GPRS support.
- Integrated features, such as digital cameras, keyboards, infrared, and Bluetooth.
- Software support, such as mobile browsers and development tools.

3. *Understanding mobile technologies:* Understanding mobile technologies on mobile devices allows engineers to identify and select the proper technologies in the system analysis phase, and to make the right design trade-offs during system design. The chapters of Part II cover the basics of different mobile technologies, such as presentation technology, media support technology, and operation platforms. In the system analysis phase, engineers need to pay more attention to the features, standards, strengths, and limitations of the provided mobile technology.

4. *Understanding mobile storage technologies:* For smart mobile client software, mobile application data needs to be stored in a persistent data store on mobile devices. It is necessary to determine the types of persistent data store technologies that are available to the selected mobile platforms, along with their key differences and limitations, before collecting and analyzing mobile data requirements. Table 9.1 shows the major features and limitations of different mobile data storage technologies on mobile devices.

5. *Understanding wireless security:* This is critical for developing wireless-based application systems. System analysts understand and document the essential wireless security requirements in network communications, mobile transactions and application sessions, and security needs from mobile device users. Chapters 11 and 12 provide the concepts, issues, and solutions of wireless security. More wireless security topics can be found in [1, 2].

9.2 Engineering System Requirements

The key in system requirement engineering is to get a clear understanding about what the customer wants.

9.2.1 Requirements Engineering Process

Software developers have realized that it is cost-effective for the tasks in the communication phase, following a requirements engineering process, to get a clear understanding about the customer's needs for the final product. A typical engineering process is given in [3], which is described in sequential steps here.

Table 9.1   Comparison of Different Mobile Storage Technologies

Name	Database Type	Major Features	Major Limitations
Windows CE database	Flat-file database	Provides a built-in database system in the form of its object store API. Object store API functions are useful to add, remove, search, and sort data.	Database resides only within the device's RAM; it cannot be placed on an auxiliary storage card. Data storage capabilities are of limited complexity and size. Database synchronization is an issue.
Symbian DBMS	Rational database	Supports multiple tables in the same database, and allow more complex schemas. Provides a subset of SQL and C++ API. Supports transactions for database changes (e.g., undone for all data changes).	Limits access to the other supported programming languages. Does not offer functionality as a regular rational DBMS (e.g., table joins). Does not support database synchronization.
J2ME RMS	J2ME MIDP RMS	The RMS is a set of Java classes for storing and retrieving simple sets of data. Each RecordStore has a collection of persistent records stored as a byte array. A set of APIs is available to open, close, insert, delete, and move data records.	Application programs need to create and maintain database. Very simple data storage mechanism limits data volume. Database synchronization capabilities are not required.
Palm database	Flat-file database	Supports two kinds of database systems: 1.record database for application data; 2. resource database for application code and user interface objects.	Application programs need to create and maintain database. Data storage capabilities are of limited complexity and size. Lacks support for data access transactions. Does not provide very robust synchronization capabilities.

- *Inception:* This is the first step, in which engineers ask a set of context-free questions given in [4]. Its purpose is to get a basic understanding about the problem, the customer who wants a solution, and the nature of the desirable solution.

- *Elicitation:* The major purpose of requirements elicitation is to communicate with the customer and users to determine what the objectives for the system are, what is to be accomplished, how the system (or product) fits into the needs of the business, and how the system is to be used by users. The detailed issues in requirements elicitation for wireless-based software systems are discussed in the following section.

- *Elaboration:* In this step, the information gathered from the customers and users in the previous steps is expanded and redefined. Its focus is to elaborate and refine the system, using the system requirements to generate a clear and complete blueprint for the final product from the end user's viewpoint.

- *Negotiation:* Some requirements gathered from the previous steps may be inconsistent, incorrect, conflicting, unrealistic, or ambiguous. During this step, engineers need to resolve these conflicts by communicating with the customer and users. All requirements should be ranked based on their priorities.

- *Specification:* In this step, all of the gathered system requirements are specified in a well-defined format to generate a preliminary system requirements document. This document can be used as a basis for a business contract between developers and a customer. It also can be used as a reference for engineers when performing system analysis and modeling. Some detailed "standard templates" for specifications are given in [5].

- *Validation:* The requirements generated from the previous steps need to be examined using customer reviews and document inspections to ensure that they are unambiguous, consistent, correct, accurate, and complete. The primary requirements validation method is known as the formal technical review, which can be found in [4, 6, 7].

- *Management:* Since it is common that the given system requirements are changed during a software development cycle, engineers need to track and maintain all changes of the given system requirements. One common practice is to use a software configuration management process to track requirement changes and document versions.

9.2.2 Requirements Elicitation for Wireless-Based Software Systems

Gathering requirements for wireless-based software systems is a challenging task because of the following considerations:

- Lack of training and understanding of the wireless domain and application problems;
- Rapid advances of wireless networking and mobile technology, and quick updates of mobile devices.

Therefore, it is important for system analysts and engineers to understand what kinds of system requirements they need to collect, and how to obtain them from the customers and users.

Table 9.2 provides system analysts with a basis for developing an inquiry list for elicitation of different types of system requirements for wireless-based application systems.

Elicitation of system requirements for network infrastructure and connectivity should be focused on finding the answers to the questions listed in the table. Based on the responses from the customer and users, the system network infrastructure should be presented with specified connectivity information (such as network name and protocol standard); required network components (such as proxy servers, gateways, and routers); as well as mobile client software and application servers.

Gathering System Data Requirements

In addition to asking questions listed in Table 9.2, we also need to pay attention to collecting special data requirements and constraints for wireless-based application systems. They can be classified into the following areas:

Table 9.2 Different Types of System Requirements for Wireless-Based Applications

Types of System Requirements	*Typical Questions for Requirements Elicitation*
Requirements for network infrastructure and connectivity	What kinds of wireless networks are needed to support the system? What is the required network connectivity between wireless network(s) and wired network(s)? What are the required communication protocols and connectivity standards in the application domain? What are the required network components (e.g., servers, gateways, routers) in the targeted network?
Requirements for mobile devices and operation platforms	For what types of mobile devices and operation platforms is the system designed? Which specific device models and platform versions must be supported by the system? What is a typical required configuration for a targeted mobile device and operation platform (e.g., memory size, screen size, and so forth)?
Requirements for system data and data storage	What kinds of data repositories are needed for the system (e.g., files systems, and database systems)? What kinds of mobile data storage are needed for the targeting mobile devices? What is the mobile data that must be stored on mobile data storage? What kinds of multimedia data are required for the system? Which data delivery mechanisms are required for different mobile data?
Requirements for mobile technologies on mobile devices	Which mobile presentation technologies will be selected to be developed for mobile user interfaces? Which network connectivity software (or APIs) will be used to support communications? Is multimedia support required for the mobile applications? If yes, what is the supporting technology? Is any synchronization capability required for supporting mobile users? If yes, what is the supporting technology? Which mobile data storage will be used to store persistent mobile data on mobile devices?
Requirements for system users and system-user interactions	Who are the users of the system? Can they be classified into different user groups (or actors)? For each user group (or actor), what are its accessible system functions and service features? What kinds of roles does each actor play? What kinds of goals do each actor have? What kinds of mobile users are supported by the system? What are the mobile user access patterns and profiles to the system?
Requirements for system application functions	What are the required system functions and service capabilities? What are the mobile accessible functions and service features? What are the business models and application workflows supported by the system? Are there any required standards for application transactions, billing, and user interfaces?
Requirements for system security and user access control	What are the network security requirements for the system? Are there end-to-end and/or peer-to-peer security requirements system communications? What are the security requirements for system application functions and transactions? What are the security needs and privacy concerns of mobile users? How are user accesses controlled, including access by mobile users?

- *Mobile data storage constraints*, such as memory size, cache size, and external storage size.

- *Mobile data delivery mechanisms* (e.g., push delivery or pull delivery). In the push delivery, mobile data is delivered to mobile users by the system without their requests. A typical example is sending notification messages to mobile users in a wireless information service system over a wireless network. In the pull delivery, mobile data is delivered to a mobile user only when requested.

- *Mobile data accessibility*. It is important to find out when, where, and how mobile users access system mobile data. For example, a mobile portal system on a subway only supports mobile users on a train to access the system through a very high-speed wireless LAN. However, a mobile portal system for wireless phone users needs to support their mobile access anywhere and at any time.

- *Mobile data synchronization*. It is necessary to determine the types of stored (or persistent) mobile data on mobile devices that must be synchronized, and the types of mechanisms that are needed to perform data synchronization.

- *Location-aware mobile data*, such as a local map, news, weather information, and product promotions based on the current location of mobile users. Location-aware mobile data always needs to be retrieved and delivered to mobile users based on their current locations.

- *User-aware mobile data*, such as a mobile address book, and a buddy list for chatting systems. User-aware mobile data must be always available to mobile users upon login.

Gathering System Functional Requirements

Gathering system functional requirements is one of the primary tasks in system requirements elicitation. The key is to communicate with the customers and users, collect their inputs about system functions and service capabilities, and classify and document them in a well-defined traceable format. System requirements can be classified into in different groups. A simple classification includes two groups: (1) function requirements, which specify the requirements about system functions and service capabilities; and (2) nonfunction requirements, which specify the requirements about required technologies, operational context, performance, reliability, availability, and scalability. Table 9.3 shows a detailed classification of system functions for wireless-based application systems.

Table 9.4 lists a set of functional requirements for an MMS-based messaging system [8] that provides each mobile user with a multimedia message mailbox, and allows them to exchange multimedia messages over a wireless network.

Gathering Wireless Security Requirements

System security is always an important concern in building wireless-based application systems. In general, security requirements can be classified into the following types:

Table 9.3   Classification of System Function Requirements

Types of Function Requirements	Descriptions
Communication capabilities and connectivity functions	They support network communications at the application level and required connectivity interfaces.
Security functions	They specify the required security solutions to meet the various system security requirements.
Mobility functions	They specify the required functional capability to support the system's mobility features.
Synchronization functions	They support the synchronization of system data, including persistent mobile data on mobile devices.
Persistent data storage functions	They maintain and access system persistent data, including mobile persistent data.
Application transaction functions	They process and support the required application transactions in the system. In a wireless payment system, a wireless payment transaction processing function must be specified.
System administration functions	They only support system administrators in performing various system administration operations.
End user application functions	They only support system functions that can be accessed by end users.

- *User-related security requirements*, which include security requirements about user authentication (e.g., checking user password); authorization (e.g., checking user PIN numbers, electronic signatures); and privileges (e.g., checking user access code) to access functional features.

- *Network-related security requirements*, which include security requirements about network firewalls and secured protocol-based communications at a specific network level. For example, a wireless Internet chatting system may only require secured end-to-end communications over a wireless network. This implies that some kind of encryption and decryption mechanisms must be used to support end-to-end communications between two points over the given network. A customer may have more strong security needs on communications over a wireless network (e.g., secured peer-to-peer communications). This suggests that an end-to-end secured application protocol is needed at the application layer to support all communications and transactions between two parties over a network.

- *Session security requirements* for certain types of user operation sessions. A typical example is to require a system to provide a secured log-on session, which implies that all transactions between a user and the system must be checked using a session ID. Another example is the required secured payment sessions in a wireless payment system, in which all payment transactions during the same payment session must be checked, based on a payment session ID.

- *Transaction security requirements* for application functions, such as payment transaction security requirements in a wireless payment system. These types of requirements usually specify the need for secured application transactions in user authentication, access control, data access, session control, and communication security.

Table 9.4　Sample List of Function Requirements for an MMS-Messaging System

End user application functions	1. Mailbox management functions for a mobile user: 　List a message mailbox index and display the first page index on the display screen. 　Select and access the current message entry from the displayed index. 　If the mailbox index has more than one page, the user can switch the current index page to the first, previous, next, and last index page. 2. User address book management functions for a mobile user: 　List a mailbox address index on the display screen. 　Select and access one address entry. 　Add, delete, and find an available entry. 3. Message management functions for a mobile user: 　Reply to a received message in his (or her) mailbox from another user. 　Send a new message to other users. 　Forward a received message to other users. 　Create a message by recording audio message, selecting graphic image file or video file, or setting a text message. 　Select and delete a message from the mailbox. 4. Mobile notification function for mobile user: 　System should be to deliver a system-generated notification message to a mobile user. 　When a new message arrives while the user is online, the user should be immediately informed about the message. 　When the user is offline, all new messages must be stored on the system. The user should be informed about these messages as soon as he is online.
System administration functions	1. User account management: 　The system administrators should be provided with functions to manage all user service accounts. The functions must allow the administrator to do the following: 　View, create, update, delete, disable, and enable mobile user service accounts. 　Track and monitor user accounts. 2. Notification management: 　The system allows administrators to create and send various notifications to mobile users. The notifications can be classified into the following types: 　Any new promotions for mobile users. 　An alert about a mobile user's mailbox limit being exceeded. 　A warning message about illegal usage of a mobile user's account. 3. System configuration: 　The system allows administrators to set up system configurations and metadata. Typical examples are: 　Mobile users' mailbox size and address book size, which should be configurable based on the user classification. 　The number of messages that can be displayed per page (which can be configured by the mobile user). 　The message size in a mailbox. 　The number of entries in address book listing page (which can be configured by the mobile user).

- *Certification security requirements* for system components, including mobile client software, wireless Web server, and application servers. These requirements specify the need for a certification solution in a system to ensure that each involved party (client or server) has been certified with a appropriate agency.

Some systems may have special security requirements relating to biometrics. For example, a wireless payment system may require a stronger form of authentication, known as biometric authentication. Biometrics provides a wide range of techniques for authenticating a mobile user, based on the user's unique physical characteristics, such as fingerprint identification, face recognition, and voice recognition.

9.3   System Analysis and Modeling

System analysis and modeling is an important phase in a software development process. Its major objective is to generate a blueprint for building a software system based on the system requirements. During this phase, engineers need to understand and analyze the system, and model the system to generate a system analysis and modeling document. This document defines the system from six different views, using well-defined models and formats, and will be used as a reference document by engineers for system construction. Figure 9.1(a) shows a procedure to perform system analysis and modeling, and Figure 9.1(b) shows the six views included in the system. To generate the high-quality outcome, formal or informal reviews can be performed to assure the quality of the generated document. The detailed review methods and guidelines can be found in [4, 6, 7].

The basic tips for generating a good system definition document are listed here.

- Apply well-known system analysis models and follow well-defined documentation standards to achieve consistency.

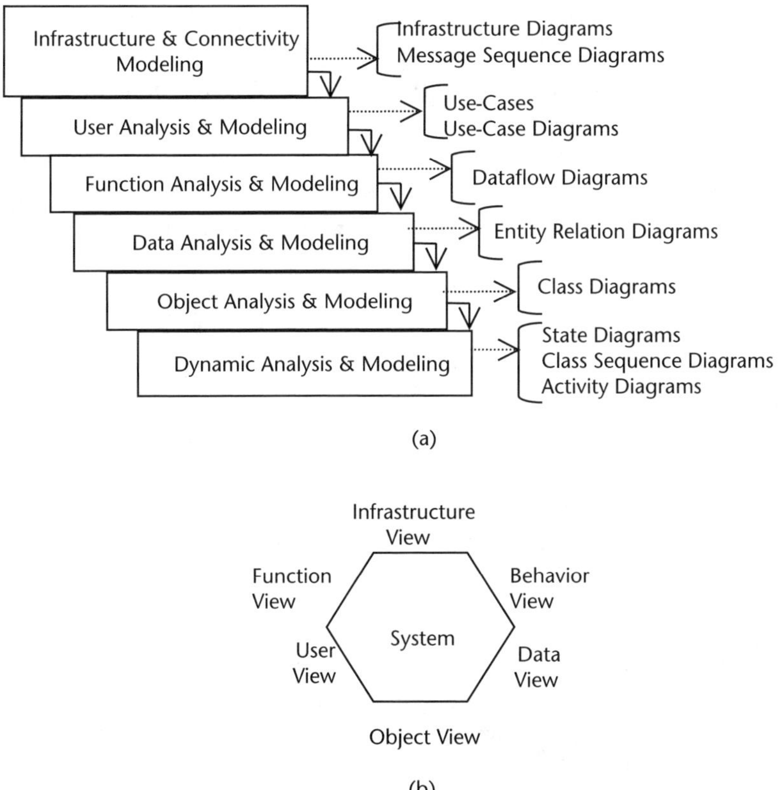

Figure 9.1   Process of system analysis and modeling: (a) six steps, and (b) six views.

- Refer back to the given system requirements to achieve traceability.
- Analyze and model the system in all six views to achieve completeness.
- Review the document to achieve accuracy.

9.3.1 System Infrastructure and Connectivity Modeling

System infrastructure and connectivity modeling can be performed in four steps:

1. Generate a system infrastructure diagram to present all servers and mobile clients in the network. Figure 9.2 presents the infrastructure of the mobile-talk system [9]. The major objective is to help students develop a wireless multimedia information service system for mobile phone users.
2. Generate a table to briefly specify the function of each server (or client). Table 9.5 lists the basic servers and clients in the mobile-talk system.
3. Specify the network connectivity protocols in the system infrastructure, using message flow diagrams. The purpose of this step is to help engineers understand network protocols, which usually are at either the session layer or the application layer, and to help them understand communication message sequences among different servers and clients over the target network. The message flow diagrams specify the protocol-based communication messages sequences. They can be derived from related protocol standards and specifications. The mobile-talk system uses the session initiation protocol (SIP) at an application layer to set up and control wireless multimedia (digital audio) call sessions between two mobile users. Figure 9.3 shows the diagrams presenting SIP-based message flows. The detailed overview of the SIP protocol is given in [10], and the related specifications and standards are given in [11, 12]. The diagrams in Figure 9.3 are derived from these SIP specifications. Figure 9.4 shows the extended SIP-based message flow for the user presence service [10], which provides messages to inform a mobile user (e.g., User A) about the presence of another

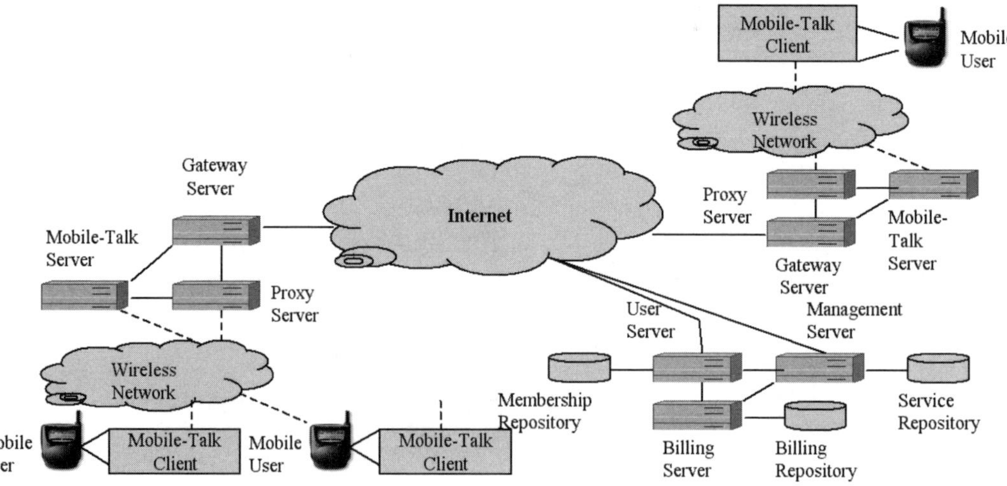

Figure 9.2 Infrastructure of the mobile-talk system.

Table 9.5 Network Component List in the Mobile-Talk System Infrastructure

Components/Systems	Descriptions
Mobile-talk client	It supports the interactions between a mobile user and the mobile-talk system on a mobile phone to access the end user functions of the mobile-talk system.
Proxy server	It is responsible for forwarding the mobile client's requests to the corresponding application server, and for forwarding the login authentication request to the authentication server. In addition, it performs mobility management, network management, and deployment services.
Mobile-talk server	It supports the two mobile clients in conducting a stream-based digital audio call over a 3G wireless network. It must set up a mobile-talk call, first using SIP-based communications with two mobile-talk clients, and then allowing two clients to talk to each other through digital audio streaming in a half-duplex mode.
Gateway	It supports the basic network gateway functions in the system, based on the stored the network Yellow Page, which keeps location information on all of the network components, and mappings between server IP addresses and mobile phone IDs. When there is a packet arrival, it looks into the Yellow Page and forwards the packet to the appropriate proxy server.
Management server	It mainly deals with the function of network management and service management. The central server stores the deployment information about all of the servers in the system for management purposes.
User service server	It maintains all user membership information and registrations, and controls mobile users' accesses by performing user authentication.

mobile user (e.g., User B) in the system, whenever User B's status changes (e.g., User B goes off-line).

4. Generate a message scenario diagram for each system feature that involves more than one server and/or client.

9.3.2 System User Analysis and Modeling

The objective of system user analysis and modeling is to analyze and specify a system's behaviors from the end user's point of view. It can be carried out in two steps: (1) identify end user groups (i.e., actors), and (2) develop use-cases. Its output is a set of use-case diagrams.

To identify various system end user groups, system analysts can first classify all system end users into different actors, based on their accessible devices, system interfaces, and function sets. Each actor should have its own goal, accessible functions, system-and-user interaction behaviors, and system interfaces (i.e., devices).

According to the system requirements of the MMS-based message system (given in Section 9.2), we can identify two system end user groups. They are online users and mobile users. Figure 9.5 shows two examples of use-case diagrams for mobile user.

A use-case is defined in [13] as a contract that describes the system's behavior under various conditions as the system responds to a request from one of its users. Pressman [4] views a use-case as a stylized story about how a role-playing end user interacts with the system under a specific set of conditions—preconditions, postconditions, and a basic operation scenario with a number of sequential operations. Considering each user group as a system's actor, engineers can develop use-cases by answering the following questions, as suggested in [14]:

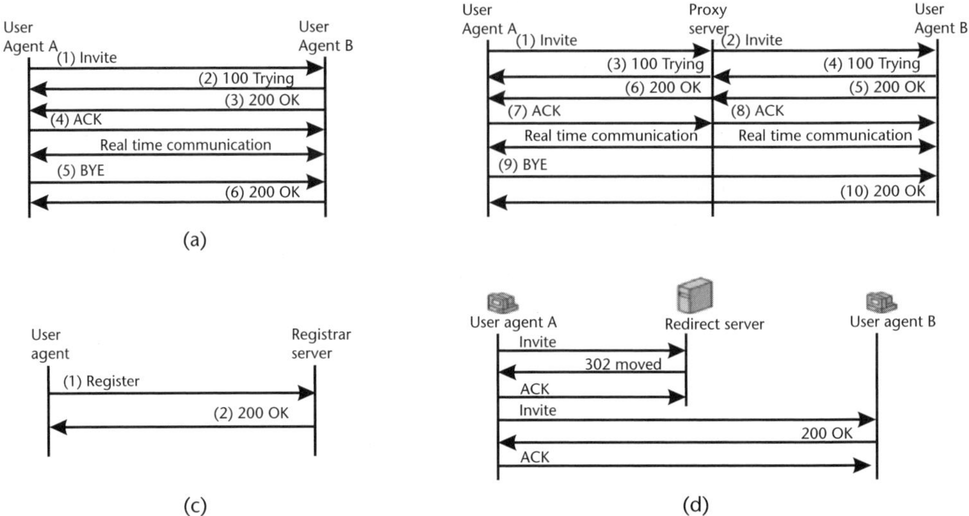

Figure 9.3   SIP-based message flows: (a) between two user agents, (b) between proxy server and user, (c) between a user agent and registrar server, and (d) between redirect server and user agents.

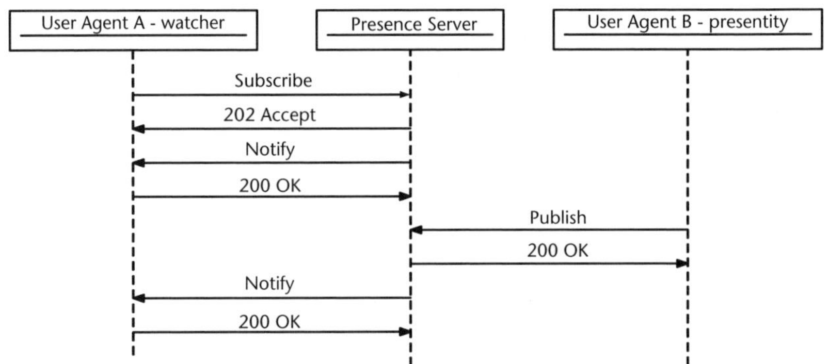

Figure 9.4   SIP-based message flows for user presence.

- Who is (are) the primary actor(s), and the second actor(s)?
- What are the actor's goals?
- What preconditions should exist?
- What major tasks or functions does the actor perform?
- What exceptions might be considered?
- What variations in the actor's interaction are possible?
- What system information will the actor acquire, produce, or change?
- Will the actor have to inform the system about changes in the external environment?
- What information does the actor desire from the system?
- Does the actor wish to be informed about unexpected changes?

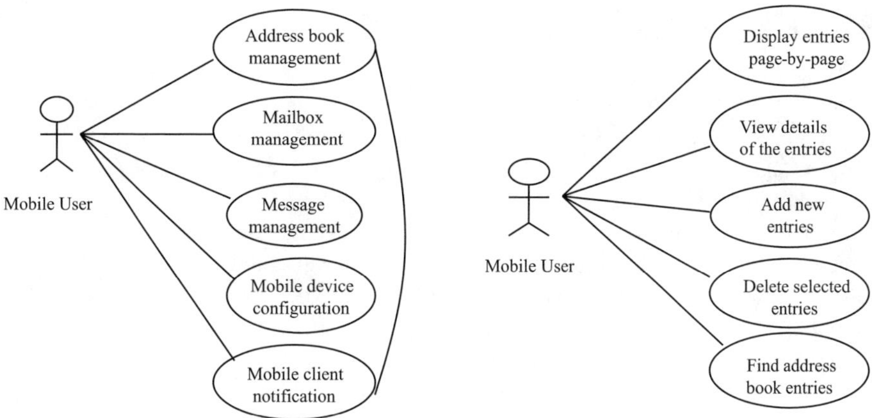

Figure 9.5   Two use-case diagrams in the MMS-based message system.

For a use-case sample for the mobile user actor in the MMS-messenger to access the address book management function, see Table 9.6.

9.3.3   System Function Analysis and Modeling

System requirements can be classified into functional and nonfunctional requirements. The nonfunctional requirements specify the required technologies, operation environment, and system performance. Typical system performance requirements specify the customers' and users' expectations for system reliability, availability resource utilization, processing speed, throughput, and scalability. Functional requirements specify system capabilities and functional features.

System function analysis and modeling only focuses on the understanding of system functions and the information flows among them. To achieve understanding, we can first partition the system into functional servers and client software, based on the system infrastructure diagram. Then, for each server (or client software), we further partition them into logic functional modules, and present them using a functional block structure diagram, as described in Table 9.7. Figure 9.6 shows an example. We can continue this functional partition process until we have gained a good understanding about the subfunctions of each module.

To analyze and specify the information flow among these function modules, we can use a well-known function modeling method, known as dataflow modeling, in which system functions are specified as functional processes, and their information flows among them at the different levels. The dataflow diagrams (DFDs) are used to present the function processes and the information flows among them at each level. The inputs and outputs of each function are modeled as the incoming and outgoing information links of its function process. For a complex function process, it may support a number of subfunctions. To clearly understand and model its low-level functions, we must partition and model the process in terms of its subfunction processes and their information flows. This operation continues until all its subfunction processes have been clearly presented and modeled.

A dataflow diagram is a graph consisting of four types of elements: (1) nodes, which stand for function processes; (2) dataflow links, which present incoming and

Table 9.6   A Use-Case Sample for Address Book Management

Use Case ID: 1

Use Case Title: Address Book Management

Primary Actor: Mobile user

Stakeholders and Interests:

1. The mobile user has created an MMS account.

2. The mobile user should be able to view, find, edit, add, and delete entries.

3. A message is shown on the mobile device screen if the above features succeed or fail.

4. Any change in the address book will be saved in the database.

Preconditions:

1. The mobile user is identified and authenticated by system.

Success Guarantee (Postconditions):

1. The address book is viewed.

2. The change to the address book (edited, added, or deleted entries) will be stored in the database and shown on mobile device.

Main Success Scenario (or Basic Flow):

1. The mobile user logs in using his or her user ID and password.

2. The mobile user sees the first n entries in the address book (n defined in configuration module, default value equals 5).

3. The mobile user views any selected entry in the address book.

4. The mobile user finds any existing entry.

5. The mobile user edits and updates any selected entry.

6. The mobile user adds new entries and saves them in the database.

7. The mobile user deletes selected entries from the database.

8. The mobile user sees a message indicating the change to the address book whenever there is a change made by Steps 3 to 7.

Extensions (or Alternative Flow):

At any time, if the system fails, it should generate an error report.

1. If the mobile user enters invalid user ID/password, the system should be able to prompt appropriate error messages.

2. When the mobile user tries to find an entry that does not exist in his or her address book, the system should be able to prompt appropriate error messages.

3. If the mobile user tries to add an entry to a full address book (number of entries = Max capacity of address book), a warning message should be shown on the screen.

Frequency of Occurrence: whenever needed/requested

outgoing dataflows among function processes; (3) data stores, which represent logical data store repositories used in the system; and (4) external entities, which refer to the external hardware and software parts that interact with the system (e.g., third party systems and hardware devices). Figure 9.7 shows a top-level DFD diagram for the application server in the mobile-talk system. Figure 9.8 shows a third-level DFD diagram that presents the supporting functions (such as handle request and handle response), and their incoming and outgoing dataflows inside a presence manager processor.

9.3.4 System Data and Object Analysis and Modeling

In this step, there are two major tasks: (1) system data analysis and modeling, and (2) system object analysis and modeling. The first task conducts system data analysis and modeling for system data objects that are stored and processed in the system.

Table 9.7   Description of the Functional Modules in the Mobile-Talk Application Server

System Function Components	*Descriptions*
Mobile talk server	It is the central controller module that dispatches requests or responses to appropriate modules for processing.
Mobile talk server configuration	The server configuration module initializes SIP stack and server database.
Mobility manager	It is responsible for storing the mobility information of the user. It keeps track of visitor information and periodically updates the database by querying user status and location.
Call manager	It manages a call that is established between users. It keeps track of the time at which the connection was established, the status of the connection, the time at which the connection was dropped, and so forth.
Presence manager	It manages presence-related requests and response processing.
Deployment service	It tracks new deployments made at this layer. It notes the new configuration of the server and dynamically changes the values.
Security manager	It manages authentication and other access rights validations. It ensures that the parties in the conversation are authorized parties. It checks the access rights for the system administrator for configuring the system.
Session manager	It manages active sessions, such as session creation, session deletion, and so forth. Each session uses a channel, and once the communication is completed, the session is deleted.
Channel manager	It manages all the communication channels, such as creation of channel, deletion of channel, and updating channel. A channel is a virtual channel established between two communicating entities, and there can be more than one call in a single channel.
SNMP processor	It manages network components. This processor informs the central server when the application server goes down, and immediately stops sending or receiving requests until another alternative is found.
SIP processor	It accepts and processes all requests related to SIP.
Application database manager	It manages the database, checking frequently for user's information, updating, deleting, and inserting into database as required; and handles database synchronization and migration with other servers.
Communication interface	It handles communication-related procedures, such as sending a packet, receiving a packet, buffer management, and queue management.

System analysts or engineers need to define all system data objects from the system data modeling, and then clarify their relationships.

Data Object Modeling

A data object is a representation of almost any composite information that must be understood and processed by the system. Each data object must have a number of data attributes or properties. For example, a user's account information record includes a set of data attributes, such as user name, account ID, user password, and so forth. However, "user password" is not a valid system data object about a user, since it has only one data attribute, and does not present other properties of a system user.

A system data object can be a system input (or output) record, an incoming (or outgoing) message, a representation of a real world object (e.g., a mobile user), an organization, or an occurrence (e.g., an event). Since a data object is only an encap-

Mobility Manager	Security Manager
Database Manager	Deployment Manager
Process Manager	Call Manager
Channel Manager	Session Manager
MobileTalkSvr	Mobile Talk Server Config
SIP-Processor	
SNMP-Processor	
Communicate Interface	

Figure 9.6 Modules of the mobile-talk application server.

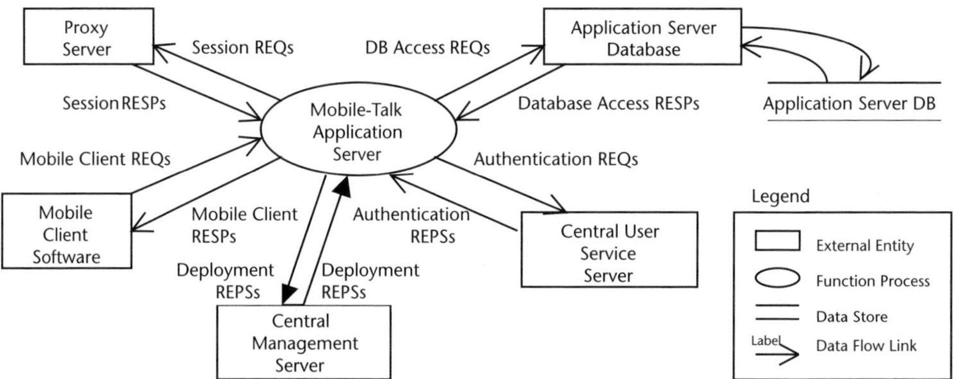

Figure 9.7 Top-level DFD for the application server in the mobile-talk system.

sulation of data attributes without the references to the operations that manipulate the data, a common way to represent a data object is by using a table. A mobile user's registration information can be represented in a table format, shown in Table 9.8.

System data object analysis and modeling has three steps:

1. Identify system data objects and object attributes.
2. Identify and analyze the relationships between system data objects.
3. Present system data objects using entity relation (ER) diagrams.

A system data object must have an object name, and each data object usually consists of three types of data attributes: (1) one or more key data attributes that are used as a key to uniquely identify an individual object instance, (2) property attributes that describe the basic properties of the data object, and (3) reference attributes that refer to another instance in another system data object. In the given mobile user registration table, user ID is the key attribute, user account ID is the reference attribute that points to user account table, and the other data fields are property attributes.

Data objects in a system are associated with one another in different ways because there are three types of data dependency relationships. We can check any

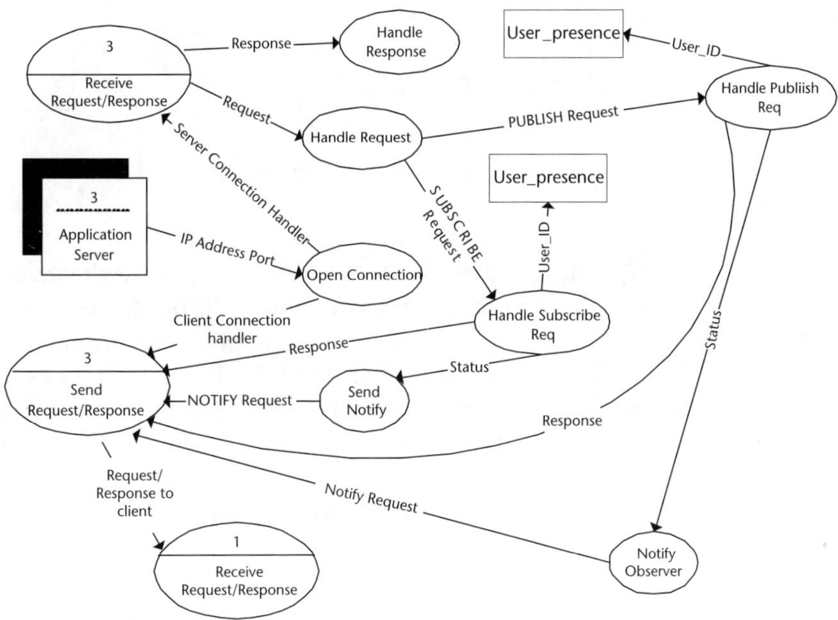

Figure 9.8   DFD for the presence management process.

Table 9.8   A Sample Data Object for User Records

User Name	User ID	Address	Mobile Phone Number	Registration Date	User Account ID
Jerry Gao	110-29-53305	One Washington Square San Jose, CA 95192-0180	408-956-3940	09-30-2005	12039401

two identified data objects to see if they are related or associated based on the given use cases and function requirements. Consider two types of data objects (mobile user and mobile service) in a an MMS system. They are related in different ways. For instance, a mobile user has a specific mobile service, such as wireless chatting, mobile commerce, and mobile portal access. A mobile user may request or cancel a mobile service. Thus, we can define the following relevant relationships between mobile service and moblie user according to the following requirements:

- *Has-service relation:* mobile user *has* a mobile service;
- *Request-service relation:* mobile user *requests* a mobile service;
- *Cancel service relation:* mobile user *cancels* a mobile service.

These data object/relationship pairs can be graphically represented in Figure 9.9(a).

Understanding data objects, attributes, and relationships in a system only provide the basis for understanding the information domain of a problem. Additional information is needed to generate a clear and accurate picture of the targeting system in its information space. After identifying the relationships between data objects, we obtain a set of objects and their relationship pairs. However, each rela-

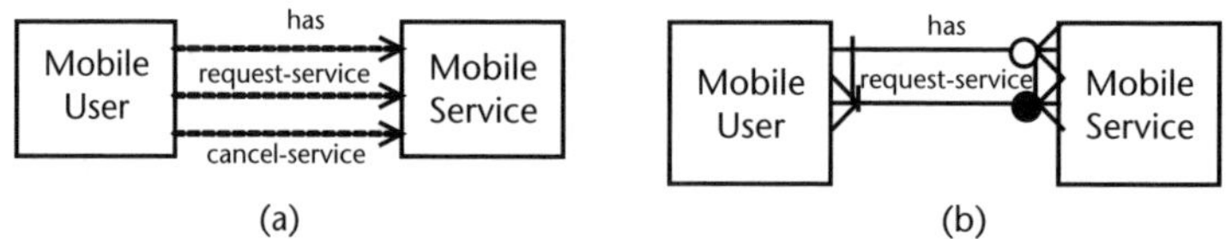

Figure 9.9   (a, b) Two views of relationships between two types of data objects.

tionship pair only indicates that object A relates to object B, which does not provide other information about each relationship (e.g., their occurrences in each relationship). We need to understand and model the number of occurrences of object A are related to the number of occurrences of object B. This is known as the cardinality concept in data object modeling. Thalheim [15] defines the cardinality of an object/relationship pair as "the speciation of occurrences of one [object] that can be related to the number of occurrences of another [object]." Usually, there are three forms of cardinality (one-to-one, one-to-many, and many-to-many). For example, a mobile user can have zero, one, or more mobile services. This suggests that the "has" relationship between mobile user and mobile service is a one-to-many relationship, because one mobile user object can relate to one or more mobile service objects.

In addition to the cardinality information, we need to specify modality to each object relation pair. The modality concept in a relationship is used to indicate whether the occurrence of one object is mandatory or optional. The modality of a relationship is 0 if there is no explicit need for the relationship to occur or if the relationship is optional. The modality is 1 if an occurrence of an object in the relationship is mandatory. Figure 9.9(b) shows that a mobile user may have zero, one, or more mobile services. A mobile user must request at least one or more mobile services. On the other hand, one mobile service may be requested by more than one mobile user.

To represent data objects and their relationships, we can use entity-relationship diagrams (ERDs), which were originally proposed by Chen [16] for the design of relational database systems, and have been extended by others. More detailed information about ERDs can be found in [15]. Figure 9.10 shows an ERD that represents the subset of data objects in the database repository of the application server for mobile-talk system [17]. This figure uses classic ERD notation, in which labeled rectangle boxes represent data objects, labeled ovals indicate data attributes of each data object, and labeled diamonds indicate relationships between data objects. This ERD shows data objects maintained in a mobile-talk's application server to support the user presence feature. They are: user_location, user_presence, session_info, app_svr_map, process_time_info, and authentication.

Object Analysis and Modeling

Object-oriented analysis and modeling is the part of object-oriented software engineering methodology. Its intent is to define and identify all classes (and the relationships and behavior associated with them) that are relevant to the application problem to be solved. Conducting object-oriented analysis and modeling allows engineers to define an object view for the system in terms of class objects and their

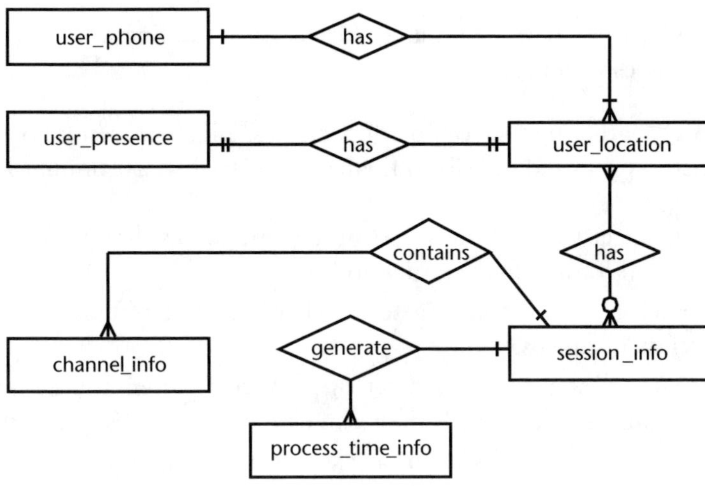

Figure 9.10   Entity relationship diagram for system data supporting the user presence feature.

relations. System data objects are the major focus in data analysis and modeling. However, classes and objects in a system are the major focus in object-oriented analysis and modeling. Here, we only focus on how to identify and define classes and model their relationships.

Conducting object analysis and modeling has four basic steps.

- *Step #1:* Identify all classes in the system components.
- *Step #2:* Identify and define class attributes and operations for each class.
- *Step #3:* Identify and define class dependency relations between classes.
    - Identify the inheritance relations between classes.
    - Identify the aggregation relations between classes.
    - Identify the association relations between classes. This is similar to the identification of various relations among data objects.
- *Step #4:* Present the object analysis results using UML-based class diagrams.

Classes may be identified by looking for different potential classes using a class categorization. Pressman [4] suggests seven types of classes:

- External entities (e.g., mobile devices, proxy servers, mobile users) that consume (or produce) information from (to) the system;
- Things (e.g., mobile transaction reports, users' buddy lists) that are part of the information domain for the problem;
- Occurrences or events (e.g., notifications) that occur in the context of system operation;
- Roles (e.g., mobile callers, system administrators) that are played by people who interact with the system;
- Organizational units (e.g., user groups, teams) that are relevant to the system;
- Places (e.g., mobile user locations) that establish the context of the problem and support the functions of the systems;

- Structures (e.g., computers, drivers) that define a class of objects or related classes of objects.

After gathering a set of potential classes, engineers can follow the selection characteristics proposed in [18] to finalize the class list, as summarized here:

- Each class must retain a set of data attributes that present information that is useful for a system to function.
- Each class must have a set of identifiable attributes that are common to all class instances.
- Each class must have a set of identifiable common operations for all class instances. These operations can change the values of the class attributes.
- Always define classes for a system's external entities that appear in the problem space and produce or consume system information.

Class attributes can be classified into the following groups:

- *Identity attributes* that uniquely identify different class instances (e.g., mobile user PIN and user ID for mobile-user class);
- *Property attributes* that describe the properties of a class (e.g., mobile user's name and age for mobile-user class);
- *Information attributes* that retain the information about an external entity (e.g., mobile device model and manufacture name are for mobile-device class);
- *State-oriented attributes* that present class object states (e.g., a mobile user's service account status is a data attribute for user-service-account class); Its data values (e.g., ACTIVE, INACTIVE, HOLD) represent the service state of the class object.
- *Occurrence attributes* that describe occurrence-related information (e.g., last-changed-date and first-sign-on-date in user-service-account class).

Class operations define the behavior of a class object. They work together to support the system functions (or subfunctions) in a problem space. Each class operation must manipulate one or more class attributes by defining, changing, or displaying their values. They can be classified into the following groups:

- *Construction operations* that define the initial status of class objects;
- *Manipulation operations* that manipulate class data attributes in some way (e.g., adding, deleting, and updating);
- *Computation operations* that perform a computation using some algorithm;
- *Inquire operations* that check the state of an object;
- *Monitor and display operations* that monitor and display an object state or the values of class data attributes.

There are three types of relationships between classes: inheritance, aggregation, and association. An inheritance relationship between two classes refers to the case in which one class inherits some properties (e.g., data attributes) and/or behaviors (e.g., operations) from another class. An aggregation relationship between two classes refers to the case in which the objects of one class become the parts of the objects of another class. An association relationship between two classes usually represents that their objects are associated with each other in the system, which is similar to the relationship in an ER diagram. Moreover, the cardinality and modality concepts in UML object-oriented modeling are the same as data object modeling using ER diagrams.

When Class A is derived from Class B, then Class A is known as a subclass of Class B. Class B is known as a superclass of Class A. This suggests that Class A objects inherit some properties (e.g., data attributes) and behaviors (e.g., operations) from Class B's objects. One common way to identify an inheritance relationship between two classes is to look for "is-a" or "subtype" relationship between two class objects, based on the given system function and data requirements. For example, a system may be required to support three types of user groups: administrator, online user, and mobile user. Based on this, we can easily identify four classes. They are: system user, online user, mobile user, and administrator, because each class has its own data properties and operation behaviors. However, we can find that system user is a superclass for the other three classes, because a "is-a" relationship exists between the system user class and the other three classes. For example, an instance of online user is an instance of system user; however, a system user may not be an online user. This implies that the objects of the system user class are a superset of the objects of the online user class. Figure 9.11 shows the inheritance relations among these four classes, using the UML notation.

The class relationships in the mobile client software for MMS-messenger are shown in Figure 9.12, and are represented using the UML notation. The rectangle represents a class. A diamond on an undirected link between two classes (message and message body) represents an aggregation relation between them. A number represents the cardinality of the relation for each class. In this case, a "1" on each side of the aggregation link represents the one-to-one binding between these two classes. This indicates that each message has only one message body. A triangle can be added onto a directed link to represent a possible inheritance relation. In this example, there are no inheritance relations among 17 classes. However, there are many association relations between classes. Directed links with a label (e.g., uses and processes) represent the association relations. The cardinality (such as one-to-one, one-to-many, and many-to-many) can be represented by using the notation "1" and "1..*". The modality of the association relation can be indicated by using either "0..1" or "0..*". Here, "0..1" means either 0 or 1 class objects are

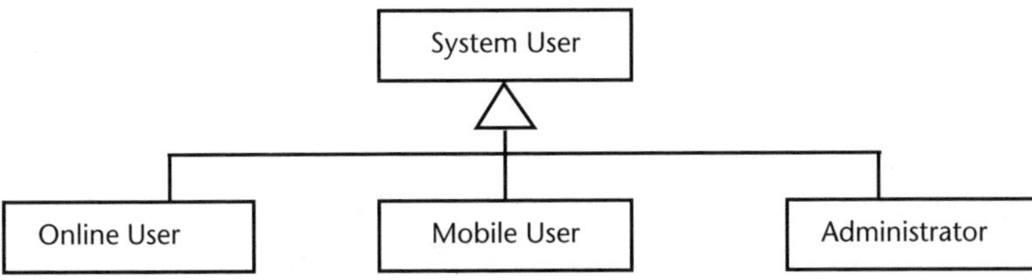

Figure 9.11   Inheritance relations among four classes: mobile user, online user, administrator, and system user.

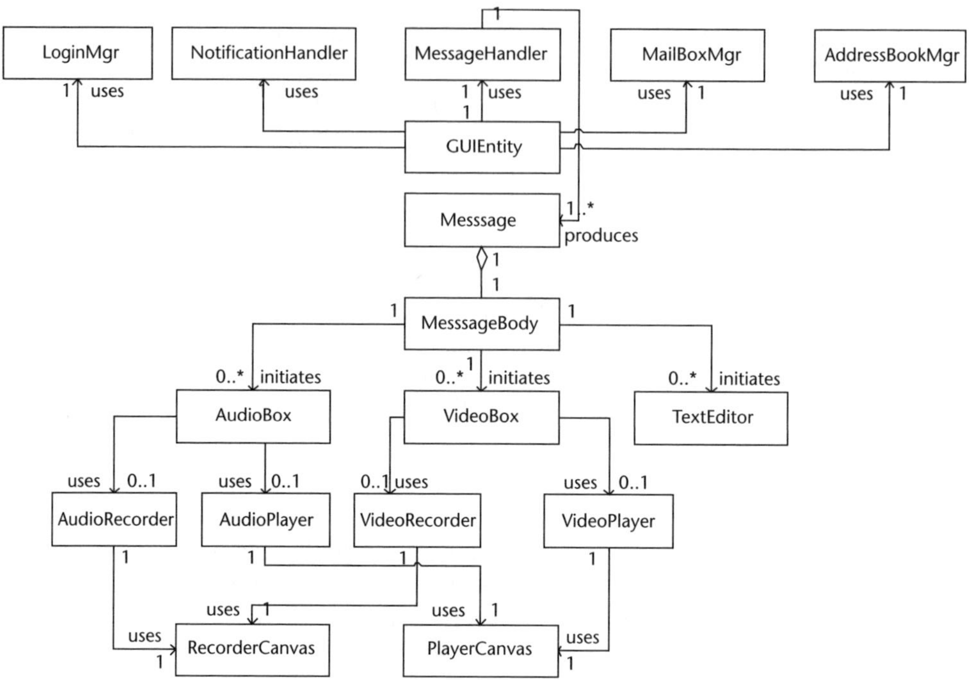

Figure 9.12 A UML class diagram of the mobile client in MMS-messenger.

participating in a relation. "0..*" means either 0 or many class objects are participating in a relation.

9.3.5 System Dynamic Behavior Analysis and Modeling

The major objective of dynamic behavior analysis and modeling for a system is to assist engineers in accurately understanding and specifying a system's state-based processes and event-driven operational behaviors. Analyzing system dynamic behaviors is very important for software systems that have a large number of state-based processes and event-driven operations. Other types of system analysis and modeling are unable to present system dynamic behaviors. The analysis of system dynamic behavior is usually conducted at two levels:

- *Component level*, in which each component's dynamic behaviors, such as component states and operations that transfer the states, are the primary modeling targets;
- *Processor level*, in which each functional process's dynamic behaviors, such as processor states and actions that change the states, are the major focus.

There are a number of well-known dynamic models that are useful in specifying system behaviors. They are: (1) state-based models, such as the finite state machine [19] and UML-based state diagrams [20]; (2) sequence-based models, such as class sequence diagrams and message sequence diagrams [20]; and (3) transaction-based models, such as Petri Nets [19]. Due to space limitations, here we only discuss the UML-based state diagrams and sequence diagrams.

UML State Diagrams

A UML state diagram consists of two sets of elements: (1) a set of state nodes, and (2) a set of state transitions. Each state node usually represents a program's active state, which could be a precondition or a postcondition of a program operation (or an event). Each diagram has only one initial state, and zero or one final state. Each state transition usually represents a program operation, a transaction, or an event that changes the program status from one state to another. It has one source state, one destination state, and its triggering event. Some transitions are more complex, since they may have a guard condition and responding actions in addition to an event. Some transitions are cyclic, which are also known as self-transitions, because their source state node is the same as their destination node. The notation of a UML-based state diagram is given in Figure 9.13.

We can identify all possible events in UML state diagram for a component or a functional processor by answering the following questions:

- What are the incoming and outgoing events (or messages) interacting with external hardware devices and software entities?
- What are the internal program operations, actions, or transactions?
- What are the internal events and messages that are processed?
- What are the interacting events and operations that interact with system users?

Based on these events, we can identify the states by asking the following two questions:

- What are the preconditions and postconditions for each event?
- Are there any business logic conditions (or constraints) for each event?

After gathering the states, we need to check each state to see if there are any guard condition and internal actions.

Figure 9.14 explains state transitions for presence handling on the application server. When the server receives a SUBSCRIBE request, it will check whether this

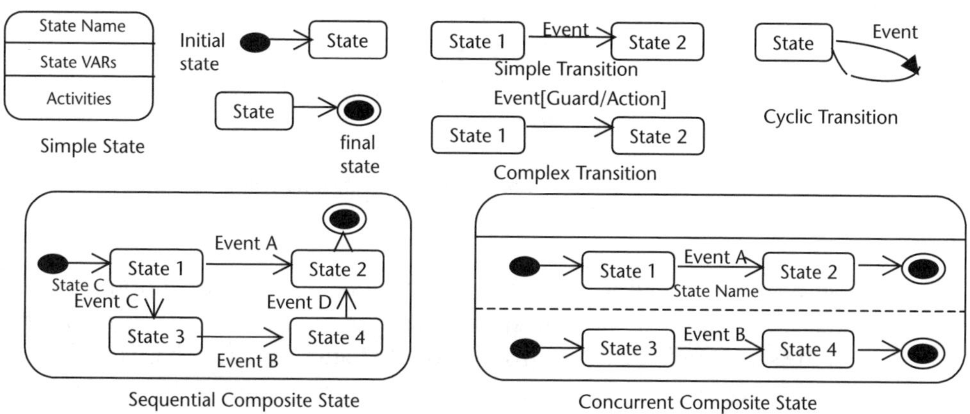

Figure 9.13 Notation of UML state diagrams.

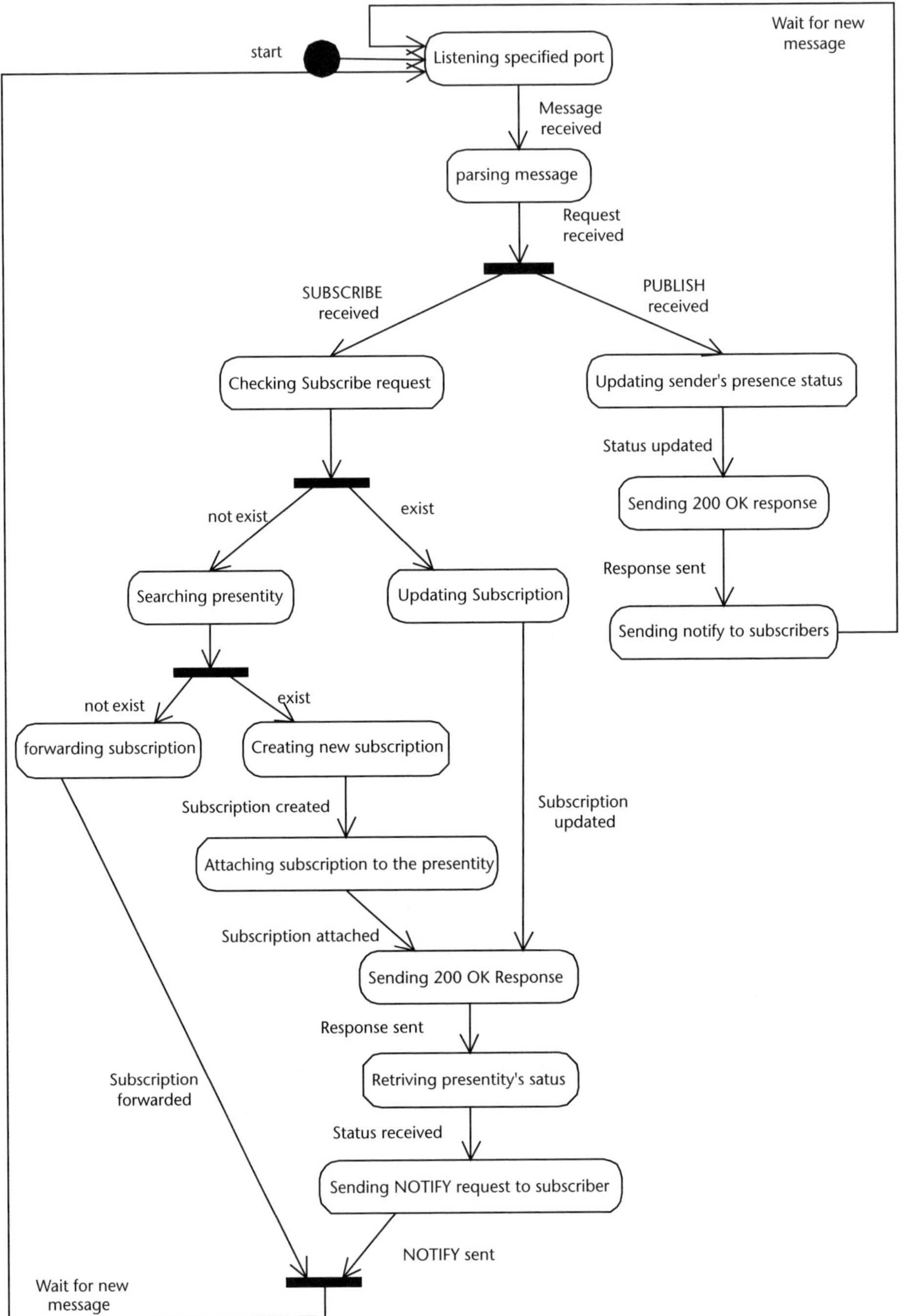

Figure 9.14 UML-based state diagram for user presence handling.

subscription is a new or existing subscription. If the subscription is new, then the server will check whether the state of user presence exists. If this state does exist, then the server will create a subscription dialogue, and attach this dialogue with the state; otherwise, it will forward the request to the router. If the subscription already exists, then the application server will update the expire information of the subscription. The server will then send a 202 accept response to the subscriber, followed by a

NOTIFY request, which contains the status of user presence information. When the server receives a PUBLISH request from the user, it will send a 200 OK, and update the user's presence information within its local database. The server will then send a NOTIFY within those subscription dialogues attached to this user.

UML Sequence Diagrams

A UML sequence diagram, which is a typical sequence-based behavior model, can be used to represent the interaction between objects in sequences. The diagram focuses on the representation of interacting messages and event sequences between objects. A UML sequence diagram consists of the following types of elements:

- Objects are represented as rectangles with a vertical timeline. Time is measured vertically (downward), and the narrow vertical rectangles represent each object's time spent in processing an activity. States may be shown along a vertical timeline.
- Event transitions are represented as directed links labeled with an event or a message. They indicate how events cause transitions from one object to another.

The notation of UML sequence diagrams is shown in Figure 9.15.

A sequence diagram can be essentially viewed as a shorthand version of the use case. It represents key class objects and the events that cause behavior to flow from one class object to another. Sequence diagrams can be used to represent events (or message flows among class objects in a component), and can be used to model the interaction messages (or events) among subsystems or packages at a higher level.

Next, we present an example of a UML sequence diagram that represents the system dynamic behaviors in the call-setup function feature in the mobile-talk system. According to the mobile-talk's system requirements, the call-setup feature is described here.

Call-Setup Feature Descriptions

When a mobile user selects one buddy from the buddy list and clicks the "Call" button to initiate a call, the request is sent to the user's mobile-talk application server. The server looks for the recipient in the server's database; if the recipient's informa-

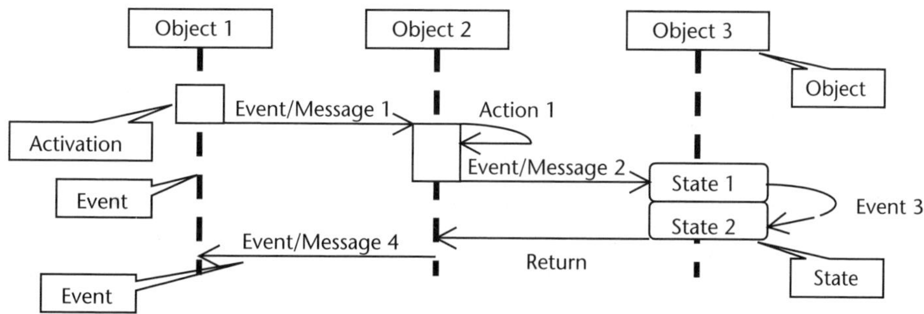

Figure 9.15   Notation of UML sequence diagrams.

tion is found, then the server sets up a call session for the calling parties. Otherwise, the server will forward the user's request to the router, and let the router forward the request to another dedicated mobile-talk application server that supports the recipient. To set up a call session, the application server uses SIP protocol to interact with the two mobile users' client software.

Figure 9.16 shows a UML-based sequence diagram that displays a call setup scenario in which only one application server is involved. This diagram shows SIP-based message interactions among two mobile users' client software and three classes (AppSvr, SessionMgr, and ChannelMgr) in a mobile-talk application server.

The corresponding event sequence is given here.

1. Client A sends an INVITE message to the Application Server (AppSvr object) to initiate a call.

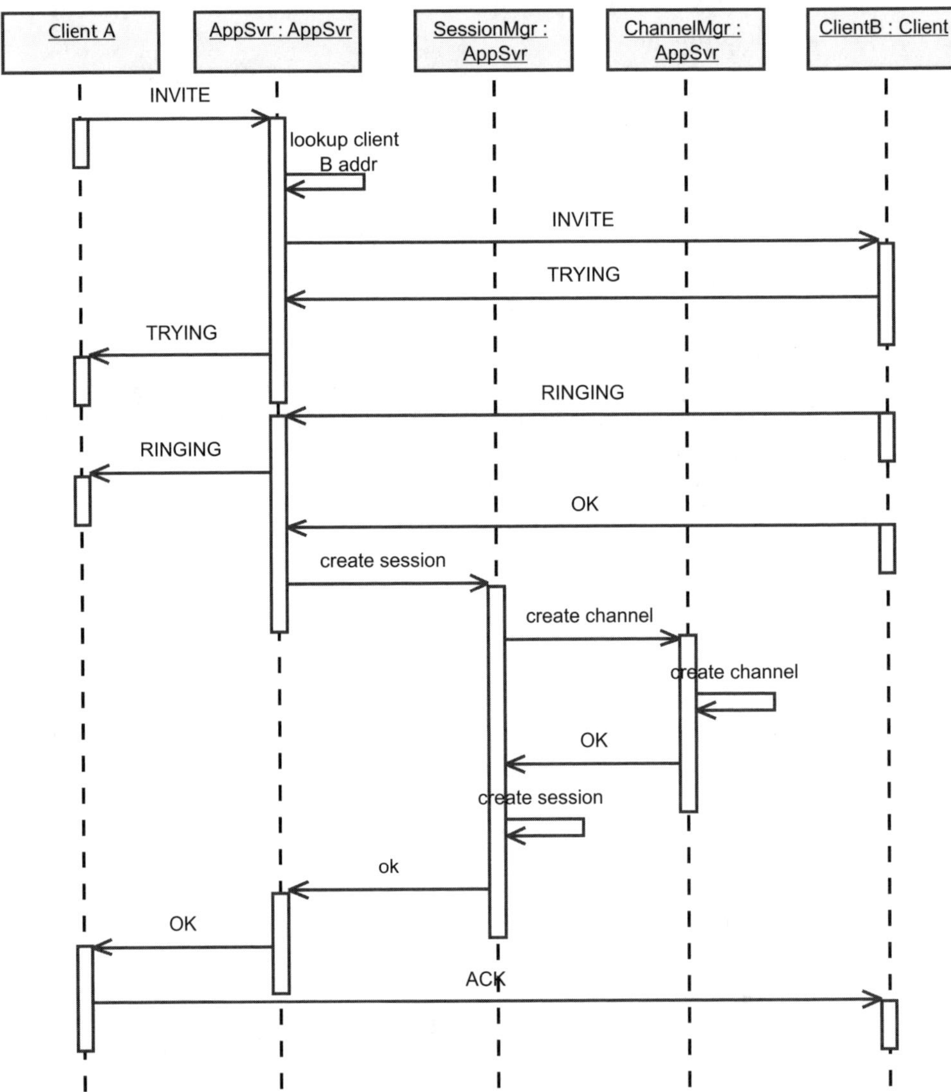

Figure 9.16 Sequence diagram for call setup involving one application server.

2. Appsvr processes the message, looks up Client B's location address, and sends an INVITE message to Client B (Client B: Client).

3. Client B responds with a TRYING message to the Application Server (AppSvr object) when it receives the INVITE message.

4. AppSvr object sends a response message (TRYING) to Client A to indicate that Client B has received the call request.

5. When Client B is ready, it sends a RINGING message to AppSvr object.

6. AppSvr object sends a RINGING message to Client A to trigger a ring tone.

7. Until AppSvr receives Client B's OK message, it starts to interact other server objects (Session Mgr and Channel Mgr) to set up a call session.

8. AppSvr sends an OK message to Client A to indicate that the call has been set up.

9. Client A sends an ACK message to Client B to indicate that the call is finally set up after receiving the OK message from AppSvr.

9.4   Summary

System requirements analysis and modeling are important phases in building wireless-based software and application systems. The objective is to generate a correct and complete blueprint of the final system. Engineers need to follow a well-defined process, communicate with the customer and end users in an effective way, and apply consistent analysis models to specify the different perspectives of the system. Since wireless applications and mobile commerce is an emerging field for most people, it is common that the customers and end users may not be able to provide clear and complete system requirements. Thus, it is a good idea to set up a pilot project to create a simple prototype of the target system using the prototyping process model, because this provides engineers with a cost-effective way to collect the detailed system requirements.

Engineers can use the following questions as a checklist when performing system analysis and modeling.

- Have we understood and specified the system infrastructure in terms of network components, application servers, and mobile clients? Have we understood and specified the network connectivity among them based on standard protocols?

- Do we have a good understanding of mobile users of a system? Who are they? What functional features can they access? When, where, and how do they interact with these functional features in the system? Have we specified the use cases for mobile users to describe various scenarios about system-and-user interactions?

- Have we gained a clear understanding of system functions by partitioning them into subfunctional modules and components? Have we understood and specified the information flows among them?

- Are we aware of the relevant standards in system network communications, mobile technologies, and domain-specific business workflows and logics?

- Do we have clear pictures of the required system data storage, and persistent mobile data and their relationships?
- Do we have clear views of all system components in terms of classes and their relationships?
- Have we specified system dynamic behaviors in terms of system states and the transitions between states? Have we modeled important operation sequences between components and class objects?
- Have we clearly specified the nonfunctional requirements of the system? These include the required mobile technologies, operation platforms, selected mobile devices, and requirements about system interoperability, performance, scalability, and reliability.
- Have we ensured that the generated requirements, specifications, and analysis models are complete, consistent, traceable, and verifiable?

References

[1] Peikari, C., and S. Fogie, *Maximum Wireless Security*, Indianapolis, IN: SAMS, 2002.

[2] Nichols, R. K., and P. C. Lekkas, *Wireless Security: Models, Threats, and Solutions*, 1st ed., New York: McGraw-Hill, 2001.

[3] Thayer, R., *Effective Requirements Practices*, Reading, MA: Addison-Wesley, 2001.

[4] Pressman, R. S., *Software Engineering: A Practitioner's Approach*, 6th ed., New York: McGraw-Hill, 2005.

[5] Somerville, I., and P. Sawyer, *Requirements Engineering*, New York: John Wiley & Sons, 1997.

[6] Gilb, T., and D. Graham, *Software Inspections*, Reading, MA: Addison-Wesley, 1993.

[7] Freedman, D. P., and G. M. Weinberg, *Handbook of Walkthroughs, Inspections, and Technical Reviews*, 3rd ed., New York: Dorset House, 1990.

[8] Johnston, A., *SIP: Understanding the Session Initiation Protocol*, 2nd ed., Norwood, MA: Artech House, 2004.

[9] Prathibha, M. G., and D. Rao, "An SIP Based Session Management in Multimedia Servers for Mobile-Talk," Master Project, San Jose State University, December 2004.

[10] RFC 3903—Session Initiation Protocol (SIP) Extension for Event State Publication, Retrieved on November 20, 2004, http://www.faqs.org/rfcs/rfc3903.html.

[11] RFC 3265—Session Initiation Protocol (SIP)-Specific Event Notification. Retrieved on November 15, 2004, http://www.faqs.org/rfcs/rfc3265.html.

[12] RFC 3261—Session Initiation Protocol (SIP), Retrieved on November 15, 2004, from http://www.ietf.org/rfc/rfc3261.txt.

[13] Cockburn, A., *Writing Effective Use-Cases*, Reading, MA: Addison-Wesley, 2001.

[14] Jacobson, I., *Object-Oriented Software Engineering*, Reading, MA: Addison-Wesley, 1992.

[15] Thalheim, B., *Entity Relationship Modeling*, New York: Springer-Verlag, 2000.

[16] Chen, P., *The Entity-Relationship Approach to Logical Database Design*, Wellesley, MA: QED Information Systems, 1991.

[17] Min, L., *Wireless Multimedia Application Server for Mobile Talk*, Master Project Report, San Jose State University, December 2004.

[18] Coad, P., and E. Yourdon, *Object-Oriented Analysis*, 2nd ed., Upper Saddle River, NJ: Prentice Hall, 1991.

[19] Davis, A. M., *Software Requirements Analysis and Specification*, Upper Saddle River, NJ: Prentice Hall, 1990.

[20] Schmuller, J. *Teach Yourself UML in 24 Hours*, 2nd ed., Indianapolis, IN: SAMS, 2002.

System Architectures for Wireless-Based Application Systems

After collecting system requirements and performing system analysis, it is time for engineers to conduct software design, address design issues, and identify appropriate solutions. System infrastructure and software architectural design are very important in software system design. We must understand different types of application architectural models for wireless-based applications, in order to obtain the best results in system infrastructure and software architecture. We also need to consider different design factors that affect the quality of system architecture, such as system complexity, performance, reliability, standardization, and security.

This chapter focuses on these topics. Section 10.1 revisits the basic concepts in software architectural design, including design process, objectives, specifications, and quality factors. Section 10.2 classifies and discusses wireless application system infrastructures based on different types of wireless networks. Section 10.3 discusses four types of wireless Internet application architectures, and examines their advantages and limitations. Section 10.4 presents smart mobile application architectures and related benefits and limitations. Section 10.5 examines two common architectures for building enterprise-oriented wireless applications. Section 10.6 summarizes the chapter.

10.1 Basic Concepts About System and Software Architectures

Before discussing system architectures for wireless-based applications, we must revisit the basic concepts, design processes, and key factors about software system architectures.

10.1.1 What Is System Architecture?

There are many different definitions of the word "architecture." *Webster's* definition is "the art or science of building." According to *Webster's* computer industry definition, "architecture" is "the manner in which the components of a computer or computer system are arranged and integrated." What is system architecture? We first look at two good definitions.

System architecture is "a description of the design and contents of a computer system. If documented, it may include information such as a detailed inventory of current hardware, software and networking capabilities; a description of long-range plans and priorities for future purchases, and a plan for upgrading and/or replacing dated equipment and software." (Retrieved at http://nces.ed.gov/pubs98/tech/glossary.asp.)

Alexander [1] defines the system architecture of information systems as follows: "Grouping data and processes into *information systems* creates the rooms of the system architecture. Arranging the data and processes for the best utility is the result of deploying an architecture. Many of the attributes of building architecture are applicable to system architecture. Form, function, best use of resources and materials, human interaction, reuse of design, longevity of the design decisions, [and] robustness of the resulting entities are all attributes of well-designed buildings and well-designed computer systems."

Wireless-based application systems include the following sets of system components:

- Wireless and wired network hardware devices, such as gateways and mobile stations;

- Mobile software components on mobile devices, including mobile browsers and mobile client software;

- Networking software supporting communications, such as gateway software;

- Application servers and supporting middleware, such as a payment system and a WAP server.

System architecture for a wireless-based application system refers to its organization and structure. This includes hardware and software components, and their network connectivity, as well as software interactions. The major objective of system architectural design is to select and define appropriate system architectures based on the given system requirements. One popular way to present network system architectures is to use a layered architectural model. Figure 10.1(a) shows a layered reference architectural model for wireless-based application systems. It consists of six layers:

- *Application layer:* This layer supports all application software to perform functions and services. At this layer, different application protocols may be implemented (e.g., a mobile payment protocol).

- *Session layer:* It supports all mobile sessions from the beginning to the end.

- *Transaction layer:* It supports all mobile transactions between two parties, a sender and a receiver.

- *Security layer:* It performs the security capability to support secured wireless communications and mobile data transfers above the transport level.

- *Datagram transport layer:* It supports wireless communications between a mobile terminal and its wireless application server.

- *Bear layer:* It refers to a bear network, in which different lower-level wireless network protocols are implemented.

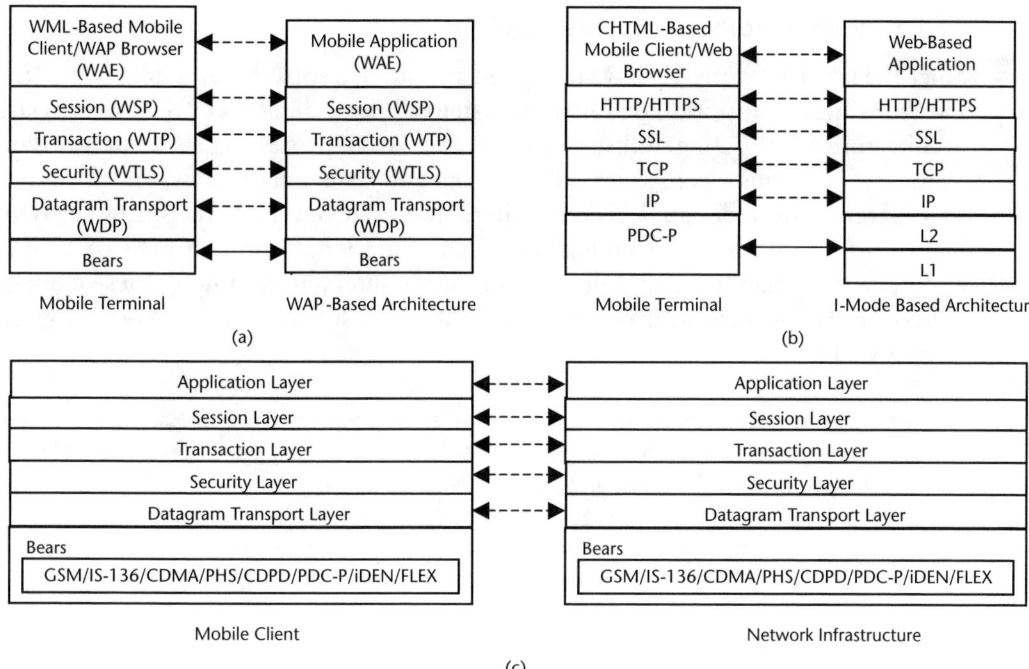

Figure 10.1   Layered system architectures for wireless-based application systems: (a) for WAP-based applications; (b) for i-mode–based applications; and (c) for wireless-based applications.

Figure 10.1(a) presents a typical layered architecture for WAP-based systems. As discussed in Chapter 3, it presents the protocol stack that supports WAP-based communications. Figure 10.1(b) displays a typical layered architecture for i-mode–based application systems, which support Internet-based communications over a package-switch network. Its details are covered in later sections. The system architectures of wireless-based applications can be specified and presented at three different levels.

- *The network level:* At this level, the underlying network infrastructure is specified and presented. It usually includes the target network structure, components, application servers, and communication protocols.

- *The system level:* At this level, the system architecture is presented using a client-and-server structure, in which mobile client software, supporting middleware components, application servers, and database servers are specified.

- *The function level:* At this level, the functional components in mobile client software and application servers are specified. If needed, the high-level interactions and interfaces between components should be specified.

According to the concept of system architecture, we understand that a software system usually is a part of a computer application system. Wireless-based application systems are the typical examples. In the following section, we revisit the basic concept of software architecture.

10.1.2   What Is Software Architecture?

Since the late 1960s, people have discussed and studied software architecture as "the structure and organization of software systems." Perry and Wolf [2] view software architecture as consisting of three parts: (1) elements—processing elements and data elements, (2) forms—constraints of elements and their interactions, and (3) rationale—architectural design decisions. Later, Shaw and Garlan define software architecture of a system as "a collection of computational components—or simply components—together with a description of the interactions among these components—the connectors" [3]. The recent definitions of software architecture are given as follows [4–6]:

> The software architecture of a program or computing system is the structure or structures of the system, which comprise software elements, the externally visible properties of those elements, and the relationships among them.
>
> Architecture is defined by the recommended practice as the fundamental organization of a system, embodied in its components, their relationships to each other and the environment, and the principles governing its design and evolution [ANSI/IEEE Std].
>
> Software architecture is the study of the large-scale structure and performance of software systems. Important aspects of a system's architecture include the division of functions among system modules, the means of communication between modules, and the representation of shared information.
>
> An architecture is the set of significant decisions about the organization of a software system, the selection of the structural elements and their interfaces by which the system is composed, together with their behavior as specific in the collaborations among those elements, the composition of these structural and behavior elements into progressively larger subsystems, and the architectural style that guides this organization.

Based on the given definitions, software architecture must address architectural design in three different aspects:

- Software organization and structure in terms of its functional partitions and components;
- System elements/components and their relationships, including interfaces, collaborations, connections, and constraints;
- Well-defined high-level architectural styles and principles.

10.1.3   Why Is Software Architecture Important?

Software architecture of a system presents its high-level structure and organization, providing customers and engineers with a structural view of software. There are three reasons why it is important.

1. Architecture is the vehicle for stakeholder communications. All stakeholders (customers, users, project managers, programmers, and testers) are concerned with different system characteristics. Software architecture provides a common language in which different concerns can be expressed,

negotiated, and resolved at a level that is intellectually manageable even for large and complex systems. Software architecture enables stakeholders to discuss their concerns and share their understanding about architectural models, functional components, and component interaction styles and regulations.

2. Architecture manifests the earliest set of design decisions and production management. Software architecture defines constraints on software implementation and production management. This includes organizational structure, technology selection, quality attributes, production management, change and evolution, and project cost and schedule.

3. Architecture is useful as a transferable and reusable model. Good software architectures usually provide a common conceptual architecture that can be shared by many projects on the same production line. With this architecture, application systems can be built with internally or externally developed components.

Defining the right architecture is crucial to the success of software design, because architectural design decisions affect many major quality factors of a final application product. Major architecturally related quality factors are: software performance, maintainability, extendibility, reusability, reliability, security, and testability.

10.1.4   Software Architecture Presentation and Specification

Software architectures can be designed, presented, and specified at four levels:

- *Meta-architecture:* It formulates a generic architectural vision that provides a fundamental base for guiding decisions during architectural design.
- *Conceptual architecture:* It presents a conceptual system partition based on functional components, responsibilities, and their interconnections.
- *Logical architecture:* It is developed based on the conceptual architecture. The logical architecture presents a logical structure view about software architecture, which is likely to be modified and refined during the cycle of a product's evolution.
- *Deployment architecture:* The deployment architecture is created for distributed or concurrent systems. It provides a physical view for system deployment because it maps the system components to the processes of the physical system.

10.1.5   Architecture Design Process

Engineers need to follow a design process to produce well-defined system architectures. Figure 10.2 shows a five-step process for system architects.

- *System architectural analysis:* In this step, system architectural requirements are collected, analyzed, and documented. For example, the related requirements include the targeted wireless network infrastructure, communication protocols and standards, and system connectivity. Other examples include nonfunctional system requirements about performance, extendibility, main-

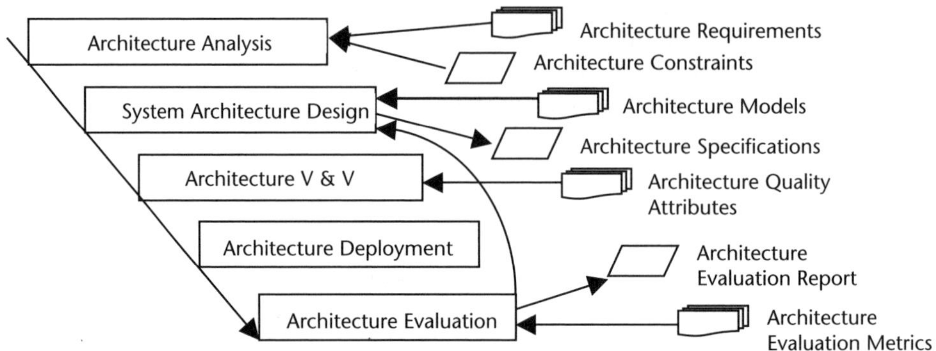

Figure 10.2 An architectural design process.

tainability, security, and reliability. The system architecture constraints need to be specified in terms of technology requirements, system interoperability, and selected standards.

- *System architectural design:* At this step, engineers need to perform four basic tasks to make architectural decisions based on the given requirements.

    - *Task #1: Select or define the system infrastructure.* For wireless-based application systems, engineers need to first determine the underlying network infrastructure (e.g., wireless LAN, GSM, GPRS, 2G or 3G networks and protocols). In many cases, wireless-based application systems may have a heterogeneous network infrastructure, such as wireless Internet. The chapters in Part III cover the basics of these wireless networks.

    - *Task #2: Select and decide well-defined system architectural models and styles.* Although many conventional architectural models have been discussed [2, 4], there is a lack of discussion about architectural models and styles for wireless-based application systems. This chapter classifies, compares, and discusses a number of architectural models and styles in wireless applications.

    - *Task #3: Partition a system into a number of subsystems and components.* A wireless-based application system usually is structured in terms of four types of components: (1) network devices and hardware components; (2) application servers, middleware, and functional components; (3) mobile devices, mobile client software, and components; and (4) system data and content repositories, such as database servers. Meanwhile, the connectivity of subsystems and the interactions between functional components need to be defined.

    - *Task #4: Present and document the system architecture using a well-defined mode.*

- *Architecture verification and validation:* The derived system architecture needs to be verified through a design review based on the architectural quality attributes. A set of architecturally related quality attributes is given in the next section.

- *Architecture deployment:* At this step, the architecture needs to be reviewed, in terms of system deployment factors, such as environment constraints.

- *Architecture evaluation:* After the system is developed and deployed, the system architecture needs to be evaluated again for the product evolution. The evaluation results are useful inputs for the architectural updates and refinements for future releases.

10.1.6 Quality Factors in Architectural Design

During architectural design, verification, and validation, engineers can use the following set of quality factors.

- *System reliability:* It is a challenge to achieve the high reliability goal, due to the poor reliability of wireless networks and mobile connectivity, as compared to online applications and conventional systems.
- *System extendibility:* System extendibility is important in the design of system architecture, due to the increase in the number of wireless applications and services in m-commerce.
- *System scalability:* System scalability of wireless-based applications is very important, due to the fast growth in the number of wireless users and the increase in the volume of mobile data transactions.
- *System performance and efficiency:* Since mobile users usually expect a very short system response time in a dynamic mobile access environment, it is very challenging to construct a system with good system performance.
- *System modifiability and maintainability:* To build a changeable and maintainable system, we need to structure the system in a way to increase its independence in the following areas:
  - Minimize the dependency of mobile client software on specific mobile devices and technologies;
  - Separate database application programs from a specific database server and its application interface;
  - Reduce the dependency of mobile client software on specific mobile layout formats, sizing, and preferences;
  - Separate application logics from specific technology middlewares;
  - Separate application logics from communication components and interfaces.
- *Standardization:* A wireless-based application system involves a number of standards in communication protocols, mobile technologies, and the application domain. Although there are many proprietary technologies and systems available, it is a good idea to design system infrastructure and architecture while considering the widely accepted industry standards in networking and mobile technologies.

Although there are many diversified wireless application systems, they share some common network infrastructures. This chapter classifies and examines system architectures of wireless applications. Section 10.2 examines different wireless-based system application infrastructures based on wireless network types. Section 10.3 discusses and compares various types of wireless Internet application system architectures. Section 10.4 explains system architectures for smart application systems, and compares their features with wireless Internet-based application systems. Finally, Section 10.5 presents vendor-oriented enterprise system infrastructures using popular wireless enterprise solutions. System architects and developers can define application system architectures based on the architectural models discussed in the following sections.

10.2   The Network-Based System Infrastructure Classification

It is natural to classify wireless-based application systems based on their underlying networks. In general, wireless networks can be classified into: satellite networks, wireless LAN networks, wide area networks, and wireless personal networks. Part III of this book has covered different types of wireless networks. Let us look the infrastructures of wireless application systems on homogeneous networks.

10.2.1   Application Systems Based on Wireless LAN Networks

Wireless LAN–based application systems usually allow mobile users to access application servers within a distance of 100m. Figure 10.3 shows a common infrastructure for a wireless LAN–based application system, which includes mobile devices, wireless access points, ports, and application servers. Over a wireless LAN network, mobile users are able to access LAN-based application servers through access points and ports. A classroom interactive learning system [7, 8] is a typical example, in which students and instructors access and share a LAN-based classroom interactive learning system for information exchanges, questions and answers, and quizzes. An Internet Web server is also needed to support the access of online class materials from the application servers. Another example is a wireless LAN–based restaurant ordering system that allows customers to order food at each table with a wireless LAN–based ordering panel [9].

10.2.2   Wireless Personal Network–Based Application Systems

As mentioned in Part III of this book, there are three major wireless personal networks: Bluetooth, Ira, and Wi-Fi networks. The application systems based on a wireless personal network only support mobile user access within 10m. They share

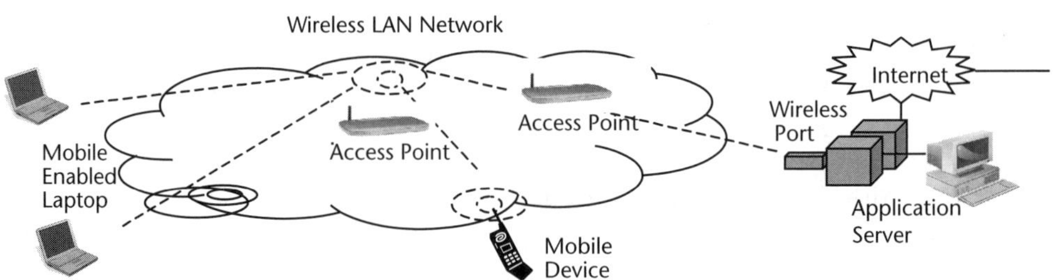

Figure 10.3   System infrastructure for a wireless LAN–based application.

common infrastructures. Figure 10.4 shows three typical system infrastructures using a Bluetooth network. Figure 10.4(a) shows two mobile devices communicating on a Bluetooth network, to exchange information or conduct peer-to-peer transactions (e.g., mobile payment transactions). In this infrastructure, each mobile device must be equipped with Bluetooth-enabled ports and drivers, as well as mobile client software. Figure 10.4(b) shows a mobile device interacting with a stationary computer (e.g., a desktop or a laptop) over a Bluetooth network, to upload and download mobile client software (e.g., games or software), or to synchronize mobile data for wireless application services. In this infrastructure, both mobile devices and stationary computer must be equipped with Bluetooth access ports and drivers and Bluetooth-enabled mobile software. Figure 10.4(c) shows a mobile device communicating with one mobile terminal (e.g., a payment station or bar code reader) over a Bluetooth network to perform transactions and validations. Meanwhile, mobile terminals interact with its application server to complete transaction processing and a validation procedure. A peer-to-peer mobile payment system in a supermarket (e.g., Wal-Mart) is a typical example. Customers can use mobile phones to make a mobile payment transactions through a mobile payment terminal, which communicates with the payment server for transaction processing and authentication. The paper in [10] describes a peer-to-peer payment system.

10.2.3   Application Systems Based on Wide Area Networks

Many wireless application systems are constructed based on wide area networks. As discussed in Part III of this book, there is a number of wide area networks, including 2G (e.g., GSM), GPRS, 3G, and 4G wireless networks. They can set up wireless application systems supporting access over a distance of 2.5 km. Figure 10.5 shows three typical examples of WAN-based system infrastructures. Figure 10.5(a) shows a GSM-based application system infrastructure, which includes mobile devices, wireless towers, base transceiver stations (BTSs), base station controllers (BSCs), mobile switch centers (MSCs), and application servers. In addition, the user registrations, authentications, service charging functions, and information services are included (i.e., HLR, VLR, SCF, EIR, and AUC). With a GSM network, many short messaging–based application systems can be built to support SMS-based advertising, marketing, payment, and customer services.

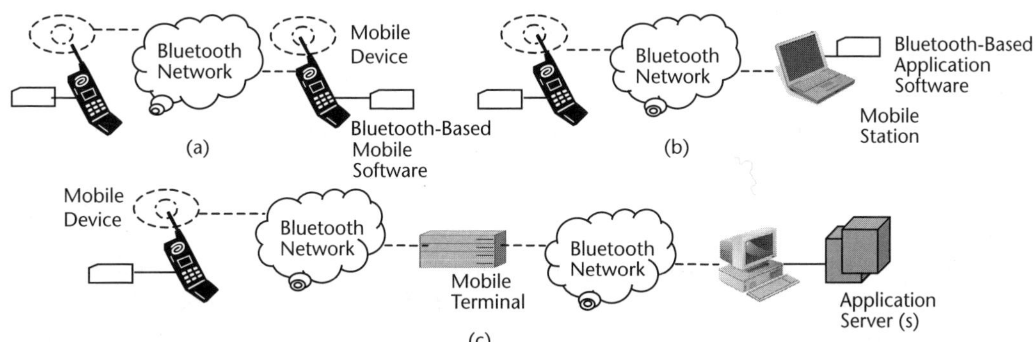

Figure 10.4   (a–c) Three typical system infrastructures for Bluetooth-based applications. See text for description of each example.

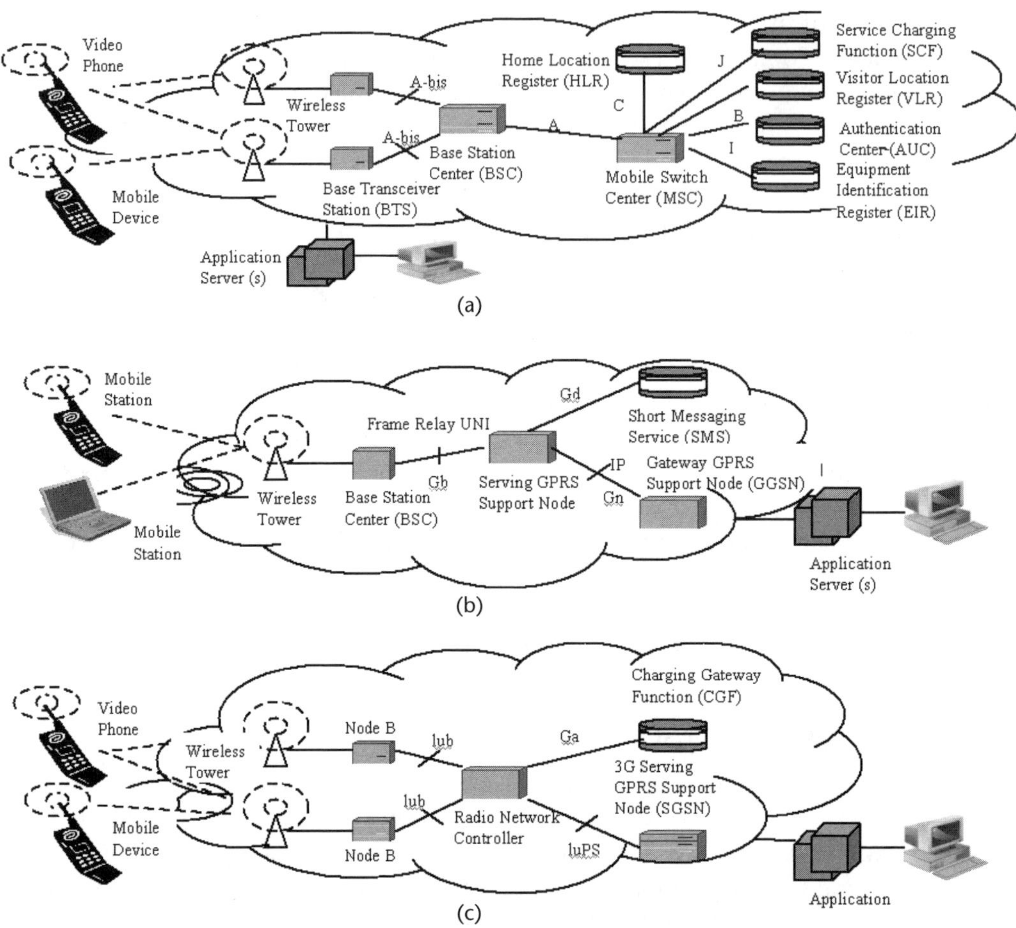

Figure 10.5 System infrastructures for WAN-based applications: (a) GSM wireless network; (b) GPRS network; and (c) 3G UMTS network.

Figure 10.5(b) shows a GPRS-based application system infrastructure, which includes mobile devices (e.g., mobile phones), wireless towers, gateway GPRS support nodes (GGSNs), BSCs, GPRS support nodes, short messaging service (SMS), and application servers. Using this system infrastructure, different wireless Internet-based portals and messaging systems can be developed to support cost-effective mobile access.

Figure 10.5(c) shows a 3G-based application system infrastructure, which includes 3G mobile devices (e.g., video phones), wireless towers, node Bs and 3G SGSNs, a radio network controller (RNC), a charging function gateway (CFG), and application servers. The distinct features of 3G-based application systems are their wireless multimedia data accesses, functions, and services. A 3G network is a wireless packet-based cellular network, on which wireless Internet voice, data, photos, audio, and video can be efficiently transferred. Typical application examples are peer-to-peer mobile video sharing, interactive gaming, and wireless multimedia learning.

10.2.4   Wireless Application Systems Based on Heterogeneous Networks

More application systems are established on heterogeneous networks, rather than on homogenous networks, for example, wired networks (e.g., PSTN and the Internet) and wireless networks. The main reason is to allow users to communicate with one another, exchange data, and share and access functions and services over different networks.

As shown in Figure 10.6, a wireless payment system infrastructure is structured over two different networks: wireless Internet and Bluetooth network. The system consists of the following components:

- *Mobile payment client software:* It interacts with mobile users to perform mobile payment operations. It supports peer-to-peer mobile payment transactions between two mobile users over a Bluetooth network, and communicates with its mobile payment server over wireless Internet for user authentication, mobile billing, and so forth.

- *Mobile payment server:* It accepts the mobile payment requests from the mobile payment client, performs security checking, and processes mobile payment transactions. It may also provide other features, such as mobile billing and account summary.

- *Merchant agent:* It provides a basic interface for merchants to connect to their internal solutions for payment and accounting.

- *Mobile payment station:* It is equipped with the hardware and software required to support two basic functions:

    - Bluetooth-based communications between a mobile payment client and a mobile payment station for mobile payment transactions;

    - wireless Internet-based communications between mobile payment clients and the payment server.

- *Wireless server and Web server:* They are the middleware bridges to the Internet and wireless networks to allow online users to set up and access account information, schedule payment transactions, and so forth.

A similar example of wireless payment systems is given in [10, 11].

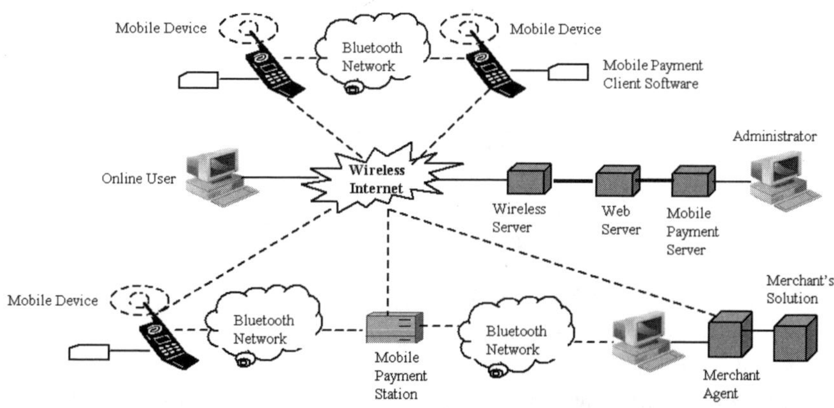

Figure 10.6   Wireless payment system infrastructure over different networks.

10.2.5 Mobile Commerce Application Systems Based on Sensor Networks

As sensor networks and pervasive computing technologies advance, pervasive commerce (p-commerce) is becoming a new research field and application paradigm in mobile commerce. According to [12], p-commerce is a new commerce paradigm based on intelligent mobile devices and ubiquitous computing. It offers the convenience of m-commerce by allowing users to shop at any time and anywhere. Mobile users are able to shop with the support of a p-commerce environment, which uses ubiquitous computing solutions to provide a seamless e-commerce experience in two aspects:

- Matching a mobile user's needs to the products offered by merchants;

- Identifying and conducting p-commerce transactions at any time and anywhere.

Figure 10.7(a) shows a smart shopping mall, called P-Mall, which consists of many smart stores. Each store has deployed a number of sensors and a store server maintaining its merchandise inventory, and sales and marketing information. P-Mall is supported by a P-Mall server, which connects to store servers, maintains the service information (e.g., store catalogs), and supports mobile commerce services. All merchandise in a store is labeled with an RFID tag identifying its electronic product catalog (EPC), which can be used to look up the product URL from its vendor. Each store server interacts with the P-Mall server by providing up-to-date information about the

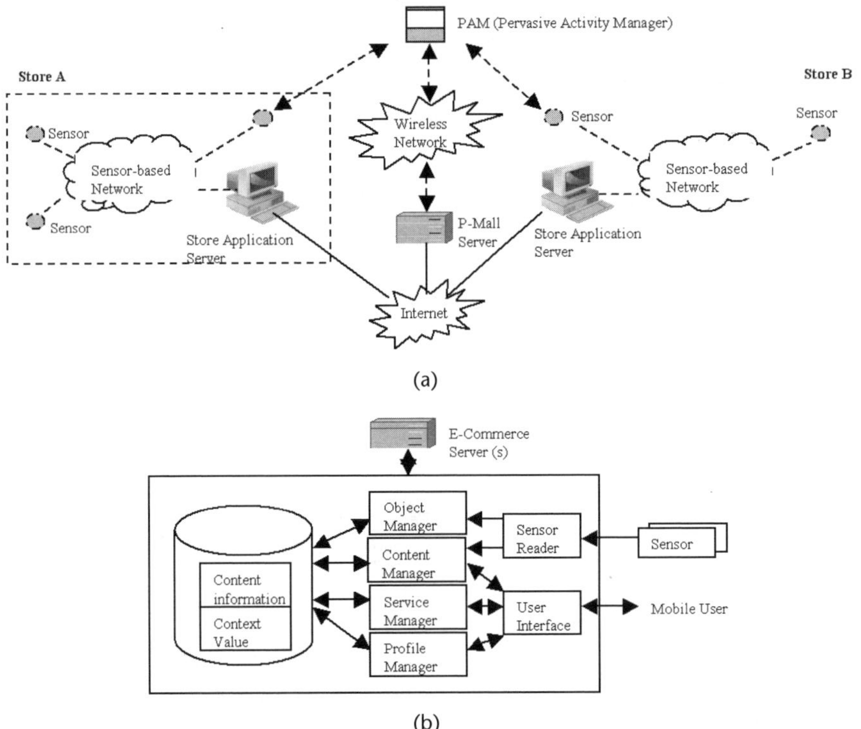

Figure 10.7 Sensor network–based application architecture: (a) P-mall architecture; and (b) PAM architecture.

store. Shoppers use a personal assistance manager (PAM) to interact with sensors and servers to easily identify their purchases and perform commerce transactions. The detailed shopping scenarios can be found in [12]. As an intelligent shopping assistant, a PAM helps a mobile user to look for preferred products and find the best results. As a user-centric program, PAM proactively conducts many context-aware operations in the background without users' explicit requests.

Figure 10.7(b) shows the architecture of a PAM given in [12], which consists of six basic components. In addition to a sensor reader and its user interface, the other four components are:

- *Context manager:* a component that manages user context;
- *Object manager:* a component that manages the configured product list;
- *Profile manager:* a component that manages user preferences;
- *Service manager:* a component that generates PAM recommendations.

According to [12], a PAM has the following major tasks:

- To automatically collect all related information about services or products identified by the user;
- To efficiently analyze the collected information;
- To systematically present the available transaction options to help users make the best decisions.

Compared to wireless Internet-based application architectures, the distinct advantage of this kind of application architecture is its independence of wireless network connectivity, so that the system reliability can be improved.

10.3 Wireless Internet-Based Application System Architectures

In the past 10 years, the Internet has been used to create diverse online systems in many different applications. It is natural to build application systems over the wireless Internet to provide mobile functions and services to global users. Jing et al. [13] review the common mobile client-server application environments and related system architectures. This section provides a detailed discussion about different types of system infrastructures for wireless Internet applications to help readers understand their strengths and limitations.

As shown in Figure 10.8, there are four types of popular wireless Internet system infrastructures: (1) i-mode–based application system architecture; (2) mobile-transparent application architecture; (3) mobile-aware application architecture; and (4) mobile-interactive application architecture.

10.3.1 The i-mode Application System Architecture

The Japanese company DoCoMo successfully launched the i-mode service in 1999, as a solution for mobile Internet services [14, 15]. The i-mode solution consists of two proprietary services: i-mail and i-mode Web services. The i-mail service is a P2P

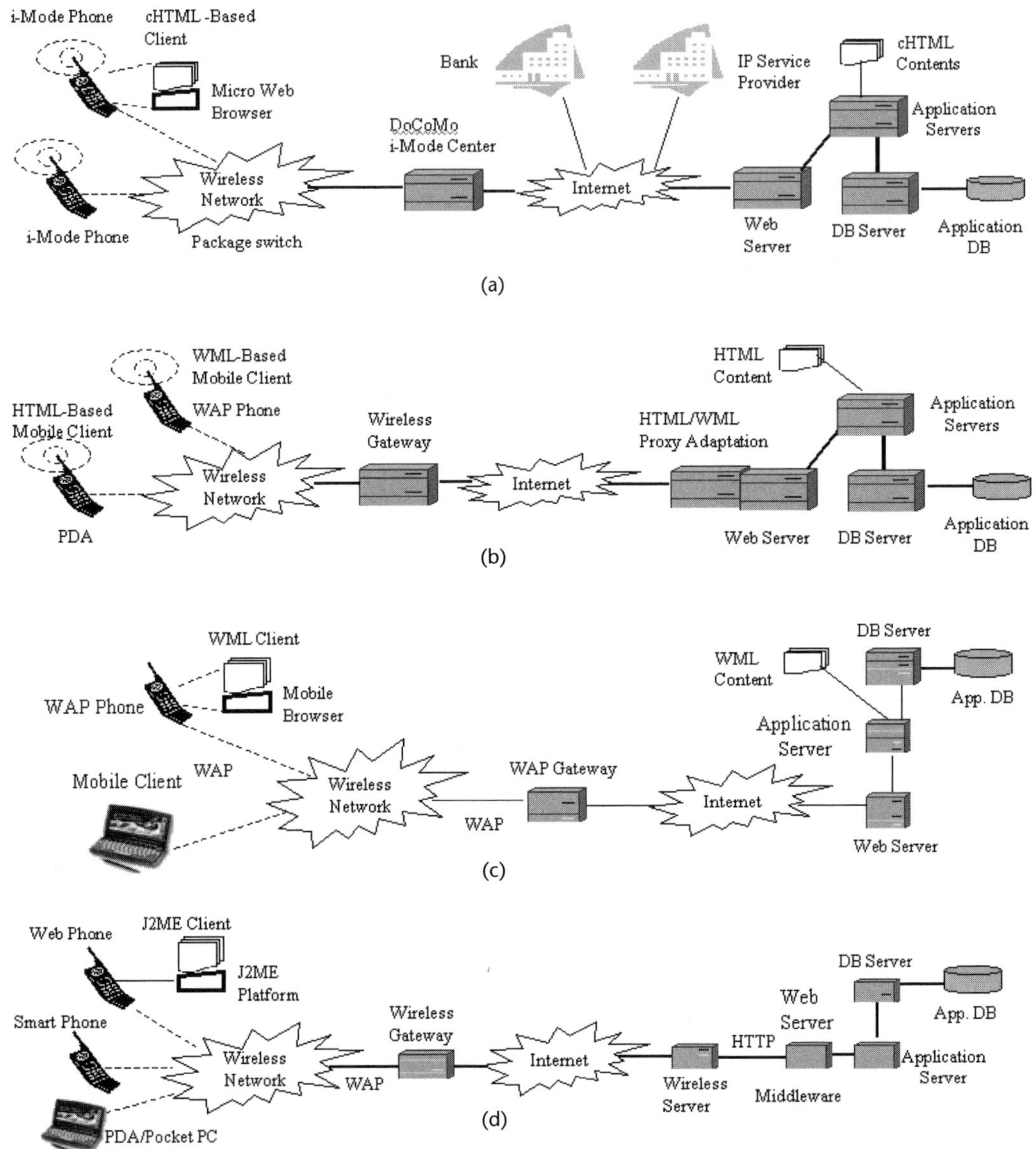

Figure 10.8 Four different types of system infrastructures on wireless Internet: (a) i-mode–based application architecture; (b) mobile-transparent application architecture; (c) mobile-aware application architecture; and (d) mobile-interactive application architecture.

messaging service that allows mobile users to send and receive short e-mail messages. The e-mail messages are sent and received by i-mail using regular e-mail message addresses. This solution creates very good interoperability between i-mail and regular e-mail servers. In the recent years, i-mail has been extended to allow users to send messages with photos and videos, which is known as i-shot.

The i-mode Web service allows the customer to access information from the Internet on their mobile devices. The information can be accessed from both registered/official and unregistered/unofficial Web sites. The i-mode service provides a cost-effective mobile presentation technology, known as compact HTML (cHTML), which is a subset of HTML 3.0 for content description [15]. All i-mode Web pages are coded using cHTML and include basic HTML markup capabilities, such as pictures, text, multicolor pages, and background sounds. On each i-mode mobile phone, a Web microbrowser is deployed to support wireless communication between the mobile phone user and the back-end Web server, known as the i-mode server.

The i-mode Web service provides an i-mode portal, which lists all official i-mode Web sites that are available to the user. A user can subscribe to the official Web sites to access the premium content. The essential capabilities of the i-mode service are listed here:

- Messaging service based on e-mail;
- Mobile Web service;
- i-mode portal listing official Web sites;
- Easy subscription and access to the official Web sites;
- Access to a great number of unofficial Web sites.

As mentioned in Chapter 4, i-mode is a proprietary protocol, and is supported by a wireless packet-switched network. It provides Internet service using personal digital cellular-packet (PDC-P) and cHTML [15]. i-mode allows application/content providers to distribute software (e.g., Java applets) to cellular phones, and allows users to download Java applets (e.g., digital games). A packet-switch network is used to support wireless data communications between i-mode phones. The TCP/IP protocol is used to support wired communications.

As shown in Figure 10.9, the Internet is used to connect the i-mode server and the Web-enabled application servers based on the TCP/IP technology. The PDC-P network includes a mobile message packet gateway (MPGW) to handle conversions between the Internet and i-mode's wireless packet-switch network. The i-mode server performs as a regular Web server and provides Internet services using PDC-P. The i-mode server is used as a middleware, where the relational markup language (RML) is used to retrieve and cache the data from different sources. Mobile data is first converted into the XML format, and then converted into the cHTML format.

Mobile applications based on the i-mode solution (e.g., its network, protocol, and Web services) have a number of benefits.

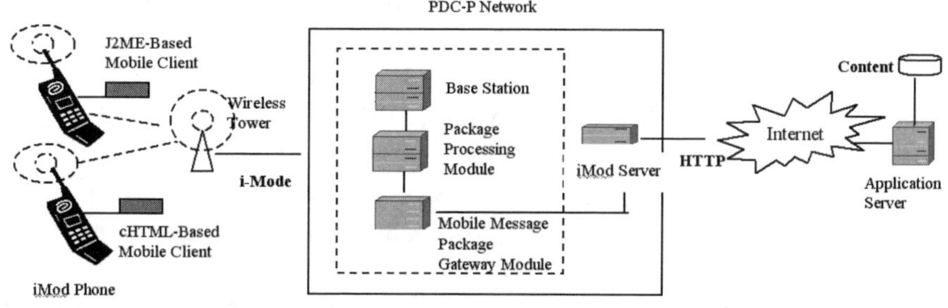

Figure 10.9   i-mode–based mobile application system infrastructure.

- The i-mode service provides a mature wireless Web service over i-mode's package-switch network, offering an efficient and low-cost mobile channel between mobile users.
- The i-mode service provides a cost-effective way to extend the existing online applications by offering wireless connectivity to Internet users.
- The cHTML-based mobile client and microbrowser provide a good mobile presentation interface to fit into low-end mobile devices, which have a small display screen, and limited memory storage and computing power.
- It is easy to provide diverse mobile messaging services to mobile users, due to i-mode's technologies and solutions, such as i-mail and i-shot. For example, they can be integrated with MMS technology to provide wireless multimedia messaging services [14].
- The i-mode service provides an end-to-end security solution in the entire mobile network, since i-mode adopted SSL/TLS protocols in March 2001.

There are two major issues in building i-mode–based mobile application systems. The first issue is the standardization. Since i-mode uses its proprietary network, protocols, and solutions, no standards and documentation are available to application vendors. Although DoCoMo has successfully deployed i-mode in Japan, it may be difficult to deploy i-mode in other markets, such as North America and Europe, due to this proprietary issue.

The second issue is its limited compatibility at the protocol level, because i-mode only supports HTTP over wireless-optimized TCP. Since 2000, i-mode has intended to solve this issue by supporting XML-based mobile presentation technology. Another concern is its acceptance by the wireless industry and phone manufacturers.

10.3.2 Mobile-Aware Application Architecture

The mobile-aware application architecture refers to the WAP-based wireless Internet application architecture, in which both mobile client software and application servers are developed using the standard WAP-based protocol and technology. As discussed in Chapter 3, WAP was designed to allow mobile users to access Internet and intranet applications using mobile devices with limited display size and connection speed. The WAP Forum was founded in June 1997 by Phone.com, Ericsson, Nokia, and Motorola. Its goal was to offer a license-free standard to the entire wireless industry, so that anyone would be able to develop WAP-based applications and services. The WAP is the result of their efforts, and it has met that goal. As an evolving standard, WAP has already built a significant degree of industry support, and is considered to be the standard for delivering and presenting wireless Internet services to the market for mobile (or handheld) devices. The WAP standardizes wireless access for mobile phones, PDAs, and pagers [16]. It works on major wireless networks, including CDMA, GSM, TDMA, and CDPD, and over circuit-switched, packet-switched, or SMS networks. The WAP-based mobile client can use any operating system, including Windows CE, Palm OS, EPOC, and JavaOS.

As shown in Figure 10.8(c), the mobile-aware application architecture also has a three-tier client-server structure.

- *Mobile client layer:* Mobile client software on this layer usually is a thin client, which provides users with a WML-based mobile interface on WAP-enabled mobile devices to interact with the application server to access mobile functions and services. Mobile client software is supported by a WAP-based microbrowser, which communicates with a WAP-based wireless server using the WAP protocol. A mobile client interacts with users based on the wireless markup language (WML), and WML Script, which is a scripting language for the WAP navigator. The client accesses WML and WML Script content using a given URL, and accesses dynamic content via CGI and Java servlets on the Internet server.

- *Middleware layer:* This layer includes two basic components. The first is the WAP gateway that is a component in a network service layer. It works as a protocol to convert between WAP protocols and common HTTP/TCP protocols used by Web servers on the Internet. As discussed in [17], a WAP gateway provides a standard interface channel between a wireless bearer network and the wired Internet. The other component is a Web server, which accepts and processes incoming mobile requests, and sends responses back to the WML-based client based on the results generated from the mobile application server.

- *Application layer:* This layer includes the application server(s) and a database server to support mobile functions and services, and WML-based mobile data and contents.

According to the WAP Forum [17], WAP-based application architecture has the following benefits:

- The architecture delivers a powerful and functional user model and is optimized for handheld wireless devices. The WML-based mobile client provides an effective user interface that is appropriate for mobile devices with limited keys and a small display screens. By using the existing Internet model as a starting point, the mobile user interface provides familiar functionality for those accustomed to the Web. The WAP specification defines a microbrowser that is the ultimate thin client, and which is able to fit in low-end mobile devices with limited memory and computing power. The use of proxy technology and compression in the network interface reduces the processing load and power consumption at mobile devices, so that an inexpensive CPU can be used for mobile handsets.

- The architecture includes standard Web proxy technology to connect the wireless domain with the Web. By using the computing resources in the WAP gateway, the WAP architecture permits the mobile devices to be simple and inexpensive. A WAP gateway usually includes two functions: (1) translations of requests from the WAP protocol stack to the WWW protocol stack (HTTP and TCP/IP); and (2) content encoding and decoding. With this infrastructure, mobile device users can browse a variety of WAP content and applications regardless of the wireless network they use. Application vendors are able to build network-independent content services and applications. The WAP gateway also decreases the response time to the mobile devices by aggregating data from different servers on the Web and caching frequently used information. The WAP gateway can also interface with other components at the network service layer, such as user information and location information repositories.

- The architecture has developed standardization for wireless Internet services, which has been accepted by the wireless industry. Today, all major phone manufacturers support WAP. This makes it easy to be deployed on various wireless networks and mobile devices.
- The architecture addresses the constraints of a wireless network. The protocol stack defined in WAP optimizes the standard Web protocols, such as HTTP, for use under low-bandwidth, high-latency conditions.
- The architecture provides a secure wireless connection, using the wireless transport layer security (WTLS) protocol. As mentioned in Chapter 3, the WTLS protocol is based on upon the industry-standard transport layer security (TLS) protocol, formerly known as the secure sockets layer (SSL). WTLS ensures data integrity, privacy, authentication, and denial-of-service protection.

The WAP specification also defines new functionalities that have not been defined by any other standard, such as a voice/data integration application interface (API), and the groundwork for wireless push functionality.

A service provider or network operator that is deploying the WAP-based solution benefits by being able to:

- Gain a whole new relationship and communication channel with their subscribers;
- Control the data connection to its subscribers;
- Gain immediate access to all WAP-enabled wireless content;
- Deploy applications such as call feature control, prepaid wireless recharge, and automated customer service;
- Choose among open standards vendors;
- Obtain the freedom to use and integrate new air interface technologies.

10.3.3 Mobile-Transparent Application Architecture

The mobile-transparent application architecture refers to another approach to set up wireless Internet application systems [13, 18]. It is also known as mobile Internet. It supports different XML-based mobile clients on diverse mobile devices, including WML, HTML, and cHTML. As shown in Figure 10.8(b), a mobile-transparent architecture is similar to a mobile-aware architecture with a three-tier client-server structure.

- *Mobile client layer:* Mobile client software in this layer provides users with a thin mobile interface that interacts with a wireless Internet application server to access mobile functions and services. The mobile client usually is supported by an HTTP-enabled Web microbrowser to communicate with a wireless Web server. A mobile client executed on a microbrowser uses a URL address to connect to a wireless Web server.

- *Middleware layer:* This layer includes a wireless Web server and middleware. The wireless Web server accepts and processes incoming HTTP-based mobile requests, and sends appropriate responses back to the client based on the results generated from the mobile application server. To accommodate the different microbrowsers and markup languages used in wireless computing, a wireless proxy adaptation module is added as a middleware. This module transforms and formats different mobile data contents into appropriate forms, based on the given information about mobile devices and platform information. As mentioned in Chapter 13, a XML application server could serve this purpose.

- *Mobile application layer:* This layer includes mobile application servers and application database repositories. Each application server provides domain-specific mobile functions and services with the middleware support. A database repository usually is implemented by a database server to store and manage domain-specific application data and mobile contents.

Compared with mobile-aware application architecture, this architecture has the advantage of supporting different types of XML-based mobile clients on different mobile devices. It has the additional performance overhead, due to the data conversion between different formats.

10.3.4   Mobile-Interactive Application Architecture

Mobile interactive architecture refers to the Java 2 Platform, Micro Edition (J2ME)–based mobile application architecture supporting the diverse needs of mobile users in a mobile client. J2ME was developed specifically to address the vast consumer market of small mobile devices, ranging from smart phones to pagers [16]. J2ME should be looked upon as a complementary technology used to further expand the usefulness of wireless access and applications, instead of looking at J2ME as a competitor to WAP.

When WAP is coupled with J2ME to construct mobile application systems, mobile users will be provided with a full-featured Java-based application environment to receive a richer mobile interaction experience. Figure 10.8(d) shows an example. Unlike the previous three architectures, this architecture is set up based on a J2ME-based mobile client. It provides an interface for mobile users that is more powerful than that provided by the WAP. With a J2ME mobile client, users will no longer be restricted to the monochromatic interfaces, but they will have full-color animated graphics and applications. A J2ME-based mobile client is executed on a J2ME virtual machine, which provides its mobile operation platform on mobile devices.

There are two choices for supporting wireless Internet connections when building J2ME-based mobile applications. One choice is to use the WAP-based approach to set up the wireless Internet connectivity between J2ME-based mobile clients and the application server. The other choice is to use a mobile transparent approach to set up wireless Internet connectivity through HTTP/TCP technology. More detailed discussions about the benefits and limitations are covered in the following section.

10.3.5 Advantages and Limitations of Different Architectures

Wireless Internet system architectures have the following advantages and features.

- Wireless Internet application systems provide the mobility to mobile users to access the Internet at any time and anywhere. All Internet-based systems can be easily extended to provide wireless accessibility and connectivity to online users.
- Minimal-to-zero software deployment is required, since there is no need of client software installation and configuration.
- Low-powered and low-cost mobile devices are needed.
- Mobile device users will find it easy to use, due to the following two reasons.
    - There is no need to upgrade mobile client software, due to Java's dynamic downloading capability.
    - Mobile users are familiar with the J2ME-based mobile user interface, due to the use of Internet-based Java applets.
- Wireless middleware can be used to easily work with the application server(s) in mobile content formatting and configuration. This accommodates different mobile platforms, microbrowsers and markup languages.
- Mobile client software provides good adaptability and interoperation ability in broad deployment to multiple mobile devices.
- No data synchronization is needed on mobile devices, which ensures that mobile users always access fresh mobile data.
- Simple mobile client software is used, since only limited data caching and functions are needed at the client side.
- Secured mobile data is stored on the server side.

Wireless Internet system architectures also have some drawbacks and issues, listed here.

- Wireless network connectivity is not reliable, due to the user's mobility during a mobile access session.
- Airtime may be costly, since wireless applications constantly need to be connected to mobile users, and system connection time is slow when circuit-switched networks (such as 1G or 2G wireless networks) are used.
- There may be a security gap between two different networks, since a wireless Internet system infrastructure involves a wireless network and Internet, while most existing security solutions only target a specific network. Thus, a total security solution is required for wireless Internet applications.
- Wireless Internet application systems may have a system performance issue due to the limited network bandwidth, dynamic network throughputs, and possible network latencies.
- System availability and fault tolerance issue may arise. Since all mobile data and applications are stored on the server, any failure of the server halts all access by mobile users.

- System and application validation could be very complex and costly. Testing the behaviors of a wireless Internet application and measuring the system performance is difficult and complex because there is a lack of available mobile testing platforms and tools for mobile client software and wireless-based applications. Today, the wireless-based system is dependent on emulator-based testing and manual operations.

10.3.6   Comparison of Different Wireless Internet Application Architectures

Section 10.3 covers four different architectures for wireless Internet applications. Table 10.1 summarizes their features in different aspects. The major points are highlighted here.

System Security

The security concerns in wireless Internet applications include: (1) communication security, (2) user identification and authentication, (3) client and server certification, and (4) security GAP and nonrepudiation.

The WAP-based mobile-aware approach provides the security for wireless communications between a WAP gateway and WAP-enabled mobile devices. The WTLS offers entity authentication, data confidentiality, and data integrity. WTLS (class 3) is used to achieve client security through client authentication and X.509 certifications [20]. To facilitate client security, future WAP phones will provide WIM [19], which implements WTLS class 3 functions. Although solutions protect the security of Internet communications between a WAP gateway to application servers, security "gaps" could appear when a session is terminated prematurely. For example, consider an intended end-to-end secure application using SSL between components X and Z, and passing through component Y. A security gap would result if the SSL session (between X and Z via Y) was broken at component Y and reestablished between Y and Z. This would result in two secure sessions: between X and Y, and between Y and Z. The message would be secure while on the network, but it would not be secure on component Y. The WAP Forum has adopted the Transport Layer E2E (end-to-end) Security Specification to address this issue [21].

The security of the i-mode mobile application architecture relies on "standard" Internet security, provided by SSL. The security protocols used in the DoCoMo network and over the air are proprietary protocols that run over SSL using its packet-switched network. According to DoCoMo [15], i-mode now provides end-to-end security "in the entire mobile network," which suggests that the security is provided both within a carrier's network and between carriers' networks. In addition, i-mode does have the ability to handle server-side authenticated SSL sessions, because i-mode phones are preconfigured with root CA keys with PKI vendors Baltimore and VeriSign [20].

System Compatibility

Since i-mode only supports i-mode phones with cHTML-based mobile clients running on a Web microbrowser, an i-mode–based mobile application has limited com-

Table 10.1   Comparison of Different Wireless Internet Application Architectures

Different Factors	Mobile Internet (Mobile Transparent)	Mobile-Interactive	i-mode Infrastructure	Mobile-Aware (WAP-Based)
Mobile presentation technology	cHTML, HTML, and WML	J2ME/HTML	cHTML	WML + WML Scripts
Mobile client platform	Microbrowser	J2ME platform	i-mode microbrowser	WML-based browser
Mobile devices	cHTML-enabled devices WML-enabled devices HTML-enabled devices	J2ME-enabled devices	i-mode phones and devices	WAP-enabled devices
Required middleware technology	Proxy adaptation middleware Wireless Web server	Wireless Web server J2EE (JSP/EJB)	i-mode server Relational markup language (RML)	WAP gateway WAP application server
Network protocol	HTTP/TCP over networks Internet protocols Underlying wireless network protocols	HTTP/TCP over networks Internet protocols Underlying wireless network protocols	Packet-switched network TCP/IP connectivity between i-mode server and the wireless Internet users i-mode's PDC-P HTTP and SSL/TLS protocols used end-to-end by i-mode devices	Lightweight directory access protocol (LDAP) WSP/WTLS/WTP/WDP WSP for back-end resources for wireless users WTLS protocol at the WAP gateway
Network connectivity	Works with major wireless networks Circuit-switched and packet-switched	Depends on underlying networks	Only i-mode network Packet-switched network	Works with major wireless networks Circuit-switched and packet-switched
Security	Internet security solutions (SSL/TLS) Non-repudiation facility Depends on microbrowser Required server security, such as certification Security GAP between wireless network and Internet	Internet security solutions (SSL/TLS) Better mobile client security due to J2ME Applet download validation by Java virtual machine Password-based user authentication with J2ME	i-mode security based on SSL/TLS protocol on the Internet Non-repudiation not facilitated Server certification No client certification Password-based user authentication	WTLS protocol for secured communications. Non-repudiation facility Password-based user identification and authentication, based on WIM [19] Existence of WAP GAP, to be addressed by future WAPs
Compatibility	Better compatibility on the mobile client side, due to conversion between XML-based forms and contents	When J2EE and J2ME is used together, good compatibility achieved on the mobile client side, because J2EE provides tools to support the dynamic formatting and conversion for XML-based forms	Compatibility problems at the protocol level i-mode only supports HTTP over wireless optimized TCP Since 2000, i-mode compatibility to WAP technology	Not compatible with other platform technologies Increased XML-based technology compatibility in the future Backward compatibility facilitated

Table 10.1   (continued)

Different Factors	Mobile Internet (Mobile Transparent)	Mobile-Interactive	i-mode Infrastructure	Mobile-Aware (WAP-Based)
Performance	Slow system performance, due to form conversion and formatting at proxy adapters	Long loading time for initial J2ME Java applet Client-server traffic reduced due to J2ME mobile clients	Efficient low-cost channel from packet-switched technology No protocol conversion on i-mode gateway	Protocol conversion at WAP gateway

patibility to other phones using different mobile technologies. In 2000, i-mode started to enhance its compatibility to the XML-based mobile technology.

A WAP-based mobile application also has a similar problem in mobile compatibility, since it only supports WAP-based mobile devices. Experts believe that XHTML-based mobile technology could be the solution for future mobile devices. The detailed discussion of XHTML is given in Chapters 4 and 15.

As shown in Table 10.1, a mobile-interactive architecture provides better mobile compatibility, because the proxy adaptation module in the middleware supports the conversion and formatting for different XML-based mobile technologies. The J2ME mobile technology has been accepted by major phone manufactures and mobile device vendors.

System Performance

System performance is always one quality factor in evaluating a mobile system architecture. Since there is no conversion for mobile data and messages between the i-mode wireless network and the internet, i-mode–based mobile applications should have a better system performance than applications that require protocol and data format conversions. Although both mobile-aware and mobile-transparent system architectures need to perform protocol conversions between wireless networks and the Internet, the WAP-based mobile-aware architecture should have a better performance, due to the following two reasons. First, the conversion is performed by a WAP gateway at the service layer. Second, the new WAP 2 protocol (see Chapter 3) has been introduced to enable HTTP over TCP/IP to be used directly to the mobile devices. This change reduces the protocol conversion work and time.

10.4   Smart Mobile Application Architectures

The smart mobile application architecture is very useful and popular for supporting smart mobile client software on high-end mobile devices, such as smart phones and PDAs. Its major focus is to provide mobile users with both online and off-line mobile application functions and services. Figure 10.10 shows the smart mobile application architecture. It includes a three-tier structure consisting of the following components.

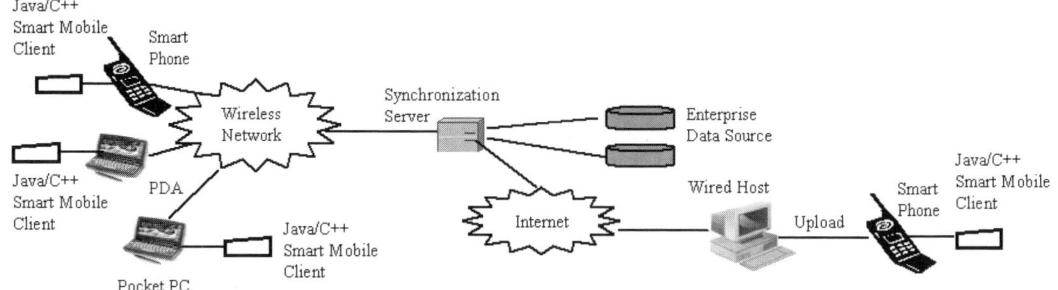

Figure 10.10 Smart mobile application architecture.

- *Smart mobile client software:* This is developed as a mobile application executed on smart mobile devices (i.e., iPods). A smart mobile client is a native executable that is a Java-based or C++-based application program deployed on high-end mobile devices. In addition to the basic application functions, a smart mobile client must include other necessary functional features:
    - A persistent data storage mechanism on mobile devices, so mobile device users can access mobile data while off-line;
    - Support for wireless or wired communication, for data synchronization with a server;
    - Ability to execute mobile smart client applications while off-line;
    - Integration with other native mobile client applications, as well as the back-end enterprise applications via a synchronization process.
- *Synchronous server:* This supports smart mobile clients by providing synchronized mobile data and contents, and data synchronization functions and services.
- *Data source store:* This usually is implemented with a database server to manage and maintain diverse mobile data and contents for each mobile device user.

10.4.1 Advantages and Limitations of Smart Mobile Application Systems

Smart mobile application systems have a number of distinct advantages. They are included here.

- The good system availability allows mobile users to access application data and functions resident on high-end mobile devices using the off-line operation mode. This eliminates the system availability issue caused by unreliable wireless network connectivity.
- High-end mobile devices with a bigger display screen provide mobile clients with a powerful operation platform and a diverse set of mobile technologies.
- Better system performance is obtained, since programs executed on mobile devices support most mobile application functions, which requires less networking communications.

- Low costs for wireless access, because less wireless connection time is needed.
- Low risk of threats to system security, because of the reduction of wireless connection time.

Smart mobile application systems also have a number of limitations. They are given here.

- Application deployment costs and overhead are limitations, because mobile client software must be reinstalled and deployed whenever new versions are generated.
- High mobile deployment costs result from multiple development cycles. Whenever mobile devices are upgraded, mobile client software (e.g., operating systems, programming languages) also must be upgraded.
- Smart mobile client software is generally more complex. Comparing with WML-based and cHTML-based thin mobile clients, a smart mobile client is much more complicated, because it supports essential mobile application functions and data. It usually is more complex and costly to be developed.
- There is limited enterprise integration at the server side.
- Mobile client security is more challenging, since mobile application software and data are stored on the mobile devices.

Table 10.2 compares smart mobile applications to wireless Internet applications.

10.5 Wireless Enterprise Application Architectures

The advance of wireless networking and mobile technology provides a great opportunity for network service carriers and application vendors to build innovative wire-

Table 10.2 Smart Mobile Applications Versus Wireless Internet Applications

Different Factors	Smart Mobile Application System Architecture	Wireless Internet Application System Architecture
System reliability	More reliable system functions and services	Unreliable system functions and services
System performance	Better system performance for mobile user accesses	Slower system performance for mobile user accesses
System security	Less risk in wireless network security, due to its off-line accesses and more mobile client security issues	More risk in network security, including network security gap and wireless network issues
Mobile user interface	Rich and better mobile user interface	Simple mobile user interface
Mobile client complexity	Complex mobile client software and higher development cost	Simple mobile client software and lower development cost
Mobile access costs	Very lost cost on mobile accesses	Higher costs on mobile accesses due wireless connectivity
Mobile client deployment	Higher cost on mobile client deployment	No need for mobile client deployment
Enterprise integration	Mobile applications are easily integrated with other applications on mobile devices	Mobile applications are easily integrated with enterprise solutions on the server side

less applications to meet the strong demand from mobile users, and allows business enterprises to set up an enterprise-oriented wireless application infrastructure. This infrastructure can be used to provide business customers and end users with a wireless connectivity to access application systems and services. It is also useful to enhance internal enterprise-based applications and workflows by offering mobile accesses and mobility to internal staff. In this section, we discuss two popular approaches and corresponding architectures for enterprises. The first uses a Microsoft-based mobile solution, and the other uses a Java-based mobile solution.

10.5.1 Wireless Enterprise Application Architecture Based on Microsoft Mobile Technology

Figure 10.11 shows the enterprise-based wireless application architecture based on Microsoft mobile technology, structured into three layers.

- *Mobile client layer:* As shown in Figure 10.10, three types of mobile clients (i.e., thin client, smart client, and thick client) can be developed and supported in this architecture. Microsoft Windows CE was developed to provide a very powerful operation platform for smart mobile devices (e.g., smart phones) and high-end mobile devices (e.g., PDAs or Pocket PCs). As discussed in Chapter 2, Window CE provides a rich set of software, including a Web server (Pocket Internet Explorer), PocketWord, and PocketOtutlook. In addition, Visual C++ and Java programming languages are also available for developers to build smart (or thick) mobile client software as a powerful mobile interface.

- *Middleware layer:* In this layer, a Microsoft Web server (IIS) is used to support HTTP communications with the Web browser on mobile devices. ASP.NET is used to support mobile controls in two areas:

  - Support of a wide range of Web-enabled mobile devices, including mobile phones, smart phones, PDAs, and Pocket PCs;

  - Support of different markup languages, such as WML, cHTML, and HTML.

  Although ASP.NET Mobile Controls do not require any client-side installation, the target device must have a client Web browser.

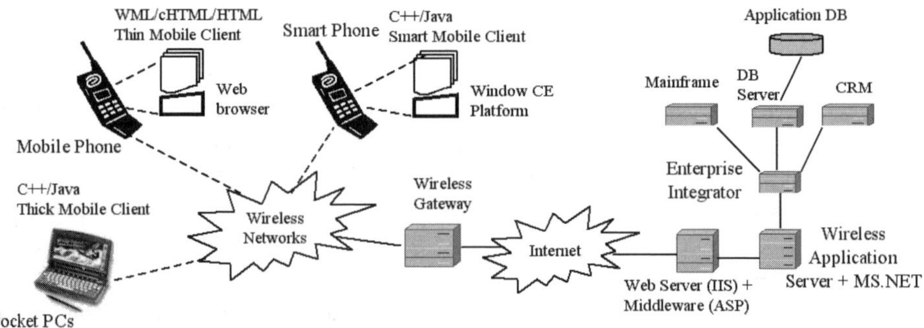

Figure 10.11 Wireless enterprise architecture based on Microsoft mobile technology.

- *Application layer:* This layer includes a wireless application server supported by Microsoft.NET technology. It usually provides the following basic functional capabilities, which allows a wireless application server to do the following:
  - Process the received mobile application requests and generate the correct responses;
  - Control and manage mobile access sessions;
  - Interact with a database server for data access and storage;
  - Transform contents between different mobile technologies (e.g., WML, cHTML, and HTML), and format mobile contents into a required form.

  - Integrate with existing applications, databases, and legacy systems.

There are four challenges concerning mobile clients and mobile devices while building wireless application systems.

- Support is necessary for different markup languages, such as HTML, cHTML, and WML.
- Mobile devices have different characteristics and configurations, such as screen size.
- Mobile devices have different network connectivity, such as GSM, GPRS, and 3G.
- Mobile devices have different capabilities, such as messaging support, notification, image display, and voice call.

To address these four challenges, Microsoft provides two solutions. The first is Microsoft Mobile Internet Toolkit, which can be used to address the challenges by isolating them from the details of wireless development. It provides the technology and tools to quickly build, deploy, and maintain sophisticated mobile applications. It also integrates with Visual Studio .NET to ensure that developers can use their existing desktop skills when building mobile enterprise applications. It can be used to add additional device support with the extensibility features. Therefore, it can quickly and easily build a single, mobile Web application that delivers appropriate markup pages for a wide variety of mobile devices. The other solution is ASP.NET, which allows creation, deployment, and execution of Web applications services. With ASP.NET, Web applications are easily built using Web forms.

10.5.2   Java-Based Wireless Enterprise Application Architectures

An alternative approach is to use Java-based technology to build enterprise-oriented wireless application infrastructure. Java-based mobile technology, developed by Sun Microsystems, has two major components that are useful to set up wireless enterprise-based application architectures:

The Java 2 Platform, Micro Edition

The Java 2 Platform, Micro Edition (J2ME) provides an effective platform on mobile devices (such as cell phones, pagers, and PDAs) to support J2ME-based mobile clients in connecting to enterprise-based application systems over a wireless Internet. The J2ME platform is a set of standard Java APIs defined for mobile devices. It provides three distinct features.

- The power and benefits of Java technology tailored for mobile and embedded devices;
- Support for both networked and disconnected applications;
- Capability for developers to write applications, dynamically download them, and leverage each device's native capabilities.

As discussed in Chapter 4, J2ME architecture defines device configurations, profiles, and optional packages as elements for building complete Java runtime environments to meet the requirements for a broad range of mobile devices and target markets. Each combination is optimized for memory, CPU processing power, and input-output capabilities of a related category of mobile devices. The result is a common Java platform that fully leverages each type of device to deliver a rich user experience.

The Java 2 Enterprise Edition

The Java 2 Enterprise Edition (J2EE) was primary designed as a cost-effective solution for enterprises to simplify problems relating to development, deployment, and management of multitier enterprise applications. J2EE is developed as a standard instead of a product, which allows application and middleware vendors to use the architecture. As shown in Figure 10.12(b), the J2EE architecture provides a framework and the corresponding services that simplify the development of Web-based applications. Developers can use the provided utilities and APIs for any Web application. Using the model view controller (MVC) pattern approach, J2EE separates the presentation layer from the business application logic layer, so that multiple views can be presented based on a shared enterprise data model. J2EE applications can be developed and deployed on any Web server, using a Java application server to provide the necessary enterprise application services, such as transactions, security, and persistence management.

Figure 10.12 shows wireless application system architecture for enterprises based on J2ME and J2EE technology. The system architecture consists of three layers.

- *Mobile presentation layer:* This layer includes a J2ME-based mobile client executed on the J2ME platform on smart phones or PDAs. The J2ME virtual machine on mobile devices can be customized based on the given configuration and profile for each type of mobile devices.
- *Middleware layer:* This layer includes a Web server and J2EE-based middleware facilities, such as enterprise Java beans (EJBs) and JSP/Java Servlets.
- *Application layer:* In this layer, J2EE application servers, legacy application systems, and corresponding databases are set up to support mobile users when they access enterprise applications and services over a wireless Internet. Mean-

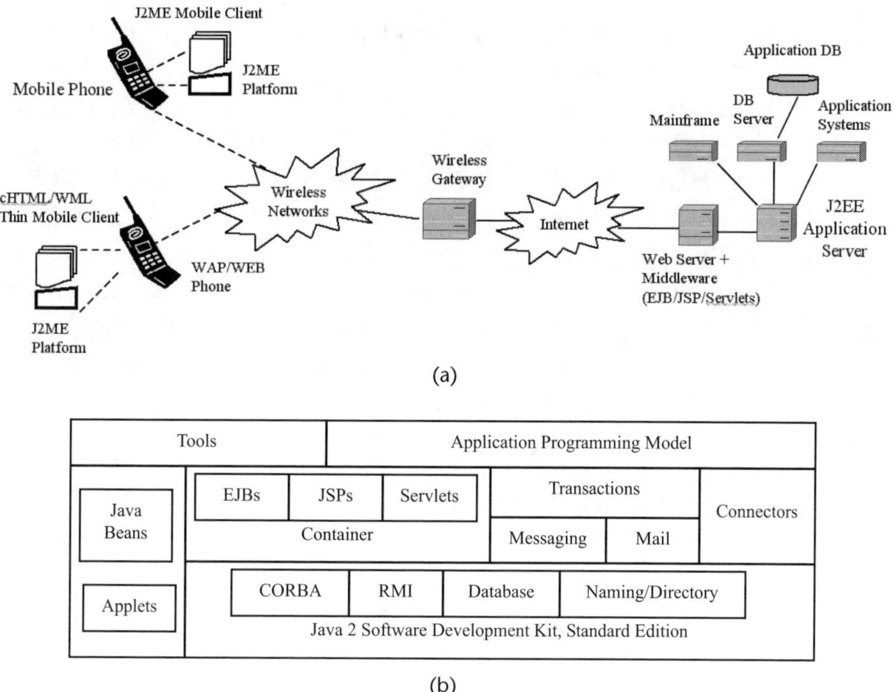

Figure 10.12   J2ME-based and J2EE-based wireless enterprise system architecture: (a) architecture for enterprises; and (b) J2EE framework.

while, the J2EE application server can be used as an integrator to connect the existing legacy applications and third-party systems.

Building enterprise-oriented wireless applications has the following advantages:

- A J2ME-based mobile client provides a powerful interactive interface that is dynamically downloaded whenever launched by a mobile user. Compared with a smart mobile client, a J2ME-based mobile client is much more easily deployed by mobile users and upgraded by application vendors.
- A J2ME-based mobile client provides a more enjoyable user interface than thin mobile clients, due to its graphic features and multiple-modal support.
- A J2ME-based mobile client can function under the connection mode and the off-line mode using the persistent data stored on mobile devices, which reduces the interactions between the mobile client and its application server. As a result, the system load at the server can be decreased, improving system performance.
- Integration to existing and legacy applications is easier, because J2EE provides a rich Java-based middleware [see Figure 10.12(b)] to support developers.

On the other hand, this approach has a number of limitations listed here [22].

- J2ME configurations and implementation releases for Java virtual machines are not identical for all mobile devices. This causes complexity in J2ME-based mobile client configuration, and reduces the interoperation ability on diverse types of mobile devices.

- Since a J2ME-based mobile client requires a certain amount of memory, it is a challenge to configure mobile devices that have limited memory size.
- Mobile users must first download the application from the J2EE server, if it is not preinstalled.
- Mobile applications using an old Java Development Kit (JDK) version usually have problems executing on new devices with an upgraded JDK version.
- Switching mobile devices to allow a user to access J2ME-based mobile applications requires special support in device profile changes and server side updates.

Since the J2EE application server has limitations in remaining current in an ever-changing mobile client environment, there is a need to extend the J2EE application server. Aligo, Inc., presents M-1 J2EE as an extension to a J2EE application server [22], which adds the following features and capabilities into the current J2EE application server.

- *Delivery suite:* The delivery suite consists of three parts: (1) presentation engine, (2) synchronization engine, and (3) messaging engine. The presentation engine automatically renders and formats the dynamic contents for each mobile device, so that the overhead of defining XSL for each mobile device can be reduced. The synchronization engine synchronizes J2ME-based mobile clients and caches the back-end data objects once the connection is established. This helps the application to function even in the disconnection mode. The messaging engine provides the messaging capabilities to every mobile device through e-mail, fax, or SMS services.
- *Processing suite:* The processing suite consists of the control and personalization engine, which detects wireless devices that access the mobile application through a user device personalization manager.
- *Integration suite:* The integration suite integrates with the back-end systems, databases, and other enterprise integration systems, such as ERP, SAP, and CRM.

As discussed in [22], using its extended server has some advantages and drawbacks. The major benefits are listed here.

- Mobile device profiles and presentation formats can be easily customized and managed, which supports diverse mobile devices and increase the flexibility of new mobile devices.
- Well-defined solutions allow developers to solve the existing issues in wireless connectivity, session management, and data synchronization.
- Multimodal session management allows users to switch devices or modes (e.g., voice to text and vice versa) without interrupting a session.

The drawbacks of M-1 J2EE are listed here.

- Since J2EE is tightly integrated with J2SE, whenever JVM is updated with a new version, this may cause a portability issue on its mobile client, requiring mobile client updating, retesting, and maintenance.
- The server handles everything from data processing to formatting and presentation, possibly causing system performance problems due to its possible network latency.
- Developers must understand all of the underlying APIs to take the full advantage of the extended J2EE application server.

10.6 Summary

System infrastructure and software architectural design are very important in constructing wireless-based application systems, because they have a strong impact on reliability, complexity, extendibility, performance, and scalability.

This chapter reviews the basic concepts of system infrastructure and software architecture, as well as the related design processes. It classifies and examines different types of wireless-based application system infrastructure and architectures, and discusses their advantages and limitations. The chapter covers the following types of application architectures.

- Network-based application architectures;
- Wireless Internet application architectures;
- Smart mobile application architectures;
- Enterprise-based mobile application architectures.

Section 10.3 focused on four types of popular wireless Internet application architectures, and compared their advantages and limitations. For example, the i-mode mobile application architecture and solution, widely used in Japan, provides a mature, low-cost approach to support cHTML-based mobile device users. Mobile-aware application architectures, popularly adopted in North America, use the standard WAP technology and WML-based mobile client to support mobile users. Section 10.4 examines the advantages and limitations of the smart mobile application architecture, and compares it with wireless Internet application architectures.

Section 10.5 discusses two popular enterprise-based mobile application architectures. The first is based on J2ME and J2EE technology, which provides an alternative approach to set up WAP-based wireless application systems. The second is based on Microsoft mobile technology, including the Window CE–based platform and its mobile-enabled program set, and .NET technology. It seems an ideal and mature solution to establish an enterprise wireless world for businesses and organizations.

It is clear that designers of wireless application system architectures should pay more attention to their impact on system interoperability, portability, standardization, deployment, and security.

References

[1] Alexander, A., *Timeless Way of Building*, New York: Oxford University Press, 1979.

[2] Perry, D. E., and A. L. Wolf, "Foundations for the Study of Software Architecture," *ACM SIGSOFT, Software Engineering Notes*, Vol. 17, No. 4, October 1992, pp. 40–52.

[3] Shaw, M., and D. Garlan, *Software Architecture: Perspectives on an Emerging Discipline*, 1st ed., Upper Saddle River, NJ: Prentice Hall, 1996.

[4] Bass, L., P. Clements, and R. Kazman, *Software Architecture in Practice*, 2nd ed., Reading, MA: Addison-Wesley, 2003.

[5] Lane, T. G, *Studying Software Architecture Through Design Spaces and Rules*, Technical report, The Computer Science Department, Carnegie Mellon University, November 1990.

[6] Jacobson, I., G. Booch, and J. Rumbaugh, *The Unified Software Development Process*, Reading, MA: Addison-Wesley, 1999.

[7] Bull, S., "User Modeling and Mobile Learning," *Proc. 9th Intl. Conf. UM 2003*, Johnstown, PA, June 22–26, 2003, pp. 383–387.

[8] Seppala, P., and H. Alamaki, "Mobile Learning and Mobility in Teacher Training," *Proc. 1st Intl. Workshop on Wireless and Mobile Technologies in Education*, New York, August 29–30, 2002, pp. 130–135.

[9] Xiang, Y., W. Zhou, and M. Chowdhury, *Toward Pervasive Computing in Restaurant*, Technical Report, retrieved on December 5, 2005, http://www.deakin.edu.au/~yxi/publications/iMenu_long_paper.pdf.

[10] Gao, J., et al., "P2P-Paid: A Peer-to-Peer Wireless Payment System," *Proc. Second Int. Workshop on Mobile Commerce and Services (WMCS 05)*, IEEE Computer Society, Munich, Germany, July 19, 2005.

[11] Gao, J., et al., "A Wireless Payment System," *Proc. Second Int. Conference on Embedded Software and Systems (ICESS 2005)*, IEEE Computer Society Press, Xi'An, China, December 16–18, 2005.

[12] Lin, K.-J., T. Yu, and C.-Y. Shih, "The Design of A Personal and Intelligent Pervasive-Commerce System Architecture," *Proc. IEEE Second Int. Workshop on Mobile Commerce and Services*, IEEE Computer Society, July 2006.

[13] Jing, J., A. Helal, and A. Elmagarmind, "Client-Server Computing in Mobile Environment," *ACM Computing Surveys*, Vol. 31, No. 2, June 1999, pp. 117–157.

[14] Visser, R. T., "Integrating I-Mode and MMS—Generating Successful I-MMS Alternatives," Master Thesis, August 2003.

[15] "DoCoMo's iMode: Toward Mobile Multimedia in 3G," presentation given at IETF Meeting, IETF 47, Plenary Session, March 2000.

[16] Fielden, T., and A. Orubeondo, "J2ME and WAP: Together Forever?" *JavaWorld*, retrieved on March 10, 2005, http://www.javaworld.com/javaworld/jw-120-2000/j2-1201-iw-j2me.html.

[17] WAP Forum, "Wireless Application Protocol," White Paper, *Wireless Internet Today*, June 2000.

[18] Kaasinen, E., et al., "Two Approaches to Bring Internet Service to WAP Devices," *Computer Networks*, Vol. 33, 2000, pp. 231–246.

[19] Wireless Application Protocol Identity Module Specification, http://www1.wapforum.org/tech/documents/WAP-198-WIM-20000218-a.pdf.

[20] Ashley, P., H. Hinton, and M. Vandenwauver, "Wired Versus Wireless Security: The Internet, WAP, and iMode for E-Commerce," *Proc. 17th Annual Computer Security Applications Conference*, 2001.

[21] WAP TM Transport Layer E2E Security Specification, http://www1.wapforum.org/tech/documents/WAP-187-TransportE2ESec-20000711-a.pdf.

[22] Hosalkaer, A., "Building Mobile Applications with J2EE, J2EE-J2ME, and J2EE Extended Application Servers," *Proc. Mid-Atlantic Student Workshop on Programming Languages and Systems (MASPLAS'02)*, Pace University, April 19, 2002.

CHAPTER 11

Wireless Security: Introduction

While wireless networking offers flexibility, portability, and connectivity, it also opens up the users to greater security risks. Hackers exploit vulnerabilities, and launch attacks at wireless networks every day, and security experts also work diligently to counter these attacks by developing security protocols and solutions. This competition between hackers and security experts will never end. This chapter provides readers with background information on security services and common security threats in wireless networks. The readers will be equipped with sufficient background to keep up with new security attacks and solutions.

A security solution generally provides the following three types of security services:

- *Confidentiality:* It provides privacy for communicating parties by ensuring that unauthorized users cannot read and interpret data transmissions.

- *Authentication:* It verifies that the identities of communicating parties are authentic. In the case of one-way authentication of a single message, it assures the recipient that the message is indeed from the claimed source. An example of one-way authentication is the wireless LAN service that verifies the identity of mobile clients. In the case of two-way authentication of ongoing message exchange, it assures two things. At the time of connection setup, it proves that the two communicating parties are indeed from the claimed sources. During message exchange, it assures that a third party cannot break in and pretend to be one of communicating parties that are already authenticated.

- *Data integrity:* It assures the recipient that the message has not been modified in transit since it was sent out by the source.

Since cryptography has been the main technology that provides these security services, we will review two major kinds of cryptography in this chapter: secret key cryptography and public key cryptography. In Section 11.1, we discuss secret key cryptography, including its basic cipher operations, major standards, block cipher modes, and how it can be used to provide security services. In Section 11.2, we study public key algorithms and how to use these algorithms to provide the security services. We then present wireless security attacks in Section 11.3. These attacks are normally developed to exploit the breeches in confidentiality, authentication, data integrity, or any combination of them. We summarize the key points in Section 11.4.

11.1   Secret Key Cryptography

Secret key cryptography, also referred to as symmetric cryptography, is the most commonly used technique. To encrypt the original message, known as plaintext, the encryption algorithm takes both the plaintext and a secret key as input, and performs various substitutions and permutations on the plaintext to produce ciphertext. Since the type of substitutions and permutations depends on the secret key, the ciphertext depends on both the plaintext and the secret key. For a given plaintext, different secret keys will produce different ciphertexts. To decrypt a ciphertext, the decryption algorithm needs to take the same secret key as that used in the input, in order to reverse the substitutions and permutations performed by encryption. Therefore, the key has to be a secret that is shared by the sender and the recipient. That is how secret (or symmetric) key cryptography takes its name.

11.1.1   Basic Operations

In this section, we will examine the two basic building blocks of secret key cryptography—substitution and permutation. The substitution technique replaces the letters of plaintext by other letters or symbols to produce ciphertext, and the permutation technique reorders the letters in plaintext to produce ciphertext letters.

Substitution

The earliest example of substitution is a simple cipher invented by Julius Caesar. To produce ciphertext, the Caesar cipher replaces each letter of the plaintext in the 26-character alphabet by the letter that is three positions down. The alphabet is wrapped around, so that X maps to A, Y maps to B, and Z maps to C.

Example 11.1
Alice wanted to send the following message to Bob, "I find this class boring." She would like to encrypt it to prevent her instructor from reading the message, so she applied the Caesar cipher. What is the ciphertext?

Answer
After applying the following mappings (i.e., I to L, F to I, N to Q, D to G, T to W, H to K, S to V, C to F, L to O, A to D, B to E, O to R, R to U, G to J), Bob received the following sentence, "L ILQG WKLV FODVV ERULQJ."

Clearly, this sentence does not make sense to the instructor if he does not know that Alice had used the Caesar cipher. However, if he knows that the Caesar cipher had been used, then it would be very easy for him to decrypt the original sentence. One way to make his guessing task more difficult would be to replace each letter in the plaintext by the letter that is k positions down, where k can be 1 to 25. In the worst case, he needs to try 25 possible substitutions to guess the original sentence.

With only 25 possible keys, the Caesar cipher is very weak in the security sense. Another cipher, called the monoalphabetic cipher, has enhanced security by allowing arbitrary substitution. In other words, A can be substituted by any of the 26 letters in the alphabet, B can be substituted by any of the remaining 25 letters, C can be substituted by 24 possible letters, and so forth. Therefore, there are 26! possible substitutions. It becomes infeasible to guess the original message by simply enumerating all the possible substitutions, since there are more than 4×10^{26} possibilities. If it took 1 μs to try each possibility, it would take 10 trillion years to try all of them. The monoalphabetic cipher is not really secure, if the attacker knows that the plaintext is English text. By measuring the relative frequencies of the individual letters and combinations of letters in the encrypted message, and comparing them to the standard distribution for English [1], he can guess the message in a much shorter time, even in tens of minutes.

There are three ways to improve the monoalphabetic cipher. The first way is to encrypt multiple letters at one time, which is known as a multiletter cipher. The second way is to use different monoalphabetic substitutions as one proceeds through the plaintext message, which is called polyalphabetic cipher. The third way is to encrypt the message with a random key that is as long as the original message, which is known as a one-time pad.

Two good examples of multiletter ciphers are the Playfair cipher and the Hill cipher. The Playfair cipher takes two letters from the plaintext as a unit and transforms it into two letters in the ciphertext. The transformation is based on a 5×5 matrix containing the 26 letters in the alphabet, with I and J counted as one letter. The construction of the matrix is dependent on the key. The encryption process takes the plaintext and this matrix as input, and generates the ciphertext based on a set of rules. Interested readers can refer to [2] for detailed discussion of these rules.

The Hill cipher is a more general cipher that takes m plaintext letters and substitutes them into m ciphertext letters, where m can be larger than two. The substitution rules are determined by a linear transformation specified by an $m \times m$ matrix. For example, if $m = 4$, then the encryption process can be described by the equation: $C_{4 \times 1} = K_{4 \times 4} P_{4 \times 1} \bmod 26$. To apply the equation, the letters are first mapped to numerical values $(0 \dots 25)$ by its alphabetic order. The plaintext message is then broken into groups of four letters, and each group is transformed into four letters using the equation.

Example 11.2
Alice wanted to send the same message as in Example 11.1 to Bob using the Hill cipher. If she used the following key, what would Bob receive?

$$K_{4 \times 4} = \begin{pmatrix} 23 & 11 & 5 & 9 \\ 2 & 2 & 16 & 71 \\ 3 & 14 & 7 & 21 \\ 23 & 47 & 4 & 9 \end{pmatrix}$$

Answer
The first four letters of the original message, "IFIN," map to vector $(8\ 5\ 8\ 13)^T$. Multiplying $K_{4 \times 4}$ with this vector, we get $(6\ 11\ 20\ 22)^T$, which maps to GLUW in the alphabet. Repeating the same procedure for the rest of the message, Alice generated the following ciphertext, "G LUWV WEWX IDGPQ NMAIOP."

The Hill cipher appears to be strong, but it can be broken if the attacker is able to collect a sufficient number of <plaintext, ciphertext> pairs to find the key matrix.

The polyalphabetic cipher is the second improvement over monoalphabetic cipher. It first chooses a set of related monoalphabetic substitution rules, and assigns each rule to a letter. The key is constructed using a sequence of letters that determines which substitution rule is used. The cipher repeatedly applies the key to the plaintext to produce the ciphertext. However, the periodic nature of the key can still be exploited by the attacker to guess the plaintext.

To improve the polyalphabetic cipher, the one-time pad (the third improvement over the monoalphabetic cipher) uses a random key that is as long as the message itself. This scheme eliminates the periodicity of the key and is theoretically unbreakable, but is difficult to use in practice. For every message to be sent, a key of equal length has to be randomly generated and distributed to both the sender and the receiver. The generation, distribution, and protection of this long key are all difficult tasks.

Permutation

Permutation reorders the plaintext letters. A simple permutation is to write the message in a matrix form, row by row, and read the letters column by column. The order of the column to be read is determined by a key.

Example 11.3
Let us revisit the same message that Alice wanted to send to Bob. First, we write the message down in a 4×4 matrix format

```
I F I N
D T H I
S C L A
S S B O
R I N G
```

Then we read the ciphertext using key (3 2 1 4). In other words, we first read the third column, followed by the second column, the first column, and then the fourth column. Therefore, the message Bob received was "IHLBNFTCSIIDSSRNIAOG."

However, the attacker can easily break this simple transposition cipher by playing the frequency game that he used to play against the monoalphabetic cipher. The transposition cipher can be made much more secure by multiple applications of permutations. This results in a much more complex transformation that is hard to analyze. Therefore, in secret key cryptography, permutations are often done more than once to enhance security.

In fact, a permutation is a special case of substitution, in which every plaintext letter is mapped to another letter taken from the plaintext set. Substitution is more general, in the sense that a plaintext letter can be mapped to another letter not in the original plaintext set. In general, secret key cryptography algorithms are built upon these two operations. Different secret key algorithms vary by how these substitutions and permutations are mixed and applied in the encryption process.

11.1.2 Block Cipher Standards

In this section, we review three symmetric encryption standards: the data encryption standard (DES), the international data encryption algorithm (IDEA), and the advanced encryption standard (AES).

DES

DES was first adopted in 1977 by the National Bureau of Standards (now known as the National Institute of Standards and Technology) as a standard for use in commercial and unclassified government applications. It survived very well for more than 20 years, until it was broken in 1998 by the Electronic Frontier Foundation (EFF) [3] using a special-purpose DES-breaking engine, EFF DES cracker.

DES is a wonderful example of a symmetric block cipher. It breaks the message into 64-bit blocks, and encrypts these blocks using a 56-bit key. The 56-bit key is obtained by stripping off the 8 parity bits from the given 64-bit key. Figure 11.1 shows the basic structure of DES.

The 64-bit input block is first transformed by an initial permutation, resulting in block (L_0, R_0). Then, (L_0, R_0) passes through round 1 to become (L_1, R_1), which then becomes (L_2, R_2) after round 2. After 16 identical rounds, (L_{16}, R_{16}) has its two

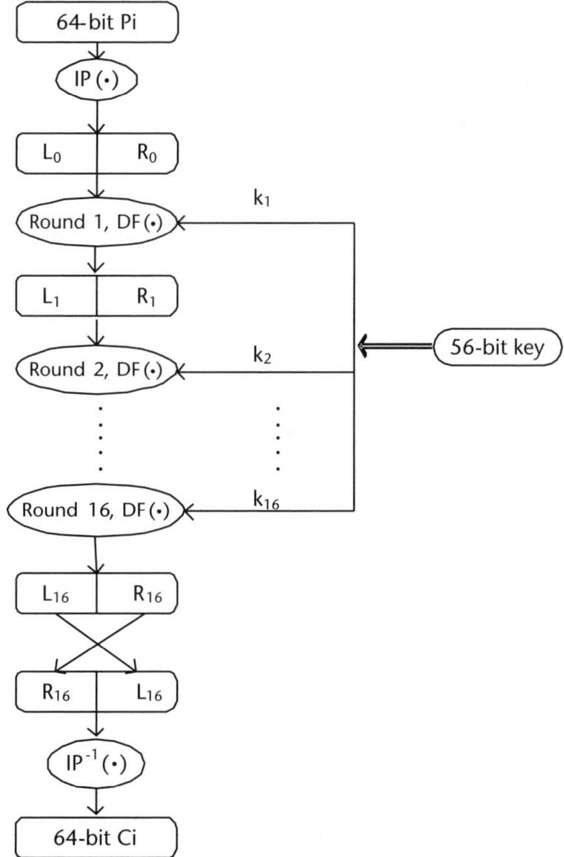

Figure 11.1 The conceptual structure of DES.

halves swapped, and (L_{16}, R_{16}) is then transformed by the inverse of the initial permutation to generate the 64-bit ciphertext block. Each round applies the same function $DF(\cdot)$ that consists of a combination of permutations, substitutions, and exclusive OR (XOR) operations with 48-bit per-round keys. These keys are generated by using the different subsets in the original 56-bit key.

DES has a nice property in which encryption and decryption follow the same algorithm shown in Figure 11.1, except they use keys in reverse order [4]. More specifically, to encrypt a plaintext block P_i, the encryption algorithm uses keys, K_1, K_2, . . ., K_{16}, to generate the cipher block C_i. To decrypt C_i, the same encryption algorithm would apply the same set of keys in reverse order, K_{16}, K_{15}, . . ., K_1, to produce P_i.

Although DES was broken in 1998, it is still possible to achieve high security by applying it multiple times using different keys. Experts estimate that triple DES is 7×10^{16} harder to crack, and will be secure for the near future.

IDEA

Xuejia Lai and James Massey of the Swiss Federal Institute of Technology developed IDEA in 1990. Its original version was called the proposed encryption standard (PES). Since its invention, it has been strengthened against various cryptographic attacks. In 1992, the revised PES was renamed IDEA. Pretty good privacy (PGP), which is widely used for privacy and authentication in e-mail applications, employs IDEA for encryption. Like DES, IDEA is also a block cipher, but it uses a longer 128-bit key to encrypt 64-bit data blocks. IDEA was patented, and found to be slow.

Figure 11.2 shows the conceptual structure of IDEA encryption. The size of input block to IDEA, P_i, is 64 bits, and the key size is 128 bits long. IDEA consists of eight identical rounds, $IF(\cdot)$, followed by a different final round, $IF'(\cdot)$. The input to each round consists of four data subblocks of 16-bit in length and a per-round key tuple. The key tuple for the first eight rounds has six subkeys, each 16 bits in length, and the key tuple for the last round has four subkeys, each 16 bits in length. In total, the algorithm needs 52 subkeys that are generated from the original 128-bit key.

The functions that are applied in these rounds are based on three fundamental operations. They are:

1. Bit-by-bit XOR of 16-bit subblocks;
2. Addition of 16-bit integers modulo 2^{16};
3. Multiplication of 16-bit integers modulo $2^{16} + 1$.

The function in the first eight rounds, $IF(\cdot)$, involves 14 steps based on these operations, while the function in the last round, $IF'(\cdot)$, involves 4 steps of such operations.

IDEA also enjoys the property in which encryption and decryption are identical except for key generations, although the relationship in keys is more complex than it is in DES [5].

There are no known cracks against the IDEA algorithm. Breaking it by brute-force enumeration of 128-bit keys requires more computing resources than are presently available.

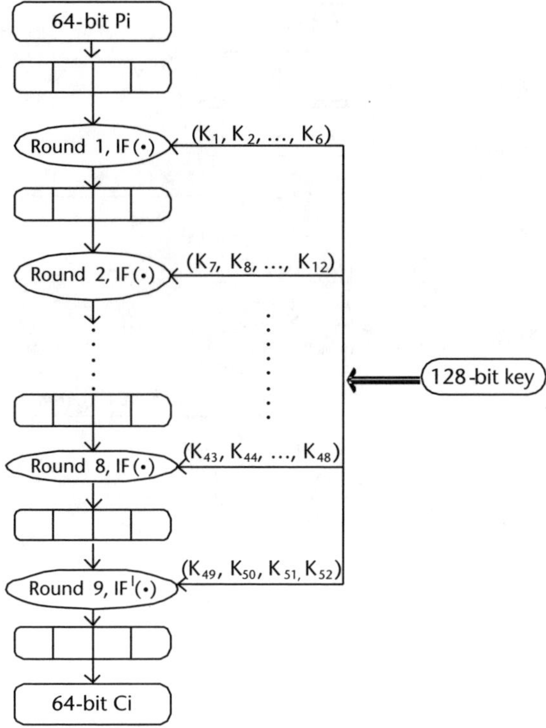

Figure 11.2   The conceptual structure of IDEA.

AES

There was a need for an improved secret key standard, since the key length of DES was too small and IDEA was protected by patent and too slow. In 1997, NIST announced a contest to select a new encryption standard, eventually choosing an algorithm developed by two Belgian cryptographers. The algorithm is called Rijndael, named after its inventors. In 2001, AES became a new secret key cryptography standard. The Rijndael algorithm is also a block cipher that encrypts data blocks of 128 bits using keys of 128, 192, and 256 bits.

AES combines the use of different key sizes into a single framework by using three parameters:

1. The block size, N_b, specifies the number of 4-byte words in the plaintext block. AES mandates the size of its data block to be 128 bits (equal to 16 bytes), so $N_b = 4$.
2. The key size, N_k, specifies the number of 4-byte words in the key. Therefore, AES-128 has $N_k = 4$; AES-192 has $N_k = 6$; and AES-256 has $N_k = 8$.
3. The number of rounds, N_r, is related to N_b and N_k in the following way: $N_r = 6 + \max(N_b, N_k)$.

Figure 11.3 illustrates the conceptual structure of AES. The input to the encryption algorithms is a single 128-bit block. This block is copied column-by-column into a state array of N_b columns. The initial state array undergoes $N_r - 1$ rounds of identical transformations, $AF(\cdot)$, and a different transformation, $AF'(\cdot)$, in the last

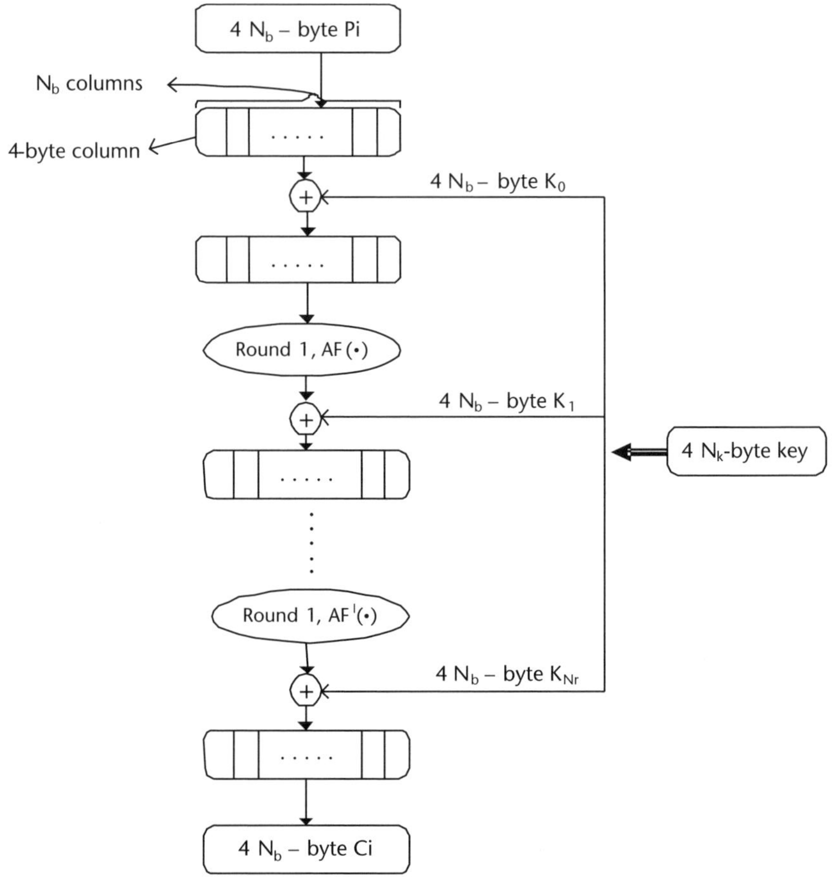

Figure 11.3   The conceptual structure of AES.

round. The state array is keyed with a subkey before the first round and after each of the N_r rounds. Like DES and IDEA, these ($N_r + 1$) subkeys are also generated from the original four N_k-byte keys.

Like DES, the AES decryption algorithm makes use of the subkeys in the reverse order. However, the decryption algorithm is not identical to the encryption algorithm.

So far we have discussed three basic algorithms to encrypt a short block of bits. To encrypt large messages of arbitrary size, several modes of block operations have been defined for DES [6]. These modes are equally applicable to IDEA and AES.

The simplest mode, known as the electronic codebook mode (ECB), breaks the plaintext message into 64-bit blocks. It independently encrypts and decrypts each of these blocks using the same key. This is a good mode to encrypt a small amount of data, such as a secret key. However, there are a number of security flaws when applying this mode to a large message. First, if a message has two identical blocks, then the ciphertext will also contain two corresponding identical blocks. It is possible for eavesdroppers to exploit these structures. Second, if eavesdroppers can figure out the block boundaries and reorganize the blocks, then the ciphertext blocks may be altered without being detected. As a result, the ECB mode is rarely used to encrypt messages, despite its simplicity.

The second mode, cipher block chaining (CBC), addresses the security problems in ECB. In CBC, instead of directly encrypting the plaintext block, it encrypts the XOR of the current plaintext block and the preceding ciphertext block. For decryption, each ciphertext block passes through the decryption algorithm, and the output is then XORed with the preceding ciphertext block to recover the original plaintext block. Like ECB, the same key is used to encrypt and decrypt every block. To encrypt the first plaintext block, we need to generate a random initialization vector (IV), which is also needed in decrypting the first block. It is easy to see that the CBC mode would produce different ciphertext blocks for identical plaintext blocks, since encryption does not solely depend on plaintext blocks. This removes the structure in ciphertext and makes cryptanalysis more difficult.

The CBC mode introduces dependencies between coded blocks. That is, the nth block cannot be decrypted until the previous $(n-1)$ blocks have been decrypted.

The third mode, the counter mode, facilitates random access to the encrypted blocks. In this mode, a counter is needed to encrypt and decrypt each block, and its value should be different for each plaintext block. In encryption, the counter passes through the encryption algorithm and then is XORed with the plaintext block to produce ciphertext block. In decryption, the same counter passes through the encryption algorithm and then is XORed with the ciphertext block to recover the plaintext block. For the counter mode to work correctly, encryption and decryption both need to agree on the algorithm of generating counter values for each block. A simple way is to randomly generate a counter value for the first block, and then increment the counter by one for each subsequent block. The counter mode has removed dependencies between blocks, and is believed to be as secure as CBC.

ECB, CBC, and the counter modes all generate encryption output in integral numbers of blocks. The other two modes, cipher feedback (CFB) and output feedback (OFB), remove the requirement of padding the message into an integral number of blocks. They can produce encryption outputs of arbitrary length, and are called stream ciphers.

11.1.3 Stream Cipher and RC4

In this section, let us turn our attention to the stream cipher. Unlike a block cipher, a stream cipher usually encrypts plaintext one byte at a time. Figure 11.4 captures the conceptual steps of a stream cipher. The sender and the receiver both share a secret

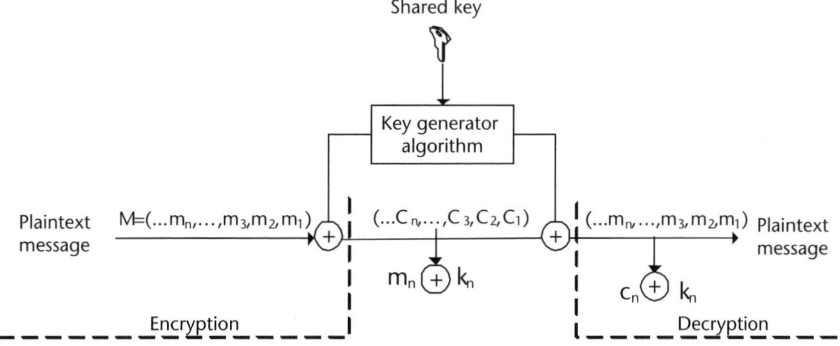

Figure 11.4 A stream cipher.

key, K, that goes through the same key stream generator algorithm to produce a sequence of key stream bytes, $(k_1, k_2, ..., k_n)$. At the sender side, the plaintext M is first broken into a stream of bytes, $(m_1, m_2, ..., m_n)$, which is then combined with the key stream one byte at a time, using the bitwise XOR operation to generate a stream of ciphertext $(c_1, c_2, ..., c_n)$, where $c_i = m_i \oplus k_i$, $i = 1, 2, ..., n$. At the receiver side, the plaintext is recovered by using XOR on the ciphertext stream and the key stream bytes.

In Figure 11.4, we see that distinct stream ciphers only differ in their key generator algorithms. RC4 is perhaps the most popular cipher among the existing stream ciphers. It is used in the secure sockets layer (SSL) protocol that forms the basis for secure communications between Web browsers and servers. It is also being used in the wired equivalent privacy (WEP) protocol in the 802.11 wireless LAN standard. In RC4, The input key, K, can vary from 1 to 256 bytes. K is used to initialize a 256-element vector $(s_0, s_1, s_2, ..., s_{255})$. The key generator algorithm in RC4 defines how to permutate this vector in order to produce the key byte k_i at each step. Detailed algorithm steps of RC4 are not covered here and can be found in [2].

11.1.4 Confidentiality

In wireless networks, there are many examples of insecure conversations. For example, a packet transmitted over radio waves can be intercepted, and unauthorized users can read a file stored in a shared server. Secret key cryptography algorithms can be very effective in providing confidentiality for transmission over insecure channels and storage on insecure media. If the sender and the receiver can agree on a shared key for use in the algorithms, then the attackers will only receive unintelligible data.

One big challenge of this classic use of cryptography is the secure establishment of a shared key between the sender and receiver. The level of confidentiality that can be provided largely depends upon the key distribution technique. Key distribution can be done in many ways, but we will discuss only two alternatives. The first method is for the sender to select a shared key and deliver it to the receiver. We need a dedicated link between the sender and the receiver for safe delivery of this shared key. In a distributed system of n hosts, where two-way communication occurs between any two hosts, we need $n(n-1)$ links. Obviously, this method is not scalable when n grows large. The second method is to rely on a central key distribution center (KDC). In this method, each party shares a secret key with the KDC. If two parties, A and B, need to communicate with each other, then the KDC will generate a shared session key S for A and B. Then, the KDC sends an encrypted S to A, using the key that it shares with A. Likewise, the KDC informs B with an encrypted S using B's key. This key distribution method prevents eavesdropping, since it encrypts the shared key before transmission. This method is also more scalable, because a distributed system of n hosts only needs n dedicated links with the KDC.

11.1.5 Authentication

In some wireless applications, it is important for each communicating party to verify the identity of the other party. For example, in a wireless payment system, the client has to ensure that the payment server is indeed the host that it trusts and to which it will send payment.

Secret key cryptography can help with this authentication process. The sender and receiver both share the knowledge of a secret key, which is not known to anybody else. If each party can prove its knowledge of the key, then it has successfully proven its identity to the other party. To do this without revealing the secret key to the third party normally involves exchanges of challenges and responses in cryptographic form. Figure 11.5 shows a basic example of challenge-response protocol. The protocol involves four messages. The first two messages convince Alice that she is talking to Bob, and the latter two messages convince Bob that he is talking to Alice. We assume Alice and Bob share a secret key K. First, Alice sends a randomly generated number, r_1, to Bob. Second, Bob sends encrypted r_1 using K back to Alice. If Alice can decode r_1 from Bob's message, then she verifies that it is Bob at the other end, since nobody else knows how to encrypt r_1 using K. Similarly, in the third and fourth steps, Bob challenges and authenticates Alice with a different random number, r_2.

This protocol is subject to some attacks, such as replay attacks. For example, if Trudy has collected many $<r_1, encrypted(r_1)>$ results by sniffing on the network, then it is possible that Trudy can answer Alice's query of r_1, and pretend to be Bob. To alleviate this problem, the random number challenges, r_1 and r_2, must be taken from a large space. There are other problems with this basic protocol [4], but it explains the general idea of the cryptographic authentication algorithm.

11.1.6 Data Integrity

Data integrity is another important concern for wireless networks. Packets of data sent over the network may be intercepted and modified in transit. For example, in the design of a secure wireless payment system, there should be a mechanism for the

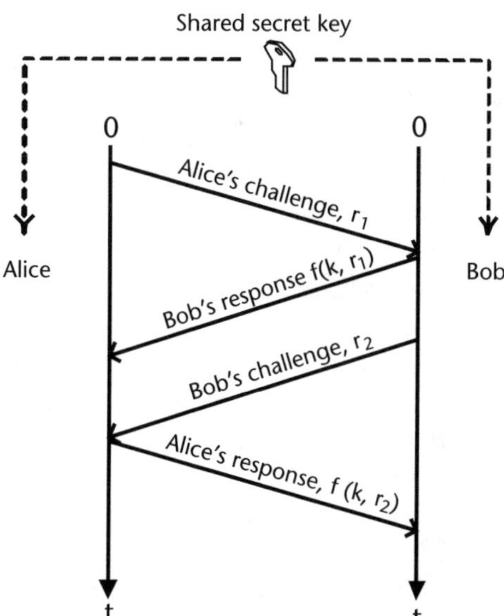

Figure 11.5 A basic challenge-response authentication protocol using secret key.

server to verify that the payment information from the client has not been changed during transmission.

A secret key scheme can be used to generate a message authentication code (MAC) that protects messages from malicious changes. MAC, a fixed-length value, is also called a message integrity code (MIC). This scheme assumes that the sender and receiver both share a common secret key, K. When the sender transmits a message to the receiver, it first calculates the MAC as a function of the message and the shared key, K. Then, the message concatenated with the MAC is transmitted to the receiver. The receiver performs the same calculation on the message, using the same secret key. If the calculated MAC matches the received MAC, then the receiver is confident that the message has not been changed. This is because any change in the message would require a new MAC, but the attacker does not know the secret key to correctly generate the new MAC. The consistency between the message and the MAC proves that the received message is unaltered from the generated message. One simple way to generate the fixed-length MAC from an arbitrary-length message is to first break the message into data blocks of MAC size, padding the last block if necessary, and then to encrypt the sum of these data blocks using the shared key. There are also more complicated ways. For these schemes to provide sufficient protection, they must make it computationally expensive to find two different messages resulting in the same MAC.

11.2 Public Key Cryptography

Whitefield Diffie and Martin Hellman invented the concept of public key cryptography in 1976 [7]. They proposed an exponential key exchange scheme to address two problems found in secret key cryptography. The first problem is that a KDC is always needed to distribute secret keys, and a compromise of the KDC could defeat the whole system. The second problem is that secret key cryptography cannot be used to digitally sign a document, so it is difficult to prove that a particular person has sent an electronic message. Since 1976, many public key algorithms have been proposed, but only a few of them remain in use. Some of these algorithms, such as the algorithm invented by Rivest, Shamir, and Adleman (RSA) [8]; the algorithm by ElGamal [9]; and the elliptic curve cryptography [10] are suitable for both encryption and digital signature. Other algorithms, such as the digital signature algorithm (DSA) [11], were only designed for digital signatures.

Public key algorithms have two keys for its operation, a private key and a public key. The private key is kept confidential by the party who generated the key, while the public key is accessible to everybody through a public repository. Determination of the private key, given the knowledge of public key, must be computationally infeasible in order for this scheme to be secure.

Figure 11.6 illustrates the basic steps for Alice to send an encrypted message to Bob, and for Bob to decrypt the message using public key cryptography. First, Alice and Bob both generate a pair of keys, a private key and a public key. Second, Alice and Bob send their public keys to a central repository. Third, Alice encrypts the message using Bob's public key, which was fetched from the central repository. Fourth,

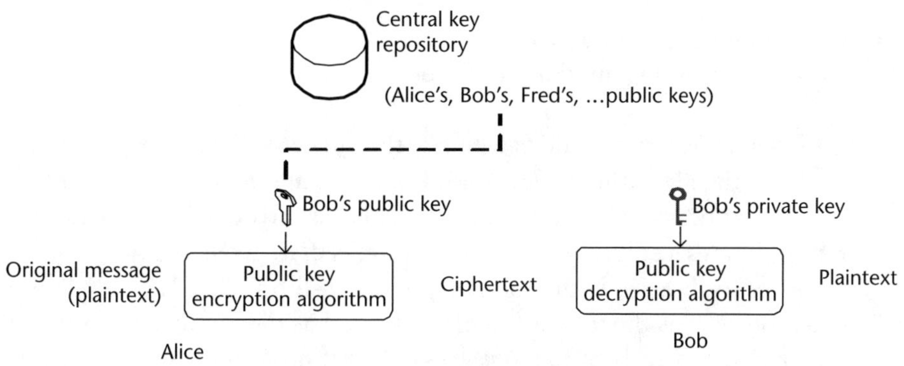

Figure 11.6   Encryption and decryption using public key cryptography.

Bob decrypts the message using his own private key. Nobody else could decrypt the message, since only Bob knew his own private key.

Figure 11.7 shows the steps for Alice to send a digitally signed document to Bob. Alice first encrypts the document using her private key, and sends the encrypted document to Bob. Bob then decrypts the document using Alice's public key, which was obtained from the central key repository. Bob is convinced that Alice sent the document, since only Alice's public key could decrypt the message.

Let us take a closer look at the type of encryption key. In Figure 11.6, Alice used Bob's public key to encrypt the message, while in Figure 11.7, Alice used her own private key to encrypt the message. Therefore, for a public key algorithm to provide both encryption and a digital signature, it must ensure that either the public key or the private key could be used for encryption, with the other one used for decryption. In this chapter, we will focus on one such algorithm, the RSA algorithm.

11.2.1   The RSA Algorithm

The RSA algorithm has been a widely accepted approach to public key cryptography. It is a block cipher in which plaintext and ciphertext are integer values. It makes use of modulo exponentiation for encryption and decryption. If the plaintext block is M and the ciphertext block is C, then encryption and decryption are done in the following way:

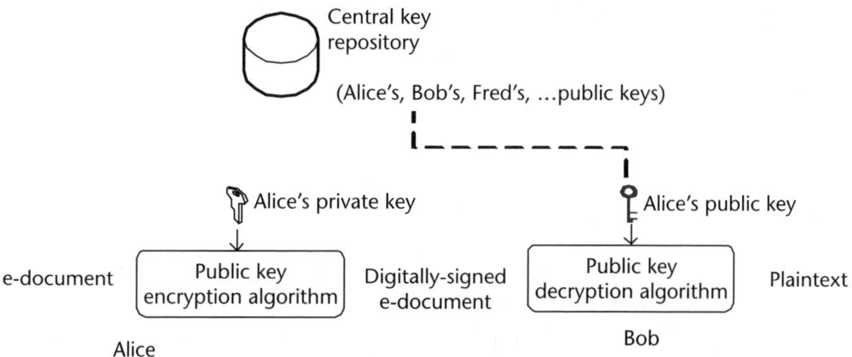

Figure 11.7   Digital signature by public key cryptography.

Encryption: $d = e^{-1} \bmod \phi(n)$,
Decryption: $M = C^d \bmod n = M^{ed} \bmod n$

Therefore, both M and C will take the integer value between 0 and $n - 1$. For this to work, the algorithm needs to find out e, d, and n, so that $M^{ed} = M \bmod n$.

This appears to be a difficult problem, but fortunately a variant of Euler's theorem facilitates a solution. The theorem states that, given two integers, $0 < M < n$, and an arbitrary integer k, the following relationship holds: $M^{k\phi(n)+1} = M \bmod n$. In this equation, $\phi(n)$ is the totient function, defined as the number of positive integers less than n and relatively prime to n. It is easy to find $\phi(n)$ for some special n. For example, $\phi(n) = n - 1$, if n is prime; and $\phi(n) = (p - 1)(q - 1)$ if n is a product of two prime numbers, p and q. The second example is used in the RSA algorithm.

If we compare the two equations,

$$M^{ed} = M \bmod n$$

$$M^{k\phi(n)+1} = M \bmod n$$

we can see that if e and d satisfy the condition $ed = k\phi(n) + 1$, or $d = e^{-1} \bmod \phi(n)$, then we have solved the puzzle.

After going over the mathematical background, let us summarize the RSA algorithm.

- Public key $<n, e>$: Choose two prime numbers p and q. Calculate $n = pq$ and to get $\phi(n) = (p - 1)(q - 1)$. Select an integer e that is relative prime to $\phi(n)$.
- Private key $<n, d>$: n and $\phi(n)$ are chosen in the above step. Here, choose $d = e^{-1} \bmod \phi(n)$.
- Encryption: $C = M^e \bmod n$ (data privacy) or $C = M^d \bmod n$ (digital signature).
- Decryption: $M = C^d \bmod n$ (data privacy) or $M = C^e \bmod n$ (digital signature).

In the RSA algorithm, both e and d can be used for encryption because multiplication of ed is commutative.

An example can help describe the various components in the RSA algorithm. We will show how to encrypt plaintext (88), and how to decrypt ciphertext (969).

1. The following steps are done to determine public and private keys.
   - Select two prime numbers, $p = 47$ and $q = 71$.
   - Calculate $n = pq = 47 \times 71 = 3,337$ and $\phi(n) = (p - 1)(q - 1) = 46 \times 70 = 3,220$.
   - Select $e = 79$ randomly from eligible numbers. Note that e should be relative prime to $\phi(n) = 3,220$, and smaller than $\phi(n)$.
   - Calculate d such that $ed = 1 \bmod \phi(n)$. The extended Euclid's algorithm [2] can be used to compute $d = 1,019$.
2. To encrypt the plaintext message $m = 88$ with $e = 79$, we need to compute:
   $c = m^e \bmod n = 88^{79} \bmod 3,337 = 88^{64+8+4+2+1} \bmod 3,337$

We have

$88^2 \bmod 3{,}337 = 1{,}070$

$88^4 \bmod 3{,}337 = (1{,}070 \times 1{,}070) \bmod 3{,}337 = 309$

$88^8 \bmod 3{,}337 = (309 \times 309) \bmod 3{,}337 = 2{,}045$

$88^{16} \bmod 3{,}337 = (2{,}045 \times 2{,}045) \bmod 3{,}337 = 764$

$88^{32} \bmod 3{,}337 = (764 \times 764) \bmod 3{,}337 = 3{,}058$

$88^{64} \bmod 3{,}337 = (3{,}058 \times 3{,}058) \bmod 3{,}337 = 1{,}090$

So

$c = 88^{79} \bmod 3{,}337 = (1{,}090 \times 2{,}045 \times 309 \times 1{,}070 \times 88) \bmod 3{,}337 = 969$

3. To decrypt the ciphertext message $c = 969$ with $d = 1{,}019$, we calculate

$m = c^d \bmod n = 969^{1{,}019} \bmod 3{,}337$

$= (969^{512} \times 969^{256} \times 969^{128} \times 969^{64} \times 969^{32} \times 969^{16} \times 969^8 \times 969^2 \times 969) \bmod 3{,}337$

$= (909 \times 380 \times 2{,}916 \times 54 \times 1{,}051 \times 948 \times 1{,}283 \times 1{,}264 \times 969) \bmod 3{,}337$

$= 88$

Up to now, nobody has been successful in breaking RSA. The assumption behind RSA's security is it is difficult to factoring a large number. If you can factor n quickly, then you can easily calculate its totient function, $\phi(n)$. Given $\phi(n)$ and the value of e obtained from the public key, it is straightforward to figure out the private key, d. The RSA algorithm then would be broken. Fortunately, even the best-known factoring methods are very slow. To factor a 512-bit number would take 30,000 MIPS-years [12].

RSA is much slower to compute than popular secret key algorithms. This is because its underlying modulo exponentiation depends on multiplication and division, processes which are much slower than the simpler bit operations (addition, exclusive OR, substitution, and shifting) of secret key algorithms. The hardware used by RSA is 1,000 times slower than that used by DES [5]. Therefore, RSA is not usually used to encrypt long messages. It is often used to encrypt a secret key, which is then used to encrypt the message.

11.2.2   Confidentiality

Public key cryptography can provide the same level of data confidentiality as secret key cryptography. Communication parties take the same steps as in Figure 11.6, in order to protect data from eavesdropping when transmitting over insecure wireless channels or to provide secure storage on insecure servers. The sender and receiver would not directly encrypt the data using public key cryptography. They would first randomly generate and distribute a secret key, using the steps shown in Figure 11.6, and then use the secret key to encrypt the data.

In Figure 11.6, when Alice needs to securely send a message to Bob, she needs to know Bob's public key. Similarly, Bob also needs to know Alice's public key when he sends a message to Alice. Knowledge of public keys is one of the biggest challenges in public key cryptography. In Figures 11.6 and 11.7, both Alice and Bob would fetch each other's public key from a central repository. It is important to ensure that the information stored in the central repository is not compromised. A bad guy, Fred, might have substituted his public key for Bob's public key in the repository. Alice would have used Fred's public key to encrypt the message, and then Fred could read the message that was intended for Bob.

A popular solution is to have a trusted certificate authority (CA) distribute public keys. The CA generates a certificate specifying the user name (Bob) and his corresponding public key, and then digitally signs the certificate using the CA's private key. Alice then retrieves Bob's public key by decrypting the certificate using CA's public key. All communication parties need to store the CA's public keys in advance, so that they can decrypt the CA's certificate. In fact, Web servers and browsers rely on public key cryptography to generate secret session keys for secure Web transactions. A Web browser's settings should contain a list of CAs.

CAs are also central trusted nodes whose compromise can destroy the security of the whole system. CAs are often maintained by highly trained security professionals. It seems to be a safer approach than spreading the functionality of CAs to individual enterprises, which may not have resources to ensure the security of the their individual public key repositories.

11.2.3   Authentication

In Figure 11.8, we show how Alice and Bob prove their identities using public key cryptography. As with its secret key counterpart, the protocol uses challenges and responses for authentication. Alice's public and private key pair is (e_a, d_a), and Bob's key pair is (e_b, d_b). For Alice to authenticate Bob, she generates a random number, r_1, encrypts r_1 using Bob's public key, e_b, and then sends this encrypted number as her challenge. Bob then sends back r_1 in plaintext. Since only Bob has the private key to decrypt r_1, Alice is assured that she is talking to Bob. Similarly, for Bob to authenticate Alice, he sends an encrypted challenge $(r_1)_{ea}$ using Alice's public key, e_a. Once he receives Alice's response, r_2, and if it is identical to the challenge he sent, then he is convinced that he is talking to Alice.

Public key cryptography is more convenient than secret key technology for authentication with multiple communication parties. When Alice needs to authenti-

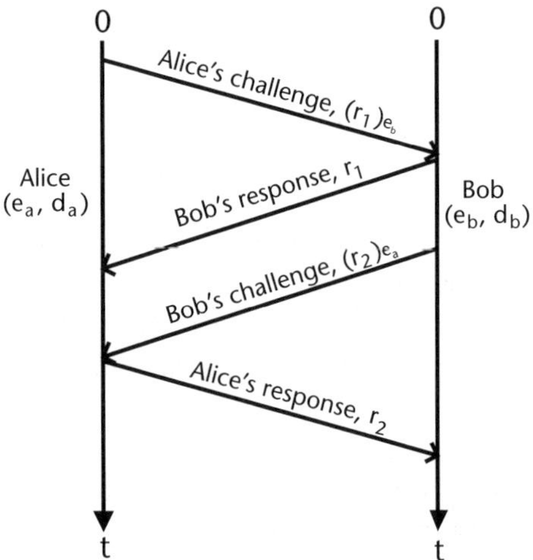

Figure 11.8   A basic challenge-response authentication protocol using public key.

cate Bob, Fred, Trudy, and so forth, using secret keys, she needs to have distinct secret keys with each of them. Alice can use the same secret key for authentication with Bob and Fred, but then Fred can claim to be Bob and vice versa, once they discover that they share the same secret key with Alice. Alice also needs to securely store all her secret keys, which is also a challenge. Alice does not need to generate or store so much secret information with public key cryptography. She only needs to fetch the other party's public key from CAs, and remember a single secret key (i.e., her private key).

11.2.4   Data Integrity

Public key cryptography can also protect data from being modified in transit by encrypting the MAC, which is a fixed-length entity generated from the original message, m. Alice accomplishes this by sending m, followed by the MAC, encrypted using her own private key. If m is modified in any way to m^o, then the decrypted MAC by Bob that is computed based on m would not match with the MAC that Bob computes from the modified message, m^o. This scheme provides a data integrity check, and proves that it Alice sent the message, since the MAC can only be generated by someone with the knowledge of Alice's private key.

This second functionality offers nonrepudiation, which is not possible with secret key–based MAC. Nonrepudiation prevents either the sender or receiver from denying that he has sent the message. Let us compare this through an example. Alice sends Bob a message followed by an encrypted MAC, which can be either a secret key–based MAC or a public key–based MAC. Alice may later want to modify the message to her advantage (e.g., to reduce the amount of money that she owes to Bob). She then sends this modified message with a modified MAC to Bob, and denies that she has sent the previous message. If the MAC is encrypted using secret key, then Bob knows that Alice has purposely made changes to the message. This is because only Alice knows the secret key, so only she can generate a new MAC that matches the modified message. However, Bob cannot prove this to other people. On the other hand, if the MAC is encrypted using public key cryptography, then Bob can prove to others that both messages and their MACs were generated and sent by Alice, because the MACs can be decrypted by Alice's public key.

To summarize, encrypting the MAC using public key cryptography can serve two important purposes. It identifies the party who generated the message, and it also proves that the message has not been altered, if the received MAC matches the computed MAC from the message.

11.3   Wireless Security Attacks

We will now describe several attack techniques targeted at wireless networks. We group them based on the dichotomy we have used in this chapter: the attacks breaking confidentiality, authentication, and data integrity. We will also discuss denial-of-service (DoS) attacks that deny the availability of network resources and services. We try to discuss these attacks in a general context, avoiding protocol-specific and platform-specific details.

11.3.1 Attacks Against Confidentiality

Confidentiality (or data privacy) is the requirement that information is not made available to unauthorized users. Attackers usually exploit several characteristics of wireless devices and wireless data transmission to break the requirement of confidentiality.

First, wireless devices are lightweight and mobile, so they can be easily stolen. The stolen device may continue to receive information to which the user does not have access. The stolen device may also have privileges to communicate with other devices in the wireless network, compromising the confidentiality of a larger network.

Second, wireless packets are transmitted in the broadcast mode when the access point is connected to a hub instead of a switch. Hubs normally broadcast all incoming data to every connected device; therefore, a laptop computer or a PocketPC device with its network interface card in promiscuous mode can monitor network traffic and gain access to unauthorized information. This problem may be somewhat alleviated if the wireless access point is connected to a switch, because the switch can be configured to prevent certain attached devices from accessing the network traffic. However, some switches return to the broadcast mode when the incoming traffic is too much to handle, so malicious attackers can still gain access to network traffic by overloading switches with heavy traffic.

Third, radio waves can travel out of the building containing the wireless devices. An interesting illustration of this fact is the 18-month "war-driving" experiment carried out by Shipley [13]. He drove around the San Francisco Bay area with a laptop and a wireless network interface card, looking for wireless networks that were accessible on the street. He found more than 9,000 visible networks, 85% of which were not configured with any security mechanism. This lack of security in existing wireless installations is confirmed by another independent study conducted in four U.S. cities: approximately 62% of the wireless networks found by driving on the streets of these four cities did not have any security feature [14]. If an attacker wants to increase the range of his eavesdropping activity, he can easily double the original range of his spiral antenna by using common items, such as PVC plumbing pipe and copper wire.

When the attacker gains access to the wireless medium, he can passively or actively eavesdrop on wireless traffic. In passive eavesdropping, he merely monitors network traffic, and may obtain valuable information about the network. First, he can determine whether there is activity in the network. If there is an important event, he can identify the increase in the amount of network traffic. He may now want to learn more characteristics about the network traffic, such as the protocol, the source, destination, size, number, and time of packet transmissions. A simple example of protocol analysis is the identification of the TCP protocol by its three-way handshake upon initialization of the connection. In TCP, the three-way handshake involves three packets. The sender first transmits an SYN packet to the receiver to indicate its intention of establishing a connection. This packet carries the sender's initial sequence number and other connection-specific parameters. The receiver then responds with an acknowledgment of the sender's SYN packet, called SYN/ACK, which also carries its own initial sequence number and other parameters. Finally, the sender responds with the acknowledgment packet (ACK) and concludes the three-way handshake. Each packet involved in the three-way handshake (SYN, SYN/ACK, and ACK) has fixed packet size, so this handshake process can be easily detected by traffic monitoring. Once the attacker is convinced that the traffic is TCP/IP traffic, he can exploit his knowledge of TCP/IP header information and launch more sophisticated attacks.

Passive eavesdropping may not help the attacker in obtaining useful information when the traffic is encrypted in other configurations. He may then actively eavesdrop on the traffic by injecting messages into the wireless network in order to obtain plaintext messages. Figure 11.9 illustrates an example of active eavesdropping by IP spoofing. The attacker intercepts an encrypted message, modifies the destination IP address of the packet to be the host under his control, and then resends the modified packet to the access point. Since the wireless access point does not perform an integrity check on the packet, it will decrypt the packet and forward the decrypted packet to the host that is controlled by the attacker. The attacker is able to collect packets in unencrypted form by this spoofing of the host.

Attackers usually do not stop with these confidentiality attacks. After collecting sufficient information through these attacks, they will analyze the traffic, find vulnerabilities, and launch more malicious attacks. They may break data integrity by injecting, dropping, or modifying data packets, and break resource availability by denial-of-service attacks.

11.3.2 Attacks Against Authentication

Authentication is the requirement that only authorized users are allowed to gain access to the network. This requirement is often defeated by techniques like the man-in-the-middle attack, rogue access point attack, and spoofed MAC address attack.

Man-in-the-Middle Attack

This is a real-time attack between two communicating parties, A and B. In Figure 11.10, A and B first employ certain authentication protocols to verify the identity of

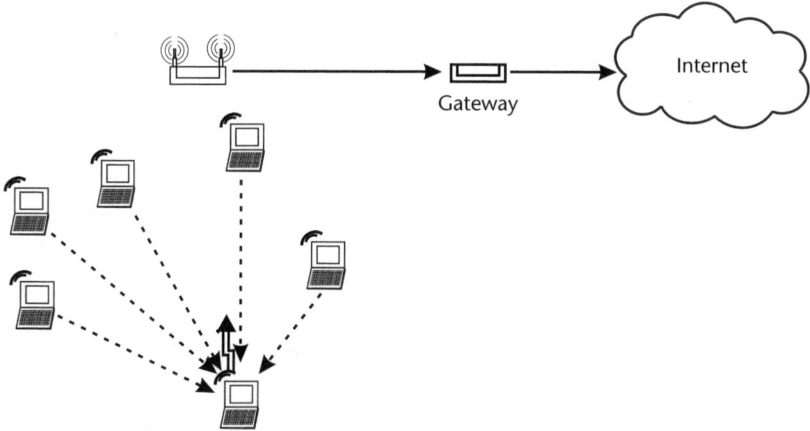

Figure 11.9 Active eavesdropping by IP spoofing.

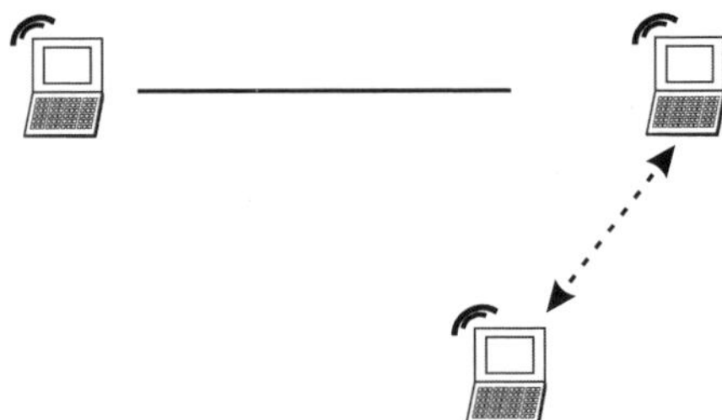

Figure 11.10   Man-in-the-middle attack.

each other. After the authenticated session is set up, an attacker, C, employs some techniques to intrude into the conversation. A (or B) now sends packets to C, which in turn forwards the packets to B (or A). However, both A and B are not aware of the fact that they are communicating with an unauthorized intruder, C.

There are many ways to implement this attack. One simple technique is address resolution protocol (ARP) spoofing [15–17]. ARP maps the network-layer IP address to the link-layer medium access control (MAC) address. Running ARP is a necessary step when A needs to send an IP packet to B. A first invokes ARP to find the MAC address of B, and then encapsulates its IP packet in a link-layer frame with B's MAC address as its destination link-layer address. The same process applies when B sends an IP packet to A. The sender usually broadcasts an "ARP request" packet to find the mapping, making the following request: If your IP address is x.y.z.w, please send your MAC. Each computer examines this request, checks to see if its own IP address is x.y.z.w, and then responds with its own MAC address in an "ARP reply" packet if its IP matches. Now suppose A sends out an ARP request packet for B's MAC address, and C responds with its own MAC address. Then, A keeps such a mapping in its own cache (i.e., B's IP address, C's MAC address). What happens if both B and C answer with their MAC address to A's ARP query? A is certainly confused as to which mapping is the correct one. In order for C to succeed in his efforts, he needs to temporarily bring down B by some denial-of-service attacks. Similarly, C can change B's ARP cache with the following mapping: A's IP address, C's MAC address. After this step, both A and B send their packets to C, not knowing that C is not the original authenticated party.

Rogue Access Points

In wireless LANs, access points (APs) are identified by the broadcasts of beacon frames. Any station that claims to be an access point and broadcasts a valid service set identifier (SSID) will appear to be a legitimate access point in an authorized network. Attackers can easily insert a rogue access point into a hidden area within the network. Most wireless LANs are set up so that access points do not need to authenticate themselves to wireless clients. As long as the rogue access point transmits a stronger signal than existing access points, wireless clients are more likely to route their traffic through the illegitimate access point. The rogue access point gains access to the network without authorization. Rogue access points often are deployed by malicious attackers and by users who take advantage of easy installation of wireless networks without notifying security administrators. Unfortunately, no good solutions exist to prevent internal users from deploying access points for their own convenience. Administrators need to educate end users about security issues, and may have to use tools like NetStumbler [18] to look for unauthorized access points.

Spoofed MAC Address

This attack exploits weak authentication of wireless clients in some wireless LANs, in which access to the network is granted or denied based on an access control list (ACL) of MAC addresses. The MAC ACL is not a strong defense mechanism by itself. Attackers can capture MAC addresses, which are transmitted in plaintext, and then can change the MAC address of their own computers to be the authenticated computers. The attackers gain unauthorized access to the network by changing the actual MAC addresses of their computers, rather than through a man-in-the-middle attack. In the man-in-the-middle attack, they change the mapping of IP addresses and MAC addresses in other computers' caches. The computers with legitimate MAC addresses also may be stolen or hacked by malicious users, who can easily gain unauthorized access to network resources by using legitimate computers.

11.3.3 Attacks Against Data Integrity

Data integrity requires that data not be modified during its transmission in a wireless medium. This requirement may be attacked by techniques such as session hijacking and message replay.

Session Hijacking

A session hijacking attack is against the integrity of an ongoing wireless session [19, 20]. Figure 11.11 illustrates the steps that the attacker performs to take away an authorized session from its legitimate owners. First, the attacker needs to successfully eavesdrop on the communications between A and B to collect sufficient information for hijacking. The information he needs may include authentication tokens, encryption keys, type of communication protocols, and so forth. This is the crucial step in launching session hijacking, since the following steps cannot proceed without sufficient knowledge about the ongoing session. Second, the attacker masquerades as A to B by crafting consistent high-level packets, in order to maintain the

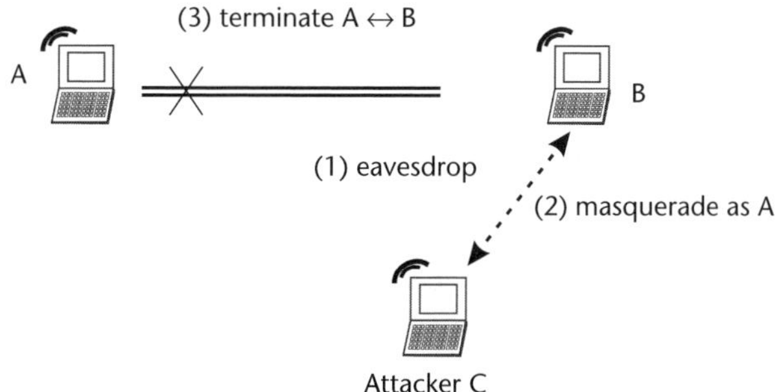

Figure 11.11   Session hijacking.

session without being detected by B. Third, the attacker removes A from the session by sending a sequence of spoofed session-terminating packets to A. After this, A thinks that it is no longer in session with B, but A is unaware that the session has been hijacked. Once the attacker owns this session, he can inject whatever packets he has crafted into the conversation, violating the data integrity of the communication.

Message Replay

Figure 11.12 displays message replay. In this attack, the intruder first eavesdrops ongoing communication sessions to collect information that might be useful in the future, such as authentication tokens, pairs of plaintext and ciphertext from encrypted sessions, pairs of challenges and responses from authentication protocols, and so forth. He may later replay the responses he has collected to pretend to be an authorized user, when the attacker sees identical authentication challenges. After successful authentication, he may insert the plaintext and ciphertext pairs from previous sessions into an ongoing session, violating the data integrity of this session. This attack is not necessarily a real-time attack, since the attacker may replay the old messages in a future session instead of the ongoing session [21–24].

11.3.4 Attacks Against Availability

Denial-of-service attacks make network resources and services unavailable to users. Signal jamming, traffic flooding, and battery draining may launch such attacks in a wireless network.

Signal Jamming

Wireless devices are susceptible to signal jamming attacks, since most wireless devices share the same 2.4-GHz ISM band with microwave ovens, cordless phones,

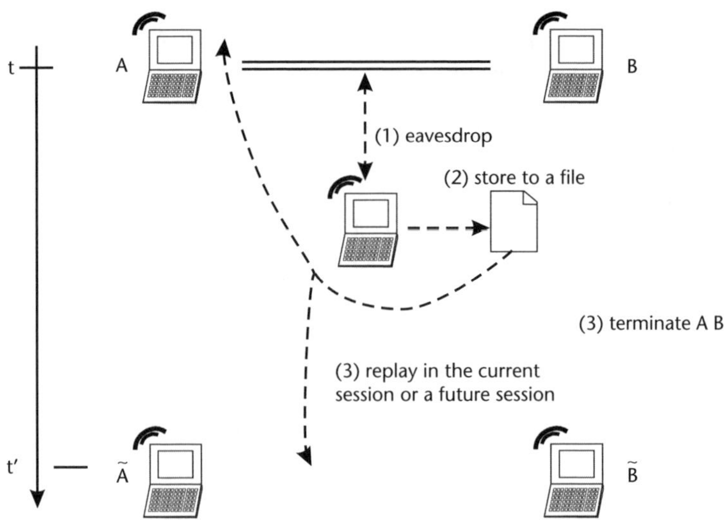

Figure 11.12 Message replay.

and other household electrical appliances. Malicious users can interfere with wireless transmissions by transmitting strong signals in the same frequency band. Normal users may also inadvertently cause disruption in wireless transmission by using the above appliances in close proximity of wireless devices.

Traffic Flooding

Wireless networks are usually connected to wired networks through gateways. Wireless networks can be easily flooded by traffic coming from the wired networks, since wireless networks have limited transmission bandwidth compared to wired networks. For example, the attacker could use a ping flooding attack from a fast Ethernet segment to overrun the capacity of a wireless access point. Traffic flooding may also be caused by legitimate users when they run applications in the wireless network that require much bandwidth, such as streaming video and downloading large files.

Battery Draining

Most portable wireless devices operate on batteries. A battery draining attack attempts to disable a device by exhausting its battery power. A malicious user may repeatedly request a device to transmit packets, which can quickly drain the victim's battery, since data transmission is the most power-consuming operation. The attacker may then pretend to be the victim device, and launch other malicious attacks.

11.4   Summary

Wireless technologies promise freedom and mobility, but also present serious security challenges. In this chapter, we give an introduction to wireless security by discussing cryptographic algorithms and common threats targeted at wireless networks.

A security algorithm or protocol often provides a subset of the following three basic security services: confidentiality service, which ensures that only the communicating parties understand the contents of transmitted messages; authentication service, which ensures that only authorized parties gain access to the network; and message integrity, which ensures that ensures the messages are not altered in transit.

Two types of cryptographic algorithms, symmetric key and public key cryptography, can be used to provide these services. In symmetric key cryptography, both communicating parties share a secret key, which is used in both encryption and decryption processes. Symmetric ciphers are basically defined by various rounds of substitution and permutation operations. We examined DES, IDEA, and AES as specific case studies, and studied how they are used to encrypt a sequence of data blocks. While we mainly focused on block ciphers, we also touched upon a very popular stream cipher, RC4, which is widely used in security protocols.

The construction of public key cryptography is based on results from number theory. Each communicating party has two keys: a private key that is kept confidential to the party, and a public key that is made available to everyone. While mathematically elegant and computationally secure, public key cryptography is rarely used to encrypt long messages because of its high computational load. It is commonly used to establish a symmetric session key for a secure transaction.

While security algorithms and protocols strive to provide confidentiality, authentication, and integrity, attackers devise schemes to compromise these services in wireless networks. Eavesdropping on a wireless medium often accomplishes attacks against confidentiality. Techniques, such as man-in-the-middle, insertion of rogue access point, and changes to MAC addresses, can launch attacks against authentication. Session hijacking and message replay cause attacks against data integrity. Attackers often take advantage of the characteristics of radio propagation and the mobility of wireless devices to launch denial-of-service attacks. It is important to evaluate defenses against these common attacks when designing a security solution for wireless networks.

References

[1]   Lewand, R., *Cryptological Mathematics*, Washington, D.C.: Mathematical Association of America, 2000.

[2]   Stallings, W., *Cryptography and Network Security: Principles and Practices*, Upper Saddle River, NJ: Prentice Hall, 2003.

[3]   Electronic Frontier Foundation, *Cracking DES: Secrets of Encryption Research, Wiretap Policies & Chip Design*, J. Gilmore, (ed.), Sebastopol, CA: O'Reilly, 1998.

[4]   Kaufman, C., R. Perlman, and M. Speciner, *Network Security: Private Communication in a Public World*, Upper Saddle River, NJ: Prentice Hall, 2002.

[5]   Rhee, M. Y., *Internet Security: Cryptographic Principles, Algorithms and Protocols*, New York: John Wiley & Sons, 2003.

[6]   National Bureau of Standards, U. S. Department of Commerce, *DES Modes of Operations*, FIPS PUB 81, 1981.

[7]   Diffie, W., and M. Hellman, "Multiuser Cryptographic Techniques," *IEEE Trans. on Information Theory*, November 1976.

[8]   Rivest, R., A. Shamir, and L. Adleman, "A Method for Obtaining Digital Signatures and Public Key Cryptosystems," *Communications of the ACM*, February 1978.

[9]   ElGamal, T., "A Public-Key Cryptosystem and a Signature Scheme Based on Discrete Logarithms," *IEEE Trans. on Information Theory*, July 1985.

[10] Koblitz, N., "Elliptic Curves Cryptosystems," *Mathematics of Computing*, No. 177, 1987, pp. 203–209.

[11] National Institute of Standards and Technology, U.S. Department of Commerce, *Digital Signature Standard*, FIPS PUB 186, 1994.

[12] Robshaw, M., *Security Estimates for 512-Bit RSA*, RSA Data Security, Inc., June 1995.

[13] Shipley, P., "Open WLANs: The Early Results of War Driving," http://www.dis.org/filez/ openlans.pdf.

[14] Ellison, C., *Exploiting and Protecting 802.11b Wireless Networks*, ExtremeTech, Inc., September 2001.

[15] Whalen, S., "An Introduction to ARP Spoofing," http://packetstormsecurity.nl/papers/ protocols/intro_to_arp_spoofing.pdf.

[16] Fleck, B., and J. Dimov, "Wireless Access Points and ARP Poisoning: Wireless Vulnerabilities that Expose the Wired Network," White Paper, http://www.cigital.com.

[17] Carey, A., "Wireless Security Vulnerabilities Continue to Surface: Cigital Identifies the Latest," White Paper, October 2001, http://www.cigital.com.

[18] NetStumbler FAQ, http://www.netstumbler.com.

[19] Skoudis, E., *Counter Hack: A Step-by-Step Guide to Computer Attacks and Effective Defenses*, Upper Saddle River, NJ: Prentice Hall, 2002.

[20] Mishra, A., and A. William, *An Initial Security Analysis of the IEEE 802.1X Standard*, Technical Report CS-TR-4328, Department of Computer Science, University of Maryland, February 2002.

[21] Borisov, N., I. Goldberg, and D. Wagner, "Intercepting Mobile Communications: the Insecurity of 802.11," *Proc. 7th Int. Conference on Mobile Computing and Networking*, Rome, Italy, July 16–21, 2001.

[22] Krishnamurthy, P., J. Kabara, and T. A. Amornkul, "Security in Wireless Residential Networks," *IEEE Trans. on Consumer Electronics*, Vol. 48, No. 1, February 2002, pp. 157–166.

[23] Moioli, F., "Security in Public Access Wireless LAN Networks," Master Thesis, Department of Teleinformatics, Royal Institute of Technology, Stockholm, Sweden, June 2000.

[24] Karygiannis, T., and L. Owens, "Wireless Network Security: 802.11, Bluetooth and Handheld Devices," National Institute of Standards and Technology Special Publication 800-48, November 2002.

Wireless Security Solutions

Wireless networks can be grouped into three categories depending on their coverage area. Cellular networks transmit voice and data in the long range; wireless local area networks (LANs) work in the medium range; while wireless personal area networks (PANs) connect small personal devices, such as laptops, PDAs, and mobile phones, in the short range.

In this chapter, we study security threats and solutions for these three types of wireless networks. For each network, we discuss its current security schemes, security threats and vulnerabilities, and common countermeasures to defend against attacks. For cellular networks and wireless LANs, we also discuss newer generation security standards that have been adopted to correct existing problems.

12.1   Security Threats and Solutions for Wireless LANs

For wireless LANs, we first inspect the initial IEEE 802.11 security standard, known collectively as wired equivalent privacy (WEP). Then, we give a classification of security threats targeted at wireless LANs. We then describe the countermeasures that a regular user can take to ensure secure access, and follow this with a discussion of the solutions and new protocols being developed by the standards body to counter current attacks. Future deployment of these more advanced protocols can enhance wireless LAN security.

12.1.1   Overview of IEEE 802.11 WEP Standard

The WEP standard provides a level of security similar to that of wired networks. It provides authentication, data encryption, and integrity protection using secret key cryptography. It is assumed that the wireless access point and the host can securely agree on a secret key using an out-of-band approach.

Authentication

WEP only requires the wireless host to authenticate to the access point but not vice versa. The authentication procedure follows a challenge-response protocol, and involves four steps.

1. The wireless host sends a message requesting authentication.
2. The access point sends a 128-bit random challenge.

3. The wireless host responds with an encrypted response, derived from encrypting the random challenge using the key it shares with the access point.

4. The access point decrypts the response and compares it with the random challenge. If both values match, then the wireless host is authenticated to use the network.

Data Encryption

WEP uses RC4 to generate key streams. RC4 accepts a 40- or 104-bit shared key, with the length depending on the configuration, and a 24-bit initialization vector (IV) to produce a stream of key values. Encryption is performed by XORing the user data with the key stream.

Proper use of the RC4 cipher requires that its input values do not repeat. However, for a given shared key, there are only 2^{24} unique input values to RC4. Furthermore, IV is transmitted in clear text, so an attacker will easily find out when a duplicate IV is sent. We will later see that attackers exploit these vulnerabilities to break the shared key.

Data Integrity

The integrity of a wireless packet is protected by a 4-byte cyclic redundancy check (CRC) value in the packet. The payload and the CRC value are then encrypted using the above procedure. The CRC is effective in detecting transmission errors such as bit flips, but it is not a cryptographic procedure. It is easy for an attacker to modify the packet content and compute a CRC over the modified packet.

12.1.2   Security Threats for Wireless LANs

This section addresses known security threats to IEEE 802.11 wireless LANs. In general, the threats either break confidentiality, authentication, message integrity, network availability, or some combinations of them. Table 12.1 gives a summary of common threats based on these four security concerns [1, 2].

Eavesdropping

The attacker monitors the wireless traffic using packet-sniffing tools and techniques that are publicly available. The attacker only needs to put his wireless interface card

Table 12.1   Common Attacks Against Wireless LANs

Threats	Confidentiality	Authentication	Integrity	Availability
Eavesdropping	✓	—	—	—
Cryptoanalysis	✓	✓	✓	—
Insertion attacks	✓	✓	—	—
Jamming	—	—	—	✓
Misconfiguration	✓	✓	✓	✓

in promiscuous mode, and install sniffing tools to collect wireless traffic. Popular sniffing tools include Ethereal [3], Kismet [4], Windump [5], tcpdump [6], sniffit [7], and dsniff [8]. This attack can be either passive or active. In passive eavesdropping, the attacker normally gains information about the session, like its source, destination, application type, and number of bytes in the session. He can also save the entire packets for later analysis. In active eavesdropping, the attacker monitors the traffic, and modifies and injects packets into the wireless session to help decode the encrypted payload. One example of active eavesdropping involves changing the destination IP address of a packet to an IP address in the attacker's control. Since the access point decrypts the packet before forwarding it to the destination, the attacker receives the packet in clear text.

Cryptoanalysis

The most effective way to defend against eavesdropping is to send wireless packets in encrypted form. WEP is the de facto data encryption standard for wireless LANs. However, WEP fails to provide strong encryption because it sends a short 24-bit IV in clear text. The IV starts from 0, and increments by 1 for each message sent. An access point running at 11 Mbps can exhaust the entire space of 2^{24} IVs within an hour. Several tools have been developed to exploit this weakness in WEP, including the following.

- *AirSnort:* It can guess a WEP encryption password in less than 1 second, given that a sufficient number of encrypted packets (e.g., 5 to 10 million) have been gathered through sniffing. The tool runs on both Windows and Linux [9].
- *WEPCrack:* It is the first open source tool that implements the attack exploiting the weakness in the RC4 stream cipher, as reported by Fluhrer, Mantin, and Shamir [10]. The tool is written in Perl, so it runs on any platform that supports Perl [11].
- *AiroPeek:* It is an advanced sniffer that captures, analyzes, and finds vulnerabilities to attack transmissions on wireless LANs. It is not an open source tool and only runs on Windows [12].
- *NetStumbler:* It can be used to detect the presence of wireless LANs, their service set identifiers (SSID), the coverage strength, and whether or not WEP is enabled on the wireless LANs. It is only available on Windows [13].

Insertion Attacks

Insertion attacks gain unauthorized access to the network through deploying wireless devices and access points. An attacker can connect to the wireless network using a client, typically a laptop or a PDA. If the access point is configured to use either the default password set by the vendor, or an easily determined password, or even no password, then the client can gain access to the network. The access is not likely to succeed if the access points are configured to use strong passwords. Rogue access points deployed by internal employees in an organization impose a more serious threat. The lack of security background and training of the normal users leaves the rogue access points extremely vulnerable to unauthorized clients.

Jamming

Devices operating in the same frequency band (2.4 GHz), such as microwaves, cordless phones, and baby monitors, may degrade the signals from wireless LAN devices. An attacker can easily jam the whole signal band using noises, resulting in denial-of-service (DoS) attacks.

Misconfiguration

Network administrators who do not understand the security consequences of different configurations may leave the door open for attackers when setting up access points. It is important that the administrators change the default configuration parameters set by vendors, since the default values are well known to the attackers. Here we provide a list of important parameters.

- *SSID:* This identifies access points for the wireless network. The wireless client needs to know the SSID in order to connect to the network. Although tools like NetStumbler are able to automatically discover SSIDs, it is more difficult for the attackers if the SSIDs are set to values that are different from the default values..
- *WEP:* The wireless network can be set in three ways: (1) without WEP, (2) with WEP with a 64-bit key, and (3) with WEP with a 128-bit key. Enabling WEP with 128-bit key can deter attackers, although WEP cannot provide total protection.
- *Client side storage:* The wireless clients normally store SSIDs and their WEP keys in their local storage. If an attacker gains access to a wireless client (e.g., through a laptop left unlocked by a user), he can steal the secrets and later use them to connect to the wireless network.
- *Configuration interfaces:* Wireless access points may be configured using different types of interfaces, like the Web, telnet, serial, or snmp. If the administrators mistakenly make the configuration interfaces publicly available, this an attacker may change the network settings to his advantage.

12.1.3   Security Countermeasures for Wireless LANs

General countermeasures to secure wireless LANs include the correct configuration of the networks and installation of proper tools. Although individual countermeasures only provide protection for some specific attacks, combining these small steps may significantly improve the security of wireless LANs. Individual countermeasures are listed here.

- Change the default administrative password for configuration interface. If the default administrative password is not changed, then an attacker may easily reconfigure the wireless LAN to allow his access into the network.
- Change SSIDs. It is important that the administrators change the default SSID that comes with an access point. SSIDs also should be changed from time to time.

- Enable WEP. Making the most out of WEP and utilizing its 128-bit encryption option may not stop a knowledgeable and determined attacker, but it will deter some novices and slow down the attack process.

- Disable SSID broadcast. Most access points periodically broadcast their SSIDs to allow new clients to associate with the network. Disabling this option makes it harder for attackers to collect SSIDs.

- Enable medium access control (MAC) address filtering. Most access points support this option by only allowing clients with authorized MAC addresses to associate with the network. This will make it harder to launch client insertion attacks.

- Use static IP addresses for new clients. Most wireless LANs use Dynamic Host Configuration Protocol (DHCP) to allocate a new IP address for an incoming client. While DHCP is easy to set up, it also makes it easy for an attacker to gain full access to the network. It is safer to use static IPs for new legitimate clients.

The following general security tools can help tighten the security of your wireless LANs.

- Set up a firewall. Like MAC address filtering, a properly configured firewall will only allow accesses from authorized clients. The firewall sets up the access rules based on IP addresses and port numbers, rather than on MAC addresses. The attackers have to be more sophisticated to apply address spoofing in order to penetrate into a firewall-protected network.

- Use virtual private network (VPN). The VPN provides a protected tunnel between two end systems, encrypting all the traffic sent between them. Setting up a VPN may be expensive, but it may be worthwhile to add an extra layer of protection to the wireless LAN.

- Install an intrusion detection system (IDS). The IDS can detect an intruder's attempts to break into the system. There are two types of intrusion detection schemes: anomaly-based and rule-based [14]. An Anomaly-based IDS watches for activities that are different from a user's or a system's normal behavior. A rule-based (or signature-based) IDS model watches for activity that is similar to known patterns of attacks. Most of the products available in the market are rule-based. Network administrators must monitor and evaluate IDS logs on a regular basis, in order to catch intrusions in a timely manner.

12.1.4   IEEE 802.11i Security Standard

The security holes in WEP have left millions of wireless LANs at risk. The standards body and the industry have worked together to define, develop, and deploy next generation solutions that address the security problems found in WEP. The most advanced solution is the IEEE 802.11i standard [15] for wireless LAN security, which enhances WEP security in encryption, authentication, and key management. The IEEE 802.11i standard is based on WiFi protected access (WPA), a quick fix to WEP problems by the WiFi Alliance, a wireless industry organization.

IEEE 802.11i has included two protocols for data protection: temporal key integrity protocol (TKIP), which is designed as an immediate fix for inherent WEP weaknesses of legacy 802.11 devices; and counter mode with cipher block chaining message authentication code protocol (CCMP), which provides improved data confidentiality for new devices. IEEE 802.11i has also chosen 802.11x to provide an effective framework for client access control [15].

Temporal Key Integrity Protocol

This protocol has fixed many known WEP weaknesses [16]. It uses a unique key for every packet, which is created by combining a shared base key, the MAC address of the transmitting station, and the sequence number of the packet; see Figure 12.1. Attackers can launch cryptoanalysis of WEP by collecting data for an extended period of time. TKIP's adoption of per-packet keys makes the cryptographic attack extremely difficult, since the mass collection of encrypted data with the same key is no longer possible.

Each TKIP packet also carries a 48-bit sequence number (also called IV). The repetition of the 24-bit IV currently allows WEP attackers to recover the shared key. Expanding the IV to 48-bit effectively thwarts such an attack, since it would take approximately 900 years to iterate through the entire IV space. Besides addressing the problem of IV collision in WEP, the 48-bit sequence number also prevents message replay attack with WEP. In TKIP, the wireless stations monitor the sequence numbers of received packets and discard those packets whose sequence numbers fall outside an acceptable range.

In addition to changing the encryption key for every packet, TKIP also generates a different shared base key whenever a wireless client associates to an access point. The base key is hashed from a session secret, a random number (nonce), and the MAC addresses of the access point and the client host. The session secret is unique with 802.1x authentication, and is transmitted securely to the client by an authentication server. This solves the WEP problem of constant reuse of the same key by everyone in the wireless LAN.

TKIP uses a new data integrity algorithm, known as message integrity code (MIC), to provide better protection against message modification and forgery. WEP uses a 4-byte integrity check value (ICV) for each packet, which can be easily cracked. An attacker can store the ICV, change encrypted packets, and update the

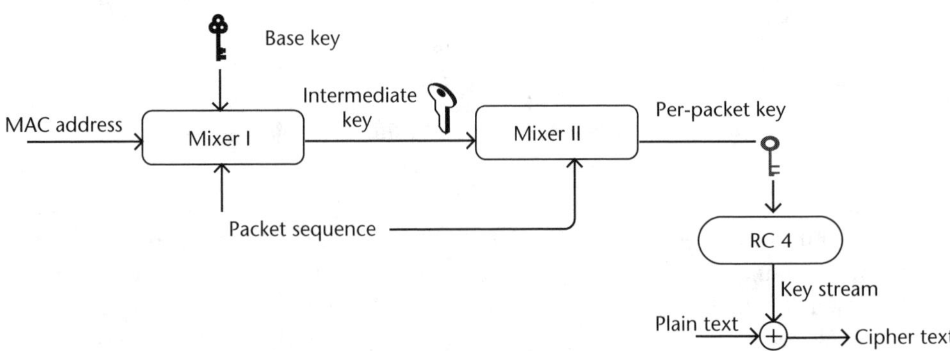

Figure 12.1   Generation of per-packet key in TKIP.

ICV without even knowing the WEP key. In contrast, TKIP uses a 12-byte ICV, in which an algorithm known as Michael creates the additional 8 bytes. Michael includes a protection mechanism that detects invalid ICV values, and automatically responds by resetting the passwords and shutting down the network. Applying a cryptographic hashing of both source and destination MAC addresses creates the MIC value. This thwarts packet-tampering attempts, since it is hard to generate correct MIC values.

Note that TKIP still uses RC4, the same cipher as WEP, which means that existing hardware can be easily upgraded to support TKIP. The WiFi Alliance expects that TKIP's advantage of easy deployment can accelerate its adoption by wireless vendors.

Counter Mode with Cipher Block Chaining Message Authentication Code Protocol

This is the preferred data encryption protocol in IEEE 802.11i. It is fundamentally different from WEP, since it uses the advanced encryption standard (AES) instead of RC4 as its cipher. The CCMP encrypts data in 128-bit blocks using the cipher block chaining mode (CBC), and provides data integrity using message authentication code (MAC). Many people have concerns about its slow AES operations, but vendors have been successful in designing hardware that can efficiently perform AES operations. The availability of efficient AES hardware and its strong cipher performance are two primary reasons for AES to replace RC4.

IEEE 802.1x

This is a port-based authentication protocol initially developed for Ethernet networks. It defines procedures for authenticating a user who connects a device into the network in an open environment. Full access to the network requires authentication by IEEE 802.1x.

The IEEE 802.11i standard has extended this protocol to wireless networks. Authentication happens when a client associates to an access point, as shown in Figure 12.2. The access point forwards the messages between the client and the remote authentication dial-in user service (RADIUS) server for authentication. Once the server has verified the client's information, it will send a message to the access point that approves the client's network access. This response message to the access point also contains a cryptographic master key. The client derives the same master key. After establishing the master key, the client and the access point also calculate a temporary key to be used for message encryption and integrity protection. Note that the RADIUS server and protocol are not part of the 802.11i standard, but they are the de facto standard for back-end authentication.

The protocol allows initial client association to the access point, in order for the client to send his authentication information. The access point will limit the client's access to the network before the authentication is complete. The client is only allowed to send packets with authentication information.

The authentication process takes place when the client initially associates to the network and during the client's network session, which verifies that the client is not compromised.

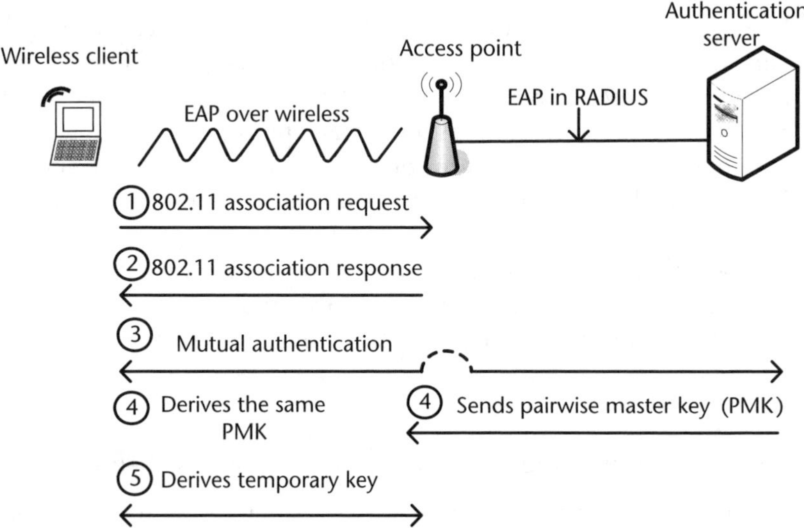

Figure 12.2   IEEE 802.1x authentication framework.

IEEE 802.1x uses extensible authentication protocol (EAP) to exchange authentication messages. EAP provides a flexible framework for vendors to plug in their own authentication modules, rather than using a fixed mechanism. This modular approach may provide better protection for possible protocol flaws.

There are several common EAP methods [17]. EAP-MD5 is a simple password-based mechanism. This is easy to implement, but is not very secure. EAP-TLS adds a transport layer security (TLS) session in EAP. This is much more secure than the simple EAP-MD5, but it requires significant setup efforts. It requires users to install certificates and enterprises to install a complete public key infrastructure (PKI). Protected EAP (PEAP), which is more secure than EAP-TLS, creates a secure authentication mechanism in a TLS tunnel.

IEEE 802.11x is very suitable for enterprises and hotspots, depending on the EAP protocols used. EAP-TLS or PEAK can provide strong client access control in enterprises that deploy a RADIUS server and PKI. EAP-MD5 can be applied to authenticate users in hotspots that are unlikely to deploy PKI. This technology is not very suitable for deployment in a small office home office (SOHO) environment. SOHO users rarely have the expertise to install and maintain a back-end RADIUS server.

Roadmap to IEEE 802.11i

IEEE approved the 802.11i standard in 2004. The WiFi Alliance developed WPA as an immediate solution to fix the WEP problems before the approval of 802.11i. WPA provides data confidentiality using TKIP, and provides authentication using IEEE 802.11x. WPA may operate in two modes. The normal full mode uses 802.11x authentication with slight modifications to work in enterprises. The preshared key (PSK) mode supports basic pass-phrase authentication in a SOHO environment.

WPA has generally corrected all the flaws in WEP, but it has its own weaknesses. The first vulnerability lies in its four-way handshake in its PSK mode. An attacker may eavesdrop the handshake messages and later use an off-line dictionary attack to guess the pass-phrase sent in the messages. Knowledge of the pass-phrase grants full access to the network. A pass-phrase with a small number of characters is more susceptible to this attack. WPA has also introduced a DoS attack. The access point will close all the connections if it receives two packets with invalid MICs within 1 second, in the WPA specification. The access point will wait for 1 minute before continuing operation. Thus, an attacker can carry out a DoS attack by sending at least two packets with wrong MICs every second.

IEEE 802.11i (also known as WPA 2) delivers security that is more robust than the security provided by WPA. It includes a stronger data encryption protocol (CCMP) for data protection, and requires completely new and more expensive hardware. Full deployment of the IEEE 802.11i protocol may come in the future.

12.2   Security Threats and Solutions for Wireless PANs

Wireless personal area networks are currently based primarily on Bluetooth technology. As described in Chapter 6, Bluetooth is an open standard for low-power and short-range wireless communications in an ad hoc network. We will review security mechanisms defined in the Bluetooth standard, common attacks, and countermeasures for the attacks. There is no ongoing effort to define a new security standard for Bluetooth networks, as there is in 802.11 networks.

12.2.1   Overview of Bluetooth Security Mechanisms

A Bluetooth device supports three security modes, but can only operate in one mode at a time. The choice of the security mode is partly based on application needs [18].

1. In the first mode, a Bluetooth device will not initiate any security procedures and will allow any other device to connect. This is essentially a nonsecure mode. Only applications that do not have security requirements use this mode.
2. In the second mode, a Bluetooth device contacts a centralized security manager for security policies at the service level. The security manager defines policies of access control for device users. It is possible to grant access to some devices while denying access to other devices.
3. In the third mode, a Bluetooth device initiates security procedures before establishing a channel with other devices. The procedures are implemented in the link-layer, and support encryption and authentication.

Since Bluetooth is a link-layer technology, we will focus on its secure mode in the link-layer.

Assume there are two Bluetooth devices, A and B, to communicate with each other. Figure 12.3 illustrates the flow of events from the initialization of the devices until the completion of their conversation. The two devices need to constantly generate many keys, including unit key, initialization key, link key, and encryption key. The key generation is dependent on three entities [19, 20]:

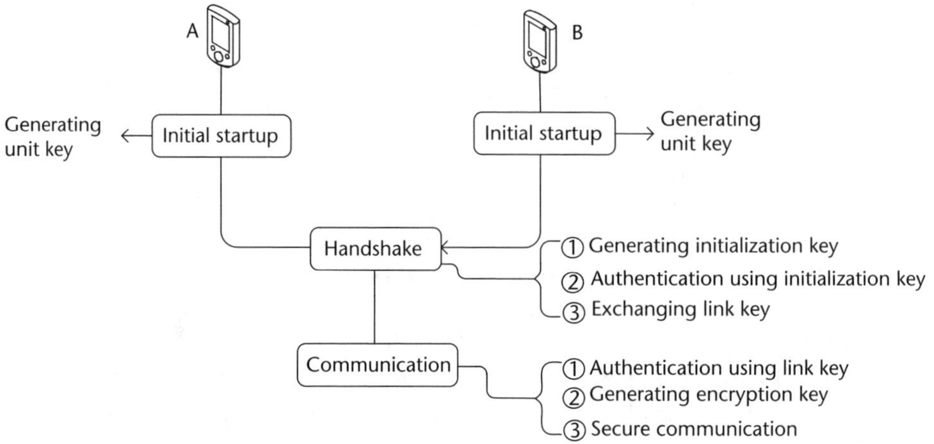

Figure 12.3   Lifetime of secure communications between device A and B.

1. 48-bit Bluetooth device address (BD_ADDR), defined by vendors, and similar to an 802.11 MAC address;
2. Private authentication key (PIN);
3. 128-bit pseudorandom number (RAND) generated by a device.

Unit Key Generation

Unit key generation is needed only once while first bringing up the device. The unit key is based on its own 48-bit device key (BD_ADDR) and a RAND. The unit key is normally stored in the device's nonvolatile memory, and very rarely needs to be regenerated.

Device Association

When device A (i.e., the claimant) tries to contact device B (i.e., the verifier) for the first time, device A needs to prove to device B that it is authorized to gain privileges in the communication that follows. It is considered to be authorized if it shares the same PIN with device B. The PIN could be a simple 4-byte value that the user enters on both device A and B, or a longer 16-byte value that is established by a key exchange protocol, such as Diffie-Hellman.

The first step in device association is to generate an initialization key, K_{init}. Device A sends a random number, IN_RAND, to device B. Then both devices compute K_{init} based on PIN, the length of PIN, and IN_RAND.

The second step is an authentication process based on challenge-response protocol, which has five steps, as shown in Figure 12.4.

1. Device A transmits its device address, BD_ADDR, to device B.
2. Device B transmits a 128-bit random number challenge, AU_RAND, to device A.
3. Device A calculates the authentication response, $SRES_A$, based on the SAFER+ algorithm, using parameters BD_ADDR, AU_RAND, and K_{init}. Device B applies the same calculation to derive $SRES_B$.

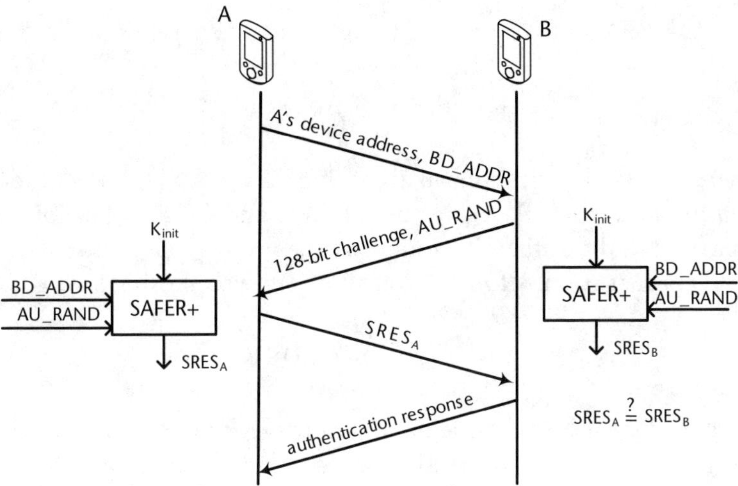

Figure 12.4 Challenge-response–based authentication procedure.

4. Device A transmits $SRES_A$ to device B.
5. Device B compares $SRES_A$ to $SRES_B$. If they are equal, then the authentication is successful. Otherwise, device A has to wait for an interval of time before the next attempt. The interval grows exponentially, to prevent possible DoS attacks by device A through a repeated use of different keys.

The third step in device association is the establishment of a link key, which will replace the initialization key in the later phases. It can be the unit key of either device A or B, if both devices have memory constraints; or a new combination key, K_{AB}, which is derived from the unit keys of A and B. In the first case, one of the devices will send its unit key (XORed with K_{init}) to the other device, which the other device can easily, since it knows K_{init}. In the second case, device A (or B) will generate a random number, LK_RAND_A (or LK_RAND_B), and then device A (or B) will derive a new key, LK_K_A (or LK_K_B), as follows.

$$LK\_K_A = E_{21} (BD\_ADDR_A, LK\_RAND_A)$$

$$LK\_K_B = E_{21} (BD\_ADDR_B, LK\_RAND_B)$$

where E_{21} is the same algorithm that is used to generate the unit key. The two devices will then exchange their random numbers, and as a result, device A can derive LK_K_B, and vice versa. The link key K_{AB} is finally set to $LK\_K_A \oplus LK\_K_B$.

Device Communication

The devices need to authenticate each other before they can communicate, following the same challenge-response protocol as in association phase. The link key, K_{AB}, will substitute the initialization key, K_{init}, in the process.

Devices A and B then need to agree on an encryption key, K_C,

$$K_C = E_3 \, (EN\_RAND_B, K_{AB}, COF)$$

where E_3 is the key generation algorithm, EN_RAND_B is the random number gener-ated by the verifier B, K_{AB} is the link key, and COF is the ciphering offset number generated in the authentication process.

Data sent from A to B (P_{data}) will be encrypted using key, K_C, as follows.

$$C_{data} = P_{data} \oplus E_0 \, (BD\_ADDR_A, clock_A, K_C)$$

In order for B to decode data, it needs to know A's device address, which it has obtained through the authentication in association phase. It also needs to know A's clock.

12.2.2 Security Threats for Bluetooth-Based Wireless PANs

Table 12.2 outlines different attacks against Bluetooth-based wireless PANs. These attacks may compromise confidentiality, authentication, integrity, and availability. We will describe each attack in more detail.

Misconfiguration

Bluetooth devices, similar to LAN devices, are also shipped with a default set of con-figurations that are intended to make initial setup more convenient. Default configu-rations often include settings that result in weak security [21].

The default setting of the first secure mode is especially dangerous. This mode does not require authentication and encryption, so that any device can connect and request information from the device.

Some devices turn on Bluetooth radio by default, disclosing their device addresses (BD_ADDR) and clocks. BD_ADDR is an important quantity in deriving the link key, which is in turn used to derive the encryption key. Revealing device addresses may imply serious threats for data confidentiality. Knowledge of the BD_ADDR and the clock of a piconet's master device enables an attacker to find out the frequency-hopping sequence for communication, thereby making it easy for the attacker to eavesdrop on the communication.

Table 12.2 Common Attacks Against Bluetooth Networks

Threats	Confidentiality	Authentication	Integrity	Availability
Misconfiguration	✓	✓	✓	✓
Device loss	✓	✓	✓	✓
Eavesdropping	✓	—	—	—
Man-in-the-middle	✓	—	✓	—
Signal jamming	—	—	—	✓
Power attacks	—	—	—	✓

Device Loss

Smaller Bluetooth-enabled mobile devices, along with their data, are easier to get lost. This imposes great security risks for the devices that have established a trusted relationship with the lost device. The lost device may be used to eavesdrop on communications of other devices, to establish communication and request confidential information from them, and to obtain the personal identification numbers of these devices.

Eavesdropping

Bluetooth networks rely on a frequency-hopping algorithm to avoid eavesdropping. The initiation of the frequency-hopping sequence derives from the device address and the clock of the master device in the piconet. Devices hop among 79 frequencies at a rate of 1,600 times per second. Frequency-hopping can be circumvented by using a listening device that can listen on all the frequencies. Listening devices are usually available through vendors for troubleshooting and diagnosis purposes. Software sniffers that "decode" frequency hopping are also available.

The ability to eavesdrop on communication is not sufficient for the attacker if the data is sent in encrypted form. It is possible for the attacker to find out the encryption key in some scenarios. If the association procedure is performed in a public area, then the attacker may capture the device address, a random number, and the PIN. These three values are used to calculate the initialization key, and subsequently, the link key and the encryption key. Even if the PIN number is not sent in the clear, the attacker may also perform a brute-force analysis of the PIN number while off-line, as long as he has captured the message sequences of the device association.

Man-in-the-Middle Attack

A device, A, which has obtained the link key for the other pair of communicating devices, B and C, can easily launch a man-in-the-middle attack. A can use the link key to pretend to be B and initiate a new communication with C, and similarly initiate a new communication with B.

An example of a man-in-the-middle attack [18, 22] that exploits the memory limitation of the victim device follows.

1. The attacker A establishes a link with the victim device B. If B has limited memory, then it doesn't have space to generate and store a new link key. B's unit key is then used as the link key for communication between A and B. Therefore, A has obtained B's unit key.

2. Another victim device C may communicate with B at a later time, using B's unit key as the link key.

3. The attacker A now can listen to the traffic between two victim devices, since it knows the link key for the communication. If A uses a spoofed address, then neither B nor C will be aware of A's presence.

4. The attacker A may also impersonate itself as B to C and as C to B, establishing new connections with both of them and manipulating their communication.

Signal Jamming

Bluetooth devices work in 2.4-GHz ISM range, similar to other 802.11 devices. Bluetooth devices are still susceptible to signal jamming, although they have built-in frequency-hopping mechanisms to minimize interference, since the 2.4-GHz band is crowded with many devices.

Power Attacks

Bluetooth-enabled devices are usually mobile devices with power constraints, functioning in various states, such as idle or active. The active state usually consumes the most power. If the attacker can keep the victim's device in the active state, then the attacker can effectively drain the victim's power [23]. One example of such an attack involves repeatedly sending service requests to a victim's device. The validation of the service keeps the victim's device active and consumes a significant amount of power.

The attacker may also make the victim's device execute a valid but power-consuming operation by passing invalid data to it. For example, when the victim's device tries to download a Web page with a gif image, the attacker may modify the image to be an animated GIF image that just repeats itself. The victim's device is kept busy displaying the animated GIF image. This is very hard to detect by device users, since the repeating animated GIF image appears like a regular GIF image.

Another attack involves the installation of a power-consuming virus program on a victim's device. The main purpose of the virus program is to consume power by repeatedly performing tasks, such as reading from memory, searching for other devices, and writing to memory. SymbOS.Cabir [24] is such a proof-of-concept worm that can spread to any Bluetooth device.

12.2.3 Security Countermeasures for Bluetooth-Based Wireless PANs

As Bluetooth technology becomes prevalent in the market over the next few years, we are likely to see more security features and solutions defined in its next generation standard. Many countermeasures can be taken to mitigate the risks from possible attacks. These countermeasures rely on an understanding of general security policies, the Bluetooth technology, and its security risks in different deployment scenarios. Examples of such countermeasures follow.

- *Proper configuration:* Many attacks result from insecure configuration settings, so the best defense is to understand the best security policy for the intended application of the Bluetooth networks. If the Bluetooth devices carry sensitive corporate information, then they need the highest security level.
- *Hide your Bluetooth:* When you go to a public place with your Bluetooth device and do not plan to use it for communication, then the best defense is to

turn off its Bluetooth feature. A related technique is to run your Bluetooth device in "nondiscoverable" mode unless a device association is needed. This will prevent the device from responding to queries from unknown devices, some of which may be hidden attackers. Both methods minimize the risks of revealing your device address and clock.

- *Strong PIN policy:* The PIN value is used to generate the initialization key in device association, so it needs to be protected from interception. The PINs can be from 1 to 16 octets in length. PIN numbers should be sufficiently long and random, which makes it more computationally challenging for the attackers to determine. A Bluetooth device has the option of storing and automatically retrieving the PIN number from the nonvolatile memory. This is a convenient but dangerous option, since the loss of a device may result in the disclosure of its PIN number, and subsequently, the compromise of its previously trusted devices. It is safer to manually enter a PIN number each time the device is initialized.

- *Protect link key:* The use of the device's unit key as its link key should be avoided whenever possible. The unit key is too weak to be used as a link key, since the link key is used in the authentication process and in the derivation of the encryption key for device communication. The link key must be deleted from the device if the association was performed only for one-time communication. Following these guidelines minimizes the risks of eavesdropping, device spoofing, and man-in-the-middle attacks.

- *Biometrics:* Corporate users should be required to provide multiple authentication tokens, such as voice, fingerprints, and passwords. Authentication in the current Bluetooth security standard is based on devices instead of users. The addition of biometric measures effectively includes user authentication, and helps to prevent lost or stolen devices from posing a security threat to the operating environment.

- *Strong security in application layer:* Applications should not depend heavily on Bluetooth devices for security, since these small devices are usually limited in power, computational capability, and memory. It is unlikely for them to implement strong security algorithms. If the communication is sensitive, then performing encryption in the application layer adds an extra layer of protection against possible eavesdropping and cracking. The deploying of a public key infrastructure can increase the security of the key exchange process in Bluetooth communications.

12.3   Security Threats and Solutions for Cellular Networks

Mobile phones are used by hundreds of millions of subscribers on a daily basis, and they communicate through cellular networks. Cellular networks have evolved through three generations with different technologies. The first generation (1G) networks used analog techniques to transmit only voice. The second generation (2G) networks were also used to transmit only voice, but they began to use digital transmission techniques. There was no worldwide standardization for 2G networks. The Global System for Mobile Communications (GSM) systems used throughout the world, except in the United States and Japan, are the most popular 2G systems. The third generation (3G) networks use advanced digital transmission schemes to achieve higher transmission rates. 3G has extended the traditional application of voice transmission to include data transmission.

We will focus our discussions on digital cellular networks. First, we will inspect security features in the popular GSM networks. Then, we will discuss the security flaws and common attacks for GSM networks. Finally, we take a brief look at security schemes in 3G networks and how they address problems in the GSM networks.

12.3.1 Overview of Security Mechanisms in GSM Cellular Networks

The GSM networks consist of three components: mobile station, base station subsystem, and network subsystem. Figure 12.5 illustrates the GSM networks' architecture.

1. The mobile station (MS) consists of the physical equipment and a smart card called the subscriber identity module (SIM). The mobile physical equipment is uniquely identified by the international mobile equipment identity (IMEI). The SIM card is used to identify a user, and provides mobility to the user. By inserting the SIM card into any GSM phone, the user can make calls or access other subscription services regardless of location.

2. The base station subsystem (BSS) consists of the base transceiver station (BTS) and the base station controller (BSC). The BTS manages radio transmission with mobile stations. In a large urban area, there will be a large number of BTSs in a BSS. The BSC manages transmission among BTSs and connects into the mobile switching center (MSC) in the network subsystem.

3. The network subsystem consists of the MSC, the home location register (HLR), the visitor location register (VLR), the equipment identity register (EIR), and the authentication center (AC). The MSC manages call establishment, routing, and connection into other networks. The HLR contains the subscription information and the current location of each user. The VLR contains information about subscribers who are currently

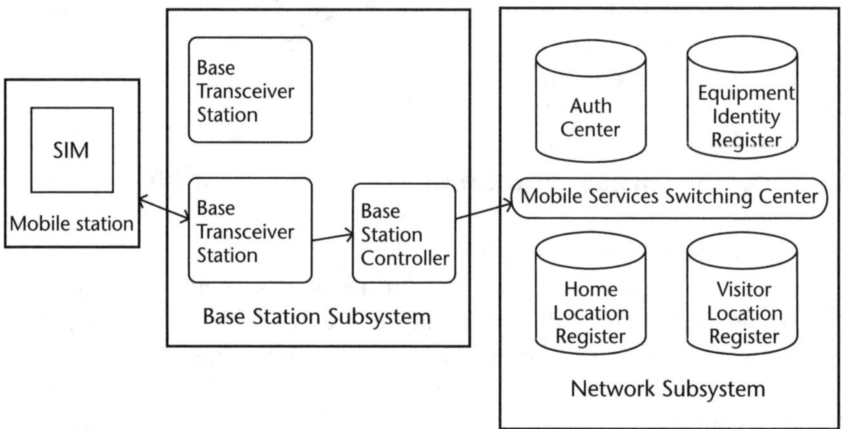

Figure 12.5 Architecture of GSM networks.

roaming in the network as visitors. The AC implements authentication and key generator algorithms, and is responsible for deciding whether or not a subscriber is authorized to access the network. The EIR keeps a list of valid mobile equipment in the network.

GSM networks have several built-in security features, including subscriber authentication, subscriber identity protection, and secure transfer of data and control information [25–28]. The GSM standard was created in secrecy, and its algorithms are not available to the public. The security algorithms have been leaked or reverse-engineered over the years.

Subscriber Authentication

This procedure checks the validity of a subscriber and decides whether or not to allow the subscriber to use the cellular network. It follows a traditional challenge-response protocol, as shown in Figure 12.6.

Each user stores a unique identity, known as an international mobile subscriber identity (IMSI), and a secret authentication key (K_i) in his SIM card. The user sends his IMSI to the VLR at the beginning of authentication, which then forwards the IMSI to the user's HLR. The HLR requests his AC to generate a 3-tuple vector, a 128-bit RAND, a signed response of the random number ($SRES_n$), and the communication key (K_c). $SRES_n$ and K_c are computed as follows.

$$SRES_n = A3\ (K_i, RAND)$$

$$K_c = A8\ (K_i, RAND)$$

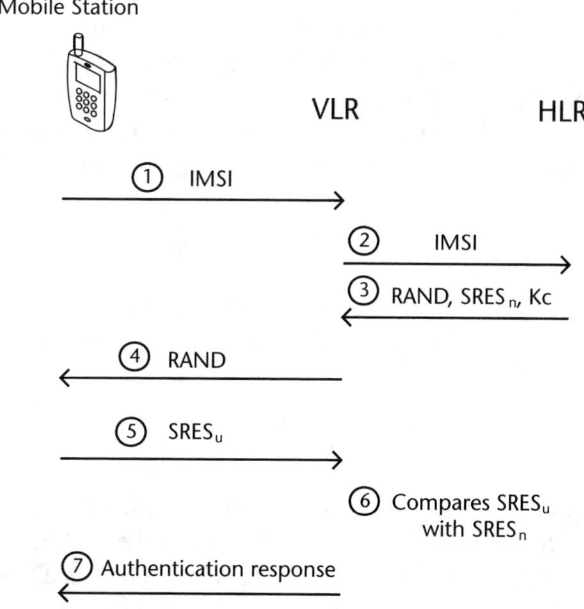

Figure 12.6 Challenge-response authentication procedures in GSM.

The HLR forwards this 3-tuple vector to the VLR. The VLR then sends the random challenge, RAND, to the user. The user applies the same A3 algorithm to compute his response, $SRES_u$, based on the shared secret key K_i, and sends $SRES_u$ back to the VLR. If $SRES_u$ is equal to $SRES_n$, then the VLR believes that the user knows the secret key K_i, making it a legitimate user.

The authentication center and mobile station both use the A3 algorithm to calculate the signed response based on the shared secret key K_i, and use the A8 algorithm to derive the communication key K_c. The A3 and A8 algorithms are one-way hash functions, and their implementations are operator-dependent. Therefore, A3 and A5 are stored in the user's SIM card.

Subscriber Identity Protection

The user's IMSI is sent in clear text during the authentication procedure. This allows eavesdroppers to obtain the identity of the GSM user. GSM networks assign a temporary identifier to protect the user's identity, which is called temporary mobile subscriber identity (TMSI) for communication after initial authentication. For every subsequent call made to the same VLR, the user must send the TMSI instead of the IMSI in its authentication transaction. The remaining steps in authentication are the same as in Figure 12.6. The mobile station stores the TMSI in its SIM card, so it is still available after the phone is switched off.

If the TMSI is sent in clear text, this defeats the purpose of protecting the mobile user's identity. TMSI is instead sent in encrypted form, using the A5 algorithm as an output of TMSI and Kc.

$$\text{Encrypted TMSI} = A5(TMSI, K_c)$$

$$K_c = A8\ (K_i, RAND)$$

Several implementations of the A5 algorithm currently exist, and the most commonly used ones are A5/1, A5/2, and A5/3. A5/1 and A5/2 were originally defined by GSM. A5/1 is widely used in the United States and Europe to provide stronger protection. A5/2 is a weaker algorithm commonly used in Asia. A5/3 was added in 2002, and it is based on the KASUMI algorithm defined by 3G standards.

Secure Transfer

Each data frame after the authentication process is encrypted with A5 using a shared key K_c. The key is generated during the challenge-response process using the A8 algorithm:

$$K_c = A8\ (K_i, RAND).$$

Figure 12.7 shows the encryption process. The cipher text is the result of adding the data frame with a key stream output by the A5 algorithm. The encryption is also seeded by the value of COUNT that is based on the TDMA frame number, ensuring that the key stream changes with each data frame.

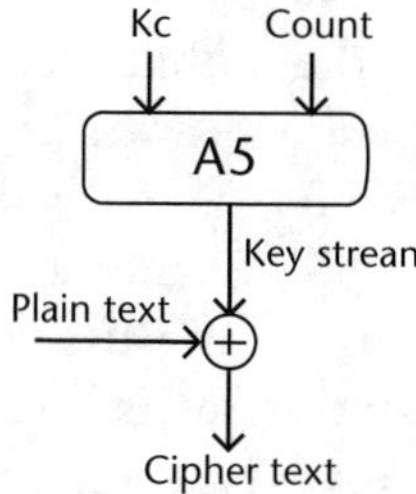

Figure 12.7   Secure data transfer.

12.3.2   Security Threats for GSM Cellular Networks

Most security experts have criticized GSM's approach of "security by obscurity." They believe that a security standard will not be trustworthy if it has not been thoroughly examined. Their criticism has been partly supported by various threats and vulnerabilities found in GSM security standards. Table 12.3 summarizes the common security threats for GSM networks and the security services that are compromised.

Impersonation of the Network

An MS in GSM networks authenticates to the network, but the network does not authenticate to the MS. It is possible for an attacker to set up a fake base station using the same mobile network code as the subscriber MS's network. The fake base station may simply send a random challenge in the authentication phase, but ignore the signed response from the MS. The MS may then end up making calls unknowingly to the fake base station, which may then forward the calls to a public telephone network. The fake base station can also choose not to activate data encryption at all, thereby reading the subscriber's calls in clear text.

Identity Caching

Identity caching refers to the attempt to retrieve a subscriber's permanent IMSI. The attacker may passively sniff the radio channel for IMSI sent in clear text. GSM introduces a temporary ID, known as TMSI, to defend against this type of eavesdropping, pages a mobile station by its TMSI after registration, and maintains a database in its VLR that maps TMSIs to IMSIs. However, if the provider encounters an error in its database and cannot match the a mobile station's TMSI to a valid

Table 12.3   Security Threats for GSM Cellular Networks

Threats	Confidentiality	Authentication	Integrity	Availability
Impersonation of the network	✓	✓	—	—
Identity caching	✓	—	—	—
SIM card cloning	✓	✓	✓	✓
MS spoofing	—	—	—	✓
Weak encryption	✓	—	—	—

IMSI, then it cannot verify the mobile station's identity. The provider will request the mobile station to send its IMSI in clear text [29].

Figure 12.8 illustrates how an attacker takes advantage of this process to retrieve a mobile station's IMSI. As mentioned earlier, the network does not need to authenticate to a mobile station. The attacker can thus imitate a legitimate base station, and page the mobile station by its TMSI. The mobile station will then establish a radio channel. The attacker can now simply send an "Identity Request" message. The mobile station interprets the message as an indication of database error, and sends its IMSI in the "Identity Response" message.

SIM Card Cloning

When an attacker can retrieve the secret key, K_i, from a victim's mobile phone, he can easily make an exact duplicate (i.e., clone), in order to make fraudulent calls billed to the victim user's account and perform various other attacks. The flaws in A3/A8 implementations have made this attack possible.

Most network providers combine A3 and A8 algorithms in a single hash function, known as COMP128, which generates a 64-bit K_c and a 32-bit SRES from the 128-bit RAND and 128-bit K_i in a single round. In COMP128, if an attacker uses carefully chosen random challenges, then he can guess the secret key, K_i, in a relatively small number of rounds. For example, a publicly available utility called Sim Scan [30] can recover K_i by using approximately 18,000 random challenges. The attacker needs physical access to the victim's SIM card to use this utility.

Even if the attacker does not have access to the victim's phone, he can still exploit the weakness in COMP128 to crack K_i by performing a chosen-plaintext attack. The attacker may:

- Pretend to be a legitimate base station and page the mobile phone by its TMSI;
- Establish radio communication with the victim's mobile phone;
- Send "Authentication Request" with $RAND_0$;
- Wait until the victim's mobile phone responds "Authentication Response" with $SRES_0$;
- Send "Authentication Request" with $RAND_1$;

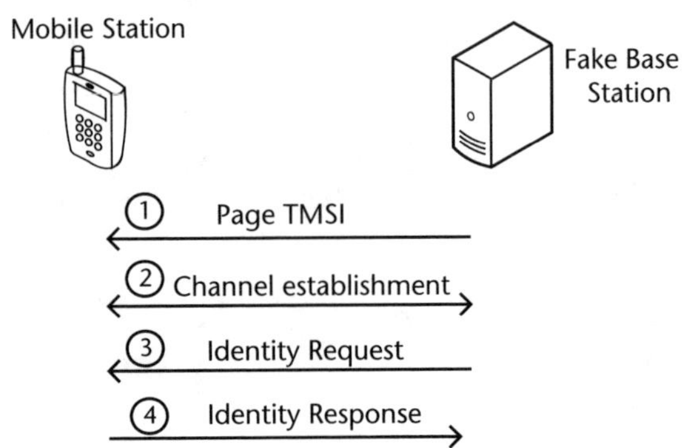

Figure 12.8 Actively retrieving a mobile station's IMSI.

- Wait until the victim's mobile phone responds "Authentication Response" with $SRES_i$;
- Repeatedly send "Authentication Request" with chosen $RAND_i$ until it has collected sufficient $<RAND_i, SRES_i>$.

The attacker needs to communicate with the mobile phone for up to 9 hours to collect sufficient information, depending on the random challenges being used. The attacker can perform the above steps in several time periods, since it not easy to continuously sustain communication for 9 hours.

Mobile Station Spoofing

Once the attacker has collected sufficient information of the victim's MS using the attacks described above, he can launch two types of spoofing attacks, leading to service disruption to the victim's MS. In the first attack, he pretends to be the victim's MS and sends a deregistration request to the network. The network deregisters the victim's MS from both the VLR and HLR, making the victim's MS unreachable. In the second attack, the attacker sends a spoofed location update request from his own area, registering this wrong area with the network under the name of the victim's MS. Any subsequent incoming calls to the victim's MS would be directed to the attacker's area, leading to denial-of-service for the victim's MS.

The above attacks can also be performed by hijacking an existing connection and modifying the signaling information. GSM networks make this possible, since signaling messages, such as location updates and deregistration, are sent in clear text.

Weak Encryption

The A5/1 data encryption algorithm was broken in 1999. Biryukov, Shamir, and Wagner [31] reported that A5/1 could be cracked in less than 1 second on a typical PC, although the attack requires large precomputed tables, approximately 2^{36} bytes or 64 GB in size. The A5/2 algorithm is intentionally weakened for export restrictions, and can be cracked using even less computation steps.

12.3.3   Security Solutions in 3G Cellular Networks

3G cellular networks support higher data transmission rates and have improved security schemes, which are designed to correct problems found in GSM networks. The 3G networks have included the following security features: mutual authentication between an MS and the network, stronger encryption of user traffic and signaling information, and integrity protection of signaling information. We will inspect these features in more detail, and discuss how they correct problems in GSM networks.

Mutual Authentication

The MS and network are both required to authenticate to each other in 3G networks. The authentication process is backward-compatible with GSM networks, and is based on challenge-response protocol and secret key cryptography. The network stores a 128-bit secret key in its authentication center, while the MS stores the same secret key in its SIM card. The way that the MS authenticates to the network is not changed, but the network needs to prove its identity to the MS. The network achieves this by sending a 128-bit authentication token (AUTN), together with a random challenge (RAND) for the mobile. The AUTN contains a sequence number encrypted using the random number, RAND, and the shared secret key, K. It also contains a MAC code, which is the expected signed response based on the sequence number. The MS will receive the RAND and AUTN, and will verify the AUTN based on the sequence number and shared secret key stored in its SIM. If the calculated XMAC is different from the MAC in AUTN, then it will send an authentication reject message to the network and terminate the connection.

This authentication procedure is apparently subject to replay attack, since the attacker can simply resend a valid AUTN to an MS to pretend to be a legitimate network. 3G phones prevent this by tracking the sequence numbers already being used by old AUTNs and rejecting duplicate sequence numbers.

Attackers impersonating as a legitimate base stations are no longer possible, since mutual authentication is mandated. Identity caching and over-the-air cracking of K_i has been made much more difficult to perform. The MS and the network agree on keys to be used for encryption and integrity protection of messages after mutual authentication. Figure 12.9 summarizes the algorithms used in the authentication and key agreement phase.

Among the five algorithms, f1, f2, and f5 are used to generate responses in the authentication phase; f3 is used to generate a ciphering key, CK, for data encryption; and f4 is used to generate an integrity key, IK, to protect signaling information. The implementation of the five algorithms is operator-specific, but it is likely that most operators will choose the draft implementation recommended by the 3G standard. The draft implementation is based on AES, and is much stronger than A5 implementations.

Data Confidentiality

The f8 algorithm, a stream cipher based on KASUMI, is used to encrypt and decrypt user data in 3G networks.

The encryption procedure shown in Figure 12.10(a) is similar to the GSM standard, but with more input parameters.

- COUNT-C is a 32-bit sequence number dependent on the data frame number.
- BEARER is a 5-bit unique ID for each radio link for an MS.

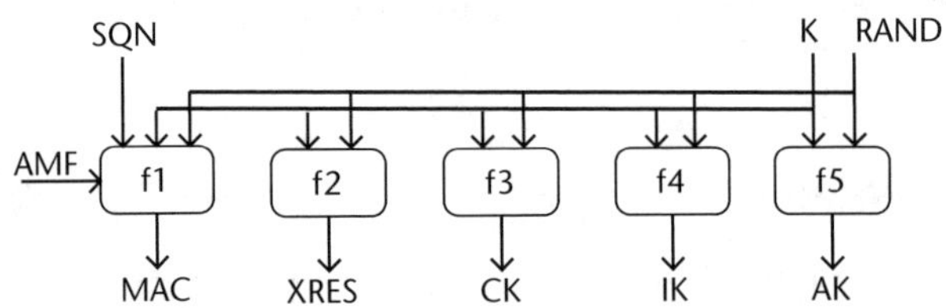

Figure 12.9 Key generation in 3G networks.

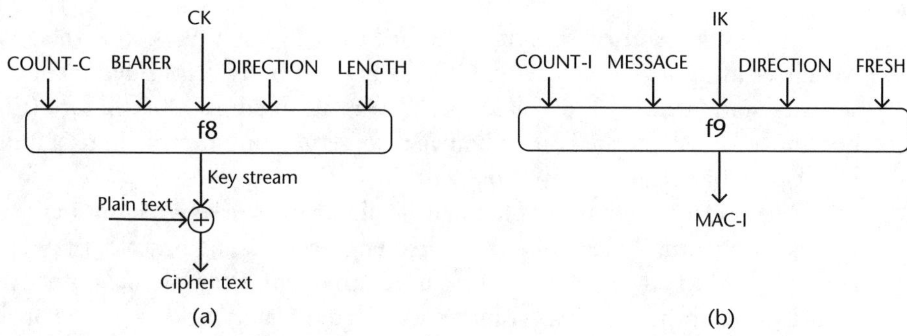

Figure 12.10 (a) Data encryption using f8, and (b) integrity protection using f9.

- DIRECTION is a 1-bit value indicating the direction of data flow.
- LENGTH shows the number of bits to be encrypted, between 1 and 20,000.
- CK is the cipher key agreed upon during authentication.

Data encryption algorithm f8 is completely different and much stronger than A5 in GSM. No attacks have been reported.

Integrity Protection

In 3G networks, an integrity value called MAC-I must be appended to the signaling message between an MS and the network. This prevents an attacker from hijacking the connection and modifying signaling message to his advantage. This makes it more difficult for the attacker to modify location update and deregistration messages to launch DoS attacks.

The integrity value, MAC-I, is generated using the f9 algorithm, which is also based on the KASUMI cipher. As shown in Figure 12.10(b), the f9 algorithm accepts MESSAGE, COUNT-I, DIRECTION, FRESH, and IK as its input. MESSAGE represents the data stream to be integrity-protected. The meanings of COUNT-I and DIRECTION are similar to the meanings in the f8 algorithm. FRESH is a 32-bit random number used for the duration of a connection, which is included to prevent an attacker from replaying the same messages in future connections. IK is the key generated using the f4 algorithm during the authentication phase.

The f8 and f9 algorithms are both known to the public and have been thoroughly scrutinized by security experts. The 3G standard strives to achieve end-to-end security in the network, which is a big improvement over GSM, since GSM networks do not provide any security on the wired network.

12.4 Summary

Wireless networks depend on radio propagation for communication, so they are especially vulnerable to certain attacks. Wireless LANs, wireless PANs, and cellular networks have all defined security mechanisms to provide authentication, data encryption, and integrity protection.

The first generation security standard for wireless LANs, known as WEP, failed to deliver all of its promises. The IEEE standard body responded by defining a new security standard, IEEE 802.11i, which was adopted in 2004. Before IEEE 802.11i became established, the WiFi Alliance also defined an intermediate standard, WPA, which was backward-compatible with WEP.

The security mechanisms for GSM cellular networks were not made public, but they were eventually leaked and reverse-engineered. The practice of achieving security by obscurity was proven to be unreliable, since people have uncovered many security vulnerabilities in its security standard. 3G networks fixed the problems with GSM networks, and invited the security community to thoroughly inspect its security mechanisms.

Wireless PANs, such as Bluetooth-based networks, have a considerably shorter range than wireless LANs and cellular networks. They cannot be attacked from long range, which has partly reduced its attacking matrix.

Achieving better security is a constant battle between the security community and underground attackers. The security industry builds better security solutions and technologies. Educating users on good security practices is also of paramount importance.

References

[1] Stripling, T., "History and Best Practices of Wireless Network Security," *GSEC Practical Assignment*, January 2003.

[2] Sacco, K., "WLAN Security: Insecurities of the WEP Protocol and How WPA Plans to Overcome Them," GSEC Practical Assignment, December 2003.

[3] Ethereal, network protocol analyzer, http://www.ethereal.com.

[4] Kismet, 802.11 wireless network sniffer and intrusion detection system, http://www.kismetwireless.net.

[5] Windump, network sniffer for Windows, http://windump.polito.it.

[6] tcpdump, UNIX network sniffer, http://www.tcpdump.org.

[7] Sniffit, network sniffer, http://reptile.rug.ac.be/~coder/sniffit/sniffit.html.

[8] dsniff, a collection of tools for network auditing and penetration testing, http://www.monkey.org/~dugsong/dsniff/.

[9] AirSnort, a wireless LAN tool to crack 802.11b keys, http://airsnort.shmoo.com/.

[10] Fluhrer, S., I. Mantin, and A. Shamir, "Weaknesses in the Key Scheduling Algorithm of RC4," *Lecture Notes in Computer Science*, Vol. 2259, 2001, pp. 1–24.

[11] WEPCrack, an open source tool for breaking 802.11 WEP secret keys, http://wepcrack.sourceforge.net/.

[12] AiroPeek, voice over wireless LAN analyzer, http://www.wildpackets.com/solutions/wireless.

[13] NetStumbler, wireless LAN wardriving tool, http://www.stumbler.net/readme/readme_0_4_0.html.

[14] Schultz, G., C. Endorf, and J. Mellander, *Intrusion Detection and Prevention*, New York: McGraw-Hill, 2003.

[15] IEEE 802.11i standard, "Medium Access Control Security Enhancements," 2004.

[16] Hanninen, T., "Wi-Fi Security," August 2003, http://www.cs.helsinki.fi/u/lamsal/teaching/autumn2003/student_final/tomi_hanninen.pdf.

[17] Porter, B., "Wireless Security's Future," *IEEE Security & Privacy*, Vol. 1, No. 4, July–August 2003, pp. 68–72.

[18] Karygiannis, T., and L. Owens, "Wireless Network Security 802.11, Bluetooth and Handheld Devices," Special Publication 800-48, NIST, November 2002, http://csrc.nist.gov/publications/nistpubs/800-47/sp800-47.pdf.

[19]   Lamm, G., et al., "Bluetooth Wireless Networks Security Features," *Proc. IEEE Workshop on Information Assurance and Security*, June 2001, pp. 265–272.

[20]   Bluetooth Special Interest Group, Bluetooth Specification version 1.2, https://www.bluetooth.org/foundry/adopters/document/Bluetooth_Core_Specification_v1.2.

[21]   Niem, T. C., "Bluetooth and its Inherent Security Issues," SANS GIAC Security Essentials Certification (GSEC) v1.4b, November 2002.

[22]   Levi, A., et al., "Relay Attacks on Bluetooth Authentication and Solutions," *Proc. 19th Int. Symp. Computer and Information Sciences*, Antalya, Turkey, October 2004, pp. 278–288.

[23]   Martin, T., et al., "Denial-of-Service Attacks on Battery-Powered Mobile Computers," *Proc. 2nd IEEE Int. Conference on Pervasive Computing and Communications*, Orlando, FL, 2004, pp. 309–318.

[24]   Ferrie, P., et al., "Symantec Security Response—SymbOS.Cabir," Symantec Incorporation, June 2004.

[25]   Howard, P., "GSM and 3G Security," Lecture Notes, Royal Holloway, University of London, November 19, 2001, http://www.isg.rhbnc.ac.uk/msc/teaching/is3/is3.shtml.

[26]   Hillebrand, F., (ed.), *GSM and UMTS: The Creation of Global Mobile Communication*, New York: John Wiley & Sons, 2002, pp. 385–406.

[27]   Brookson, C., "GSM Security: A Description of the Reasons for Security and Techniques," *Proc. IEE Colloquium on Security and Cryptography Applications to Radio Systems*, 1994, pp. 1–4.

[28]   Boman, K., et al., "UMTS Security," *Electronics & Communication Engineering J.*, Vol. 14, No. 5, October 2002, pp. 191–204.

[29]   Quirke, J., "Security in the GSM System," http://www.gsm-security.net/gsm-security-papers.shtml.

[30]   Kaljevic, D., Sim Scan 2.0, http://users.net.yu/~dejan/.

[31]   Biryukov, A., A. Shamir, and D. Wagner, "Real Time Cryptanalysis of A5/1 on a PC," *Lecture Notes in Computer Science*, Vol. 1978, 2000, pp. 1–18.

CHAPTER 13

Design of Mobile Client Software

Mobile client software refers to a program that executes on a mobile platform and interacts with mobile users to support their access to mobile application data, functions, and services. Mobile client software has three basic common functions. They are: (1) accepting and processing the inputs and requests from mobile users; (2) communicating with wireless-based servers, applications, and service systems over a wireless network (or wireless Internet); and (3) displaying the system outputs and responses to mobile users. Some mobile client software, such as smart client software, allows mobile users to access local device resources and applications to obtain a rich mobile experience.

Since mobile client software is an important part of wireless-based application systems, developing reliable, efficient, and user-friendly mobile client software is a key to the success of these products. Mobile client developers must understand more than the implementation mobile client software using existing mobile technologies, such as WML and J2ME. They also need to understand mobile users' expectations, deal with new technical challenges, and address the issues in the design of mobile client software. This chapter focuses on this mobile client software design.

Section 13.1 discusses the basics about mobile client software, including mobile users' expectations, mobile client features and requirements, and design and technical challenges. Section 13.2 classifies and discusses different types of mobile client software, including architecture styles (e.g., thin client, smart client, and thick client) and enabled technologies. Section 13.3 presents a mobile client development processes, and covers the design guidelines and principles. Section 13.4 addresses some special issues in designing mobile client software (e.g., support of reliable transactions and unified user interfaces). Section 13.5 provides mobile client design examples. Finally, Section 13.6 gives a summary.

13.1 Developing Mobile Client Software on Mobile Devices

Most development engineers have good knowledge and experience in developing Window-based and/or Unix-based application software and user interfaces. Many of them design and implement Web-based mobile client software, using Web technologies, such as Java, J2ME, JSP, and ASP. However, this is not enough for them to develop reliable and efficient mobile client software.

13.1.1   Understanding Mobile Accesses and Mobile Users' Expectations

It is difficult to design a user-friendly mobile interface for mobile device users without understanding user needs and expectations. Unlike online and conventional user accesses to application systems, user accesses through mobile client software to wireless-based application systems have several special features.

- Diverse mobile access platforms;
- Highly dynamic access environments (e.g., at a subway station or an airport);
- Very short access time;
- Limited accessible system resources (e.g., small screen display, limited storage space, and limited computing power);
- Location-aware and profile-based accessible content and features.

Most mobile device users are people with very limited computing knowledge and operation experience; for example, elementary school kids and retired seniors. They expect a very user-friendly interface. Mobile client software must meet the following expectations and requirements.

- Ease of use for end-users with little or no computer knowledge and experience;
- Reliable access over wireless networks with a limited network bandwidth and unreliable network connectivity;
- Short response time, good system performance, and efficient transactions;
- Mobile personalization to meet end-users preferences and presentation formats;
- Integration with multiple input/output media formats, including text, short messages, audio, and video.

13.1.2   Enable Mobile Technologies

There are a number of mobile operating platforms and presentation technologies on mobile devices, as discussed in Part II. They can be classified into five types, based on supporting platforms.

The first type is browser-based, written in a mobile markup language, and executed on minibrowsers on mobile devices. WML, HDML, cHTML, and XHTML are typical mobile markup languages. They are available on low-powered mobile devices, such as cell phones.

The second type is J2ME-based, which refers to mobile clients developed using J2ME technology, and is executable on the micro-Java virtual machine on mobile devices. J2ME-based mobile clients are useful to support system-and-user interactions in wireless applications. Table 13.1 provides a detailed comparison of different mobile presentation technologies on mobile phones.

The third type is Windows-based, which refers to mobile clients developed using Microsoft solutions. They are executable in the Windows CE and .NET environments. Since Microsoft has a rich set of software and enterprise-oriented solutions in the Windows CE and .NET environments, people believe it is easy to develop smart or thick mobile software on Windows CE–enabled mobile devices.

Table 13.1   Comparison of Wireless Mobile Presentation Technologies on Mobile Phones

Languages	Text Support	Image and Graphics Support	Audio Support	Video Support	Popularity on Mobile Devices
WML	Supports multiple text fonts and formats.	Supports black, white, and color images. No graphics.	No audio support.	No video support.	Available on almost every mobile phone.
J2ME	Supports multiple text fonts and formats, and text formatting.	Supports black, white, and color images. No graphics.	Supports audio recording, playback and streaming on the local device, as well as on the Internet.	Supports video capture, playback and streaming on the local device, as well as on the Internet.	Available on J2ME-enabled mobile devices.
HDML	Supports multiple text fonts and formats.	Supports images with the bmp format. No graphics.	No audio support.	No video support	Available on Openwave browsers only.
i-mode/ cHTML	Supports text without multiple fonts and style sheets.	Supports images, with limitations: (e.g., no JPEG images, no image maps, no more than two colors). No graphics.	Supports audio retrievals from the Internet.	Supports video retrievals from the Internet.	Presentable on most minibrowsers. Popularly used in Japan.
xHTML	Supports multiple text fonts and formats, and text formatting.	Supports images with different formats and 3D graphics.	Supports audio with different formats.	Supports video capture, playback and streaming on the local device, as well as on the Internet.	Industry merging trends.

The fourth type is Palm OS-based, which is only executable on PDAs. Palm OS is an earlier platform for PDAs, which was selected in 2000, due to its availability. Users found that Palm OS was not suitable for wireless enterprise applications, due to the limitations of its earlier versions. Later versions of Palm OS (e.g., Palm OS 5.0, 6.0, and i705) have added many new enterprise features and wireless capability (e.g., multitasking capability, and mobile database technology).

The fifth type of mobile platform is Symbian OS, which is created by Symbian. Most people believe Symbian OS is an ideal platform to develop mobile client software for mobile phones, such as smart phones, because it was originally designed for wireless-based application systems. One of its major advantages is its good network connectivity support.

13.1.3   Design Issues and Challenges

Engineers have encountered a number of design issues in creating mobile client software [1–3]. They are listed as follows:

- Development of a user-friendly mobile client with good mobile experience;
- Delivery of a consistent and unified user interface on diverse mobile devices;

- Support of reliable mobile transactions and services, coping with unreliable network connections, battery shortages, and abandoned client sessions;
- Conversion of existing Web contents into mobile contents.

Engineers also face several technical challenges along with the design issues. The rapid changes in mobile devices and mobile technologies increases the difficulty of providing a consistent mobile client interface to end users, and increases the project costs in system maintenance and testing for different mobile devices. Building mobile client software using the limited computing power and system resources to meet users' expectation in system performance, usability, and reliability, is another technical challenge.

13.2 Classification of Mobile Clients and Architecture Styles

Engineers need to understand different types of mobile clients before creating mobile client software. This section classifies and compares different types of mobile clients in terms of their strengths and limitations. It also covers the concepts of unimodal and multimodal mobile clients, discusses their special features and applications, and examines several special issues relating to mobile client development.

13.2.1 Classification of Mobile Clients

There are a number of ways to classify different mobile clients based on client complexity, enabled technologies, and architecture styles [4]. Figure 13.1 shows the classification of various mobile clients into three groups, based on their complexity, operation models, and dependency of network connectivity. These are:

- Thin mobile clients;
- Smart mobile clients;
- Thick mobile clients.

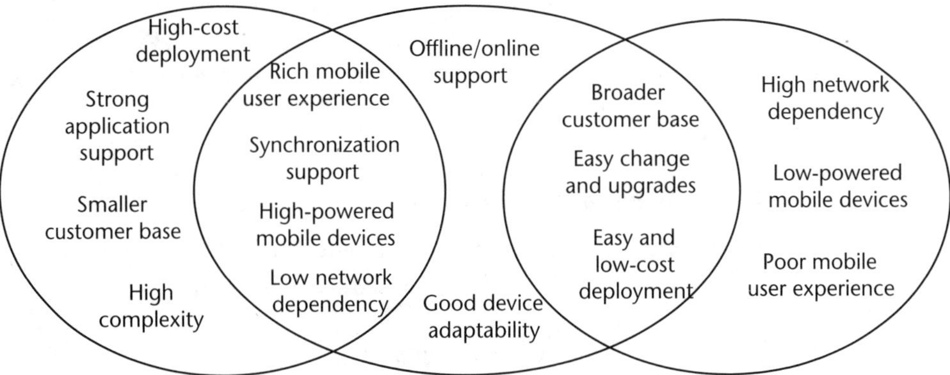

Figure 13.1   The major features of three types of mobile clients.

Thin Mobile Client

A thin mobile client refers to mobile software designed for low-powered mobile devices. It provides a thin user interface between mobile users and a central stationary server to retrieve data, and access functions and services. A thin mobile client has several special features listed here.

- It is suitable for operation on low-powered mobile devices, such as mobile phones.
- It is presented with a script-based mobile markup language, such as WML, cHTML, and XHTML.
- It is provided with a very limited set of functions, and is highly dependent on a wireless-based application server to support functional services.
- It has no need of installation and synchronization.

There are a number of advantages to using thin mobile clients.

- Low-powered mobile devices are needed. Using a thin mobile client reduces heavy computing processes on mobile devices by keeping these processes at the server side.
- It is easy to leverage to existing online Web-based HTML pages to mobile pages.
- It is easy to deploy, execute, and upgrade mobile clients on different types of mobile devices.
- It is easy to construct, due to its limited complexity.

A thin mobile client has its disadvantages and limitations, listed here.

- A thin client is usually implemented using a script-based mobile markup language, which makes its mobile user interface very primitive, without rich graphics and multimedia features.
- It is highly dependent on wireless network connectivity. This increases network traffic, slows down system performance, and raises the technical challenge of supporting reliable application transactions.
- A thin mobile client executed on a mobile browser has very limited access to the local resources on mobile devices.
- A thin mobile client may not be able to deploy on different mobile devices.

Smart Mobile Client

A smart mobile client refers to mobile application software designed for smart mobile devices. It uses local computing processors and resources to provide certain application functions and services without the support from its application server. A typical example is a downloaded digital game on smart phones. A smart client has a number of special features listed here.

- It is highly dependent on local resources (e.g., XML Web Services and .NET Framework for Pocket PCs) to support mobile computing processes on mobile devices.
- It supports users with two application operation modes: standalone and off-line.
- It requires mobile client installation and updates on mobile devices through synchronization capability.
- It provides rich mobile user interfaces supporting mobile data with text and multimedia data.

A smart client also has a set of different advantages and limitations. Its major advantages are given here.

- Its off-line support and rich user interface features improve mobile access.
- Its weak network connectivity reduces network traffic, improves system performance, and increases user accessibility of the system.
- Battery life on mobile devices is extending by avoiding wireless transmission and reception.

Similar to a thin client, a smart client also has its limitations, which are listed here.

- A smart client cannot be remotely downloaded to mobile devices. It needs to be installed or reinstalled on mobile devices whenever it is updated.
- Mobile users need to pay more money for their higher-powered mobile devices.
- Smart client software usually is more complex and costly to develop.
- Users have increased upgrade and reinstallation work.
- It is difficult and costly to construct smart client software to fit into diverse mobile devices and platforms.

Thick Mobile Client

A thick mobile client refers to complex mobile software, highly dependent on the local resources of a mobile platform, which supports users in accessing enterprise-oriented wireless applications. A thick mobile client, designed for high-powered mobile devices (e.g., PDAs and Pocket PCs), has a number of following features.

- A thick client must be installed and deployed on mobile devices. Thus, it is costly to upgrade on mobile devices.
- Thick clients usually are developed to support certain groups of mobile users to perform specific enterprise applications, unlike smart clients that are developed for public mobile users.
- Thick clients offer mobile users with complex off-line applications that are executable on a certain platform.

- Thick mobile clients provide mobile users with the synchronization capability for mobile data and program synchronization.

A thick mobile client has several strengths listed here.

- It supports complex off-line applications that provide seamless integration of enterprise systems with the existing business applications, such as SCM, ERP, and so forth.
- It offers an open standard mobile platform to access wireless data, multimedia, and Web services channels.
- It provides users with a rich mobile experience based on its strong graphics and multimedia features.
- It has reliable mobile accessibility and efficient mobile transaction processing due to its independence of network connectivity.

The weaknesses and limitations of a thick mobile client are listed here.

- It requires high-end mobile devices and powerful operation platforms.
- It has a rich and extensive graphic user interface, which is costly to develop, upgrade, and maintain.
- Developers of thick mobile clients need more knowledge about mobile devices and operation platforms, including wireless communication and programming, application interfaces, and enterprise frameworks.

Table 13.2 compares the three types of mobile clients in terms of their features, strengths, and limitations.

Table 13.2   Comparison of Thick, Thin, and Smart Mobile Clients

	Client Types		
Feature	*Thick Client*	*Thin Client*	*Smart Client*
Development complexity and cost	High	Low	Moderate
Network connectivity	Weak and reliable	Strong and unreliable	Weak and reliable
Online/off-line operations	Mostly off-line	Mostly online	Both online and off-line
Client deployment	Synchronization	Dynamic download	Dynamic download or synchronization
Enable presentation technology	C++/Java	cHTML/WML/ XHTML/J2ME	J2ME/Java/C++
Local resource usage	High	None	Moderate
Maintenance efforts	Low	High	Moderate
Mobile user experience	Very good	Primitive	Good
GUI features	Rich	Poor, without graphics	Basic, with simple graphics
Performance	High	Low	Moderate
Mobile devices	High-powered	Low-powered	Smart or high-powered

13.2.2 Unimodal Versus Multimodal Mobile User Interfaces

Another way to classify mobile client software is based on its models supporting user inputs and outputs. Mobile clients can generally be classified into unimodal mobile interfaces and multimodal mobile interfaces.

A unimodal mobile interface only supports one single channel to input data (e.g., keypad), and one single channel for generating output data (e.g., display screen). The currently available mobile interfaces are almost exclusively unimodal interfaces. Since a unimodal mobile interface only supports one single channel for inputs and outputs, it must be simple and easy to develop and implement. A unimodal user interface provides users with only a very primitive mobile access experience due to its limited input and output channel.

A multimodal mobile interface provides a much richer user interface with multiple channels for user inputs and outputs. For example, a user may see the displayed outputs on a small display screen, and receive the supplemented information with audio at the same time. The user can input data using a phone keypad, and input digital voice information using speech recognition technology.

A multimodal mobile interface has several advantages when compared to a unimodal mobile interface. Mobile users can interact with a mobile device and its mobile client using multiple operation models. Different operation modes and input/output channels clearly are better suited to different dynamic mobile environments. Consider a movie reservation system for theaters. Booking and purchasing information can be easily entered into the system with audio inputs. Seat selection can be easily made through a screen-based user interface.

Another advantage is its flexibility of allowing mobile users to dynamically select and switch the proper operation modal and input/output channel, according to a given dynamic mobile environment. It also has error prevention capability, because errors that occur in one operation mode can be corrected using another mode. The support of multimodal operations increases the complexity of mobile client software and its development costs.

Multimodal mobile interfaces can be further classified into two types: sequential and simultaneous multimodal mobile interfaces.

Sequential multimodal mobile interfaces allow a mobile user to select and use only one channel to enter input data and one channel to generate output data. Some service providers claim to provide "multimodal" access to e-mail, in which access to e-mail occurs through a graphical user interface (GUI) on a wireless personal digital assistance (PDA), or through an interactive voice response (IVR) telephone system. There is no link between the modes. A user chooses one mode and completes the transaction in the same mode.

Simultaneous multimodal mobile interfaces allow a mobile user to interact with the mobile device and client interface using more than one channel at the same time. For example, a user could specify the required dates in a hotel booking system using voice input, while at the same time select a hotel room using a screen-based user interface. One problem in the simultaneous multimodal interface is conflicting user responses (or inputs). This can arise from two sources. First, the conflict is caused by the user's conflicting inputs. Second, the system misunderstands one of the inputs from the user. There are different strategies for dealing with conflicting inputs: (1) Accept the input from the most reliable mode; (2) accept the most recent input;

(3) accept the original input; and (4) reject the conflicting inputs, and confirm with the mobile user.

Building mobile client software supporting multimodal interactions meets several major challenges in wireless-based application systems.

- It deals with the mobile device limitations in keypads and screen display to allow users to select a suitable mode to support user inputs and outputs.

- It enables mobile users to perform multitasking in a dynamic mobile environment.

- It provides a better environment for mobile users, so that they can change their mobile interaction modes at any time due to changing circumstances. For example, when a mobile user entered a high ambient noise area (e.g., a railway station), the speech recognition mode becomes more error-prone. A text-based user interaction provides the user with a more effective interaction channel.

- It enhances the mobile access experience by allowing mobile users to control their interaction modes of mobile devices.

- It offers mobile users with input confirmation and error correction, especially compared to a voice-only mobile client interface.

Constructing mobile client software with multimodal user interfaces can be implemented in two ways: thin client multimodal interface and thick client multimodal interface.

A thin client multimodal interface only provides the basic functions supporting the multimodal interactions and applications. For example, an audio client interface supports input speech encoder and audio playback decoder. It interacts with a stationary voice server to support audio-oriented multimodal interactions.

A voice server usually consists of a speech recognizer and a speech synthesizer to perform speech recognition and generation with the support from a Web-based audio-content server. Figure 13.2(a) shows an example of the basic components of a thin client multimodal interface on mobile devices and its supporting servers in a wired network.

The thin client implementation is generally more flexible, since it does not require very powerful mobile devices with rich system resources. It can deliver multimodality to low-power mobile devices, and easily support a multimodal application solution on diverse mobile devices. It must deal with unreliable network connections, since it is highly dependent on the network connectivity.

An alternative approach is to move the voice server and multimodal application interaction processor to mobile devices, as shown in Figure 13.2(b). This is known as thick client multimodal interface. It is generally more appropriate for high-power mobile devices, because it requires more computing power and system resources to support the voice server and multimodal application controller on the mobile client. This approach has several advantages. First, the network traffic is low due to its weak network connectivity. Next, the latency is low after the application is downloaded on mobile devices. Finally, the speech recognition can be speaker-dependent.

However, this approach has its weaknesses. It is only suitable for high-power mobile devices, and it is not suitable for applications that require large recognition

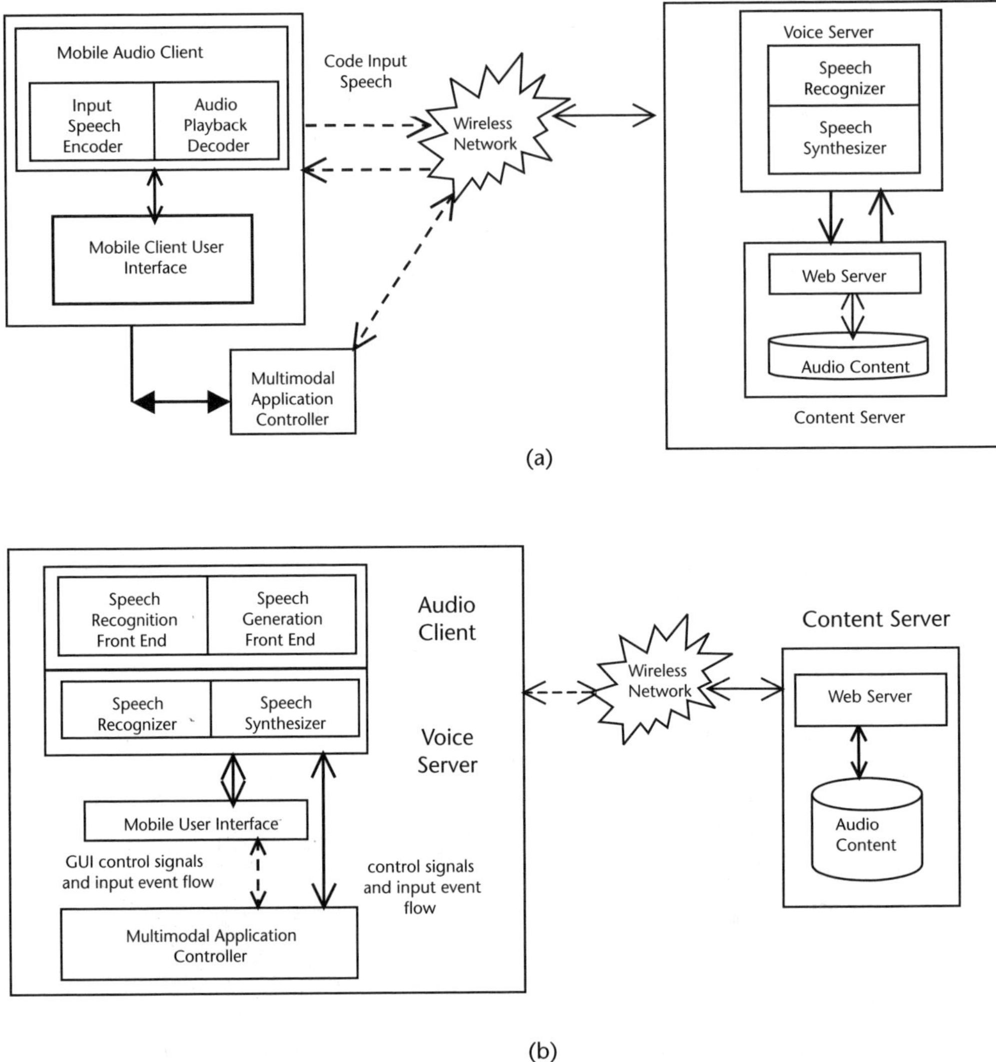

Figure 13.2   Different types of multimodal mobile clients: (a) thin client structure for a multimodal mobile user interface, and (b) thick client structure for a multimodal mobile user interface.

grammars (e.g., name and address lookups). Applications may need more downloading time because multimodal applications must be passed over the underlying wireless network.

13.3   Design for Mobile Client Software

Engineers need to understand wireless programming languages (e.g., J2ME, WML, and XHTML), need to know the basic development process for mobile client software, and need to understand the essential design principles and guidelines. This section discusses the basic steps in the development of mobile client software, then covers various design principles and guidelines.

13.3.1 Mobile Client Development Process and Methods

Figure 13.3 shows a rational development process for mobile client software. It consists of the following steps.

- *Step #1: Mobile client architecture design:* The focus of this step is to select an appropriate mobile client architecture model (e.g., thin client), based on the given system requirements. The decision to support different types of input/output modes (e.g., unimodal and multimodal) should be made. The architecture style and structure of mobile client software must be defined in terms of its function components and interactions. Mobile client design also involves wireless network connectivity (e.g., WAP, HTTP, SIP, and MMS) and related interface technologies.

- *Step #2: Mobile client technology selection:* Engineers need to finalize the required mobile platforms and technologies, including mobile platforms, mobile browsers, and presentation technologies, at this step. Mobile-enabled frameworks and enterprise-oriented application interfaces should be selected when developing thick or smart mobile clients. Some service-oriented application interfaces and frameworks may be selected to support service features, such as location detection, service discovery, language, and multimedia support.

- *Step #3: Mobile client logic design:* The primary task of this step is to conduct logic structure design by identifying classes and objects, as well as their attributes and relationships in mobile client software. This step is very useful for thick or smart mobile client software, but may not be required for creating a WML-based mobile client. Section 13.5 provides more detailed examples.

- *Step #4: Mobile client data design:* This step is required only when a mobile data repository is needed on mobile devices to store mobile data. Thick or smart mobile client software usually requires a mobile database to support

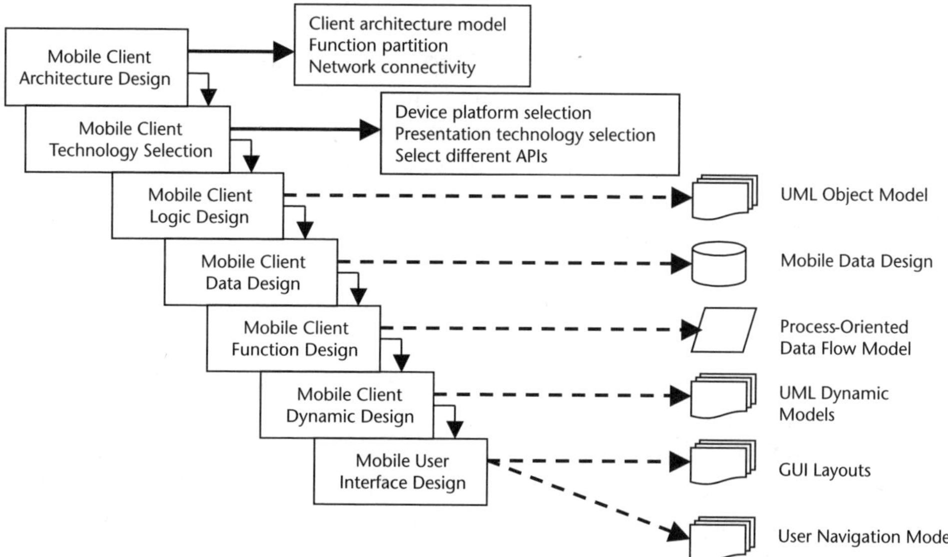

Figure 13.3 A development procedure for mobile client software.

mobile application functions. All mobile data objects, including their attributes and relationships, need to be identified in order to perform mobile data design. The corresponding mobile database technology and supporting application interfaces should be selected and finalized. Data cache management decisions also need to be finalized, and appropriate data-hoarding algorithms should be defined or selected.

- *Step #5: Mobile client function design:* The function design includes function partitioning, functional dataflow modeling, and multithreading. The output of this step includes a functional structure and data flow diagram presenting the functional processes and their inputs and outputs.

- *Step #6: Mobile client dynamic design:* The dynamic design focuses on dynamic behaviors of mobile client software, including dynamic states and transitions between states. Two major outputs are generated from this step: (1) the UML-based state diagrams presenting protocol-based communication processes, and (2) class sequence diagrams presenting the internal interactions between class objects. This step is useful and important for all types of mobile client software. Section 13.5 provides examples.

- *Step #7: Mobile user interface design:* This step focuses on the design of the mobile graphic user interface, including interface layouts, styles, and operation sequences. The outputs of this step include interface layouts and mobile user interaction models, known as mobile user navigation diagrams. Section 13.5 provides some detailed examples.

13.3.2 Design Principles, Guidelines, and Tips

A number of published articles [5–8] have discussed various design guidelines and tips for building wireless Internet mobile clients. A number of design principles are listed as follows.

- Provide the consistency of a mobile client interface on diverse mobile devices [9]. It is important to keep this consistency of a mobile client on different devices by offering the same "look and feel" to mobile users; for example, by using a well-defined consistent presentation style, formats, and elements. Using consistent button names, option labels, colors, and fonts is another way to maintain the consistency of mobile clients. It is important to offer the consistent inputs/outputs modes for mobile devices to avoid features that are unique to specific devices and platforms, unless developing thick mobile clients.

- Offer informative feedback to mobile users, such as alerts, confirmations, and error messages.

- Enhance mobile experience by design for mobile users. Due to the short access time, mobile users expect their mobile access to be more interactive, user-friendly, convenient, and enjoyable. More attention should be paid to providing interactive user-friendly dialogues with mobile users. One effective method is to provide frequent mobile users with shortcuts and hotkeys to reduce the number of system-and-user interactions.

- Provide users with reversal of actions and support error recovery [2] to cope with network disconnection and abandon sessions. The mobile client should offer error prevention, handling, and recovery capabilities.
- Design for multiple and dynamic contexts [3]. A mobile client interface should be easily accessed in various dynamic environmental conditions (e.g., brightness, noise levels, and weather), in diverse mobile environments.
- Design for "top-down" user-and-system interactions [10]. Very limited information can be presented on each screen, due to its small display size. A better way of presenting mobile contents is to partition it using a multilevel hierarchical mechanism [10].

Some design guidelines are needed to help engineers to design high-quality mobile user interfaces. They are listed here.

1. Consider network throughput while designing mobile clients. The network throughput is always an issue when more users are competing for limited resources. The apparent throughput for a user (i.e., the amount of data that appears to be exchanged) is opposed by the network throughput, which can be measured by calculating the average amount of data exchanged over time. It is always a good idea to control network throughput by controlling mobile page length through appropriate formatting. Grouping and classifying mobile input/output data is a useful way to present mobile information (input/output) in smaller data units. Finally, using images judiciously can increase the perception of high throughput.
2. Carefully select and use fonts for mobile clients. Low-power mobile devices have very limited resources, including fonts. It is a good idea to stick to the default font on mobile devices, and avoid using many fonts. Engineers need to be aware that fonts with the same name may appear different from device to device..
3. Avoid scrolling in mobile clients. Mobile users may need scrolling if the page content is longer than the display screen size. Most mobile users do not like to scroll mobile contents in a dynamic mobile environment [6]. Therefore, it is a good idea to design mobile pages small enough so that they are displayable on a small screen without scrolling.
4. Minimize mobile text inputs. It is important to minimize user inputs in different ways. One way is to allow a user to select from a list, an option menu, or a check box. Another way is to provide the default values for the given text fields (e.g., the current date). Option buttons can be used to replace input test fields. Supporting mobile audio inputs is another method of reducing text inputs.
5. Control and reduce the use of images. Images are rarely used for mobile branding and wireless advertising. Keep necessary images or icons as small as possible, and present them clearly on a monochromatic device. Images are supported only in some mobile Web browsers. Ensure that images are created with high-contrast colors that are also viewable in black and white.
6. Select and use concise text string and labels. Sentences should be simple and concise. Long sentences and subordinate clauses should be avoided. Never

use a long word unless necessary. Carefully select concise labels for function buttons and options, and keep all system messages short and simple.

7. Use colors sparingly. Most mobile devices do not have a color screen. It is a good idea to use color sparingly. When a color user interface is available, use default colors and minimize the use of colors in each page.

8. Define simple tables with few columns and rows. Mobile devices can only display simple tables with few columns and rows, due to the limited screen size. It is important to avoid nested tables. Good table examples are weather forecasts and stock quotes. Convert a complex table is into a multilevel simple table with indexes.

9. Create buttons effectively. Creating buttons using the screen spaces provides an opportunity for the wireless content developers to emulate a device's native interface. It is important to avoid using the specific features and keys (or buttons) from mobile devices, and keep consistent button style and labels.

10. Find current locations for mobile users. Location detection capability is very helpful to mobile users. Only some mobile devices and service vendors provide this feature. One approach is to design and implement an automatic positioning mechanism in the system. Another approach is to provide mobile users with a navigation aid, such as maps and directions, and optimize user input for location finding.

There are a number of design tips for mobile client developers.

- *Tip #1: Pay attention to the limits of mobile devices.* The largest block of information should not exceed 1,200 bytes, due to the limited bandwidth and memory of devices. The maximum screen resolution typically is 95×45 pixels. A mobile user has keys 1 through 9, and letters for inputting selections. Other available keys vary from device to device. It may be a good idea to avoid special keys offered by certain mobile devices.

- *Tip #2: Limit and control graphic display.* It is important to restrict the size of graphics. Some browsers will not display an image if it is larger than the screen size. WAP requires the conversion of graphics to WBMP, a monochrome bitmap image.

- *Tip #3: Provide concise and consistent text and pay attention to word length.* Many mobile devices (e.g., mobile phones) only provide 3 to 6 lines of text, including possibly 12 to 20 characters. Most mobile devices support text wrapping, although certain devices may truncate sentences when they are too long.

- *Tip #4: Provide basic navigation methods and avoid scrolling through a list.* Mobile clients, unlike Web browsers, cannot rely on minibrowsers to provide navigation, even basic navigation such as moving forward and backward.

- *Tip #5: Offer classified contents with guidelines.* Mobile content should be classified into groups with simple and effective content guidelines to help users quickly narrow their search and find the required content. A mobile portal should be grouped based on a well-defined classification scheme, so that diverse mobile contents, such as entertainment, news, and weather forecasting, can be effectively grouped.

- *Tip #6: Support mobile users with search.* Supporting mobile users with an effective search function is necessary because it is impossible to shift through pages to find the right information. Different search methods must be provided to mobile users. For example, search-by-alphabet can be provided to allow users to jump to specific letters. Search-by-zip-code can be provided to search location-based information, such as a map.
- *Tip #7: Provide all options.* It is possible that mobile users will want to view all options in a list. This feature can be provided to mobile users by limiting the amount of information in each page, and including a an option that allows the user to navigate to the next page. Users also may switch to a specific page by entering page number on a keypad.
- *Tip #8: Effectively support mobile users to input.* Make sure to effectively use the keys on the mobile devices to support mobile users to input information. For example, develop mobile client software that allows users to use the device keys as numbers when zip codes are requested.
- *Tip #9: Chunk information into decks (small screen pages).* Use multiple cards in separate decks, if the text (or list) of links exceeds the recommended 1,200 bytes. Provide a link on each card that allows the user to move to the next card or to another deck. Provide a link on the last card that allows the user to return to the initial screen.

13.4   Mobile Client Design Issues and Solutions

Engineers encounter a number of design challenges, as mentioned in Section 13.1. This section discusses the challenges by providing possible solutions.

13.4.1   Design for Reliable Mobile Client Software

The design of reliable mobile client software is one of the major challenges in a wireless-based application system design, due to the unreliability of wireless network connectivity. Designing mobile client software requires attention to this issue, particularly when mobile clients are highly dependent on wireless network connectivity. Typical examples are application systems that involve connection-based wireless communications and thin mobile client software.

Engineers need to design for reliability in three areas. The first is to design for reliable communications between mobile clients and application servers, which is essential and critical for all wireless-based application systems. The second is to design for reliability of mobile client software that resides on mobile devices. This is very important for developing thick and smart mobile clients. The third is to design for reliable application transactions, which is important when long life application transactions are required.

Engineers need to understand underlying protocols, and perform protocol analysis and dynamic modeling for message communications, in order to design for reliable communications in wireless-based application systems. They can use a

UML-based state diagram to present protocol-based communication states and transitions for mobile client software. We need to consider the following cases for each type of incoming message, their corresponding responses, and handling processes (as shown in Figure 13.4):

1. Incorrect or unrecognized incoming messages;
2. Truncated messages;
3. Timeouts for receiving messages;
4. Network connection failures, which can be identified by message sending failures;
5. Server connection failures, which only can be recognized by repeated timeout events.

A NAK message should be sent to the server to cope with the first three cases. For the last two cases, mobile users must be altered if their usage is affected. The cached mobile data should be used to support mobile access, if it is available. Some dynamic transaction data may need to be stored on a mobile database for subsequent recovery.

Engineers must consider how to deal with the following abnormal cases to enhance the reliability of mobile client software, particularly for smart or thick mobile clients.

- Out of memory;
- Out of battery;
- Abandonment of application sessions;
- Failure to access external devices, such as a digital camera and a flash card;
- Failure to create or access mobile data in a data repository on mobile devices;
- Network disconnection.

The support of reliable transactions between an application server and its mobile clients is a challenge in the design of mobile client software. The reliability of

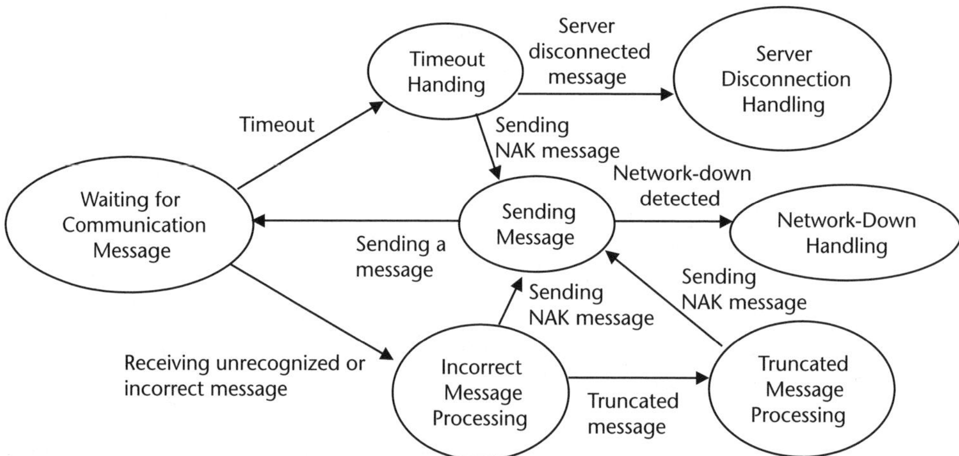

Figure 13.4 Dealing with abnormal states in mobile communications.

application transactions in wireless-based systems depends on the following two factors.

- *The connection modes between mobile clients and the server:* There are two connection models: connectionless and connection modes. In the connectionless mode, each mobile transaction is implemented as a request and its response. The application server processes each request from a mobile client as a stateless message. In the connection mode, a mobile client communicates with its application server over a wireless network to perform state-based mobile transactions during a service session. The application server using a state-based processor often processes mobile transactions during a mobile access session.

- *Duration of mobile transactions:* Mobile transactions can be classified into either short-lived or long-lived transactions. Short-lived transactions usually last for just a few minutes. Typical applications using short-lived transactions are wireless messaging systems; for example, the SMS-based messaging service application. Long-lived transactions usually last for up to a number of hours. Workflow management, cooperative editing, and digital games are typical application examples using long-lived transactions.

Supporting long-lived transactions is difficult in wireless-based applications using low-end mobile devices. There are two basic strategies in dealing with the abnormal terminations of mobile clients. The first is to simply restart a new transaction when the original mobile client is abnormally terminated, and cancel the incomplete mobile transaction. Since this abandons the original transaction with a new transaction with a different transaction identifier, it causes the problems for mobile transactions that require atomic operations, consistent transaction identifiers, and durable property. The other strategy is to create an alternative mobile client to play as an alternating client host to provide reliable processing of transactions. This approach cannot be used for all scenarios, as noted in [11]. However, it is useful for applications in which a mobile client host accompanies but does not replace the fixed client host. One example given in [11] is the sales management and support system, which supports online accesses from a fixed host and supports a traveling salesman's mobile accesses. Buchholz et al. [11] discuss three approaches supporting reliable transactions in multiple mobile clients using TXAgent (a program in the fixed network) to manage transactions between a wireless-based application server and its mobile clients. Figure 13.5 shows its architecture.

- *Approach #1:* A mobile client, rather than a TXAgent, holds a copy of the execution states of each transaction. Consequently, a failure of the mobile client leads to a loss of the execution state and the transaction must be rolled back.

- *Approach #2:* A TXAgent keeps the execution state of the transaction in combination with stateless mobile clients, while mobile clients only support use-and-system interactions. The TXAgent keeps all transaction states for each mobile client. One TXAgent object is created for each mobile client to support and manage its transaction states. This approach may result in the exchange of many messages between a client and the TXAgent, and a delay in

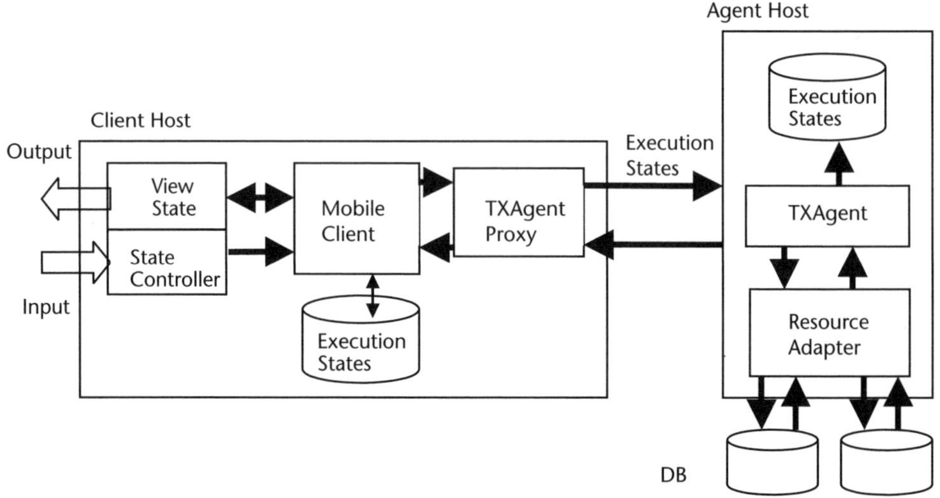

Figure 13.5   The basic architecture of a TXAgent and a mobile client. (*From:* [11]. © 2000 IEEE Computer Society. Reprinted with permission.)

the interactivity, because all interactions of a mobile transaction will be propagated to the TXAgent.

- *Approach #3*: Replicated copies of the execution state of the transactions are kept by a mobile client and the TXAgent. This is a hybrid approach that combines the advantages of the previous approaches. A mobile client processes transactions and keeps the execution state of the transactions. The TXAgent also holds a replica of this execution state. A failure of the mobile client does not necessarily lead to a rollback, because a new client may resume the transaction, using the TXAgent's copy of the execution state. The copy of the execution state of the transaction kept by the TXAgent is not updated every time that the mobile client's copy of the execution state changes. The TXAgent's copy updates only at certain points in time, called checkpoints. This approach reduces the number of messages exchanged between a mobile client and the TXAgent. A new client cannot resume the transaction of a failed client at the current execution state, but rather at the state of the last checkpoint.

Table 13.3 displays a comparative summary of the three approaches supporting reliable transactions.

13.4.2   Design Unified Mobile Client Interfaces

Many mobile devices provide a mobile browser to display Web content pages on a small display screen, although each device may have its own browsing format and presentation. This causes a design issue for a thin mobile client, because it is a time-consuming and tedious job to write mobile pages with a specific browsing format for each device.

Yang et al. [12] provided an answer to the issue of designing for a unified mobile user interface. The basic idea is to build a systematic process and tool to provide an XML-based unified user interface to support different mobile devices, by converting the content pages to different presentation formats on diverse mobile devices.

Table 13.3   Comparison of Different Approaches Supporting Reliable Transactions

Feature	Client Types			
	One Migrating Client	Approach #1	Approach #2	Approach #3
Commitment control	By the client after migration onto a fixed host	By the TXAgent on the fixed network	By the TXAgent on the fixed network	By the TXAgent on the fixed network
Coping with exception during operation calls	Transaction rollback	Get the result from the TXAgent (e.g., by call repetition)	Exceptions caused by faults on the fixed network only; then transaction rollback	Get the result from the TXAgent (e.g., by call repetition)
Local processing during disconnection	Possible (until the next data resource access)	Possible (until the next data resource access)	Impossible	Possible (until the next checkpoint)
Transfer of the execution of transaction states	By migration	Direct transfer between clients	No need to transfer (execution state is kept by the TXAgent)	Indirect transfer between clients
Coping with failures of the client	Transaction rollback	Transaction rollback	Resume the transaction with a new client in the current execution state	Resume the transaction with a new client in the execution state saved at the last checkpoint

Source: [11].

As shown in Figure 13.6, this solution has the following features:

- Use of the characteristics of XML to separate content and presentation;
- Use of a server to save the profiles of Web-enabled mobile devices, which record device capability and functionality;
- Use of a document generator (or converter) to convert XML documents to a content page in HTML/WML, and other types of markup languages based on the given mobile device's profile.

The major benefit of this solution is to deliver unified-content documents and cost reduction on page generation for different mobile browsers. The delivery of mobile content pages usually is slower than in nonunified content generation systems, since all mobile content pages are statically or dynamically generated by XML-based conversions.

The key to this solution is to use XML to handle the data in mobile content pages. The XSLT style sheets extract and format the mobile data for presentation. Different style sheets must be used for different devices, because it is very difficult to construct a single XSLT style sheet to produce a single WML for different WML-enabled wireless devices and microbrowsers. Different XSLT style sheets must be manually constructed for PDAs or mobile phones [13]. Whenever a new wireless application system is set up, various WMLs have to be generated from these different XSLT style sheets by the XSLT engine. The Servlets, JSPs, or ASPs can identify the mobile device client during the run time by detecting two HTTP header

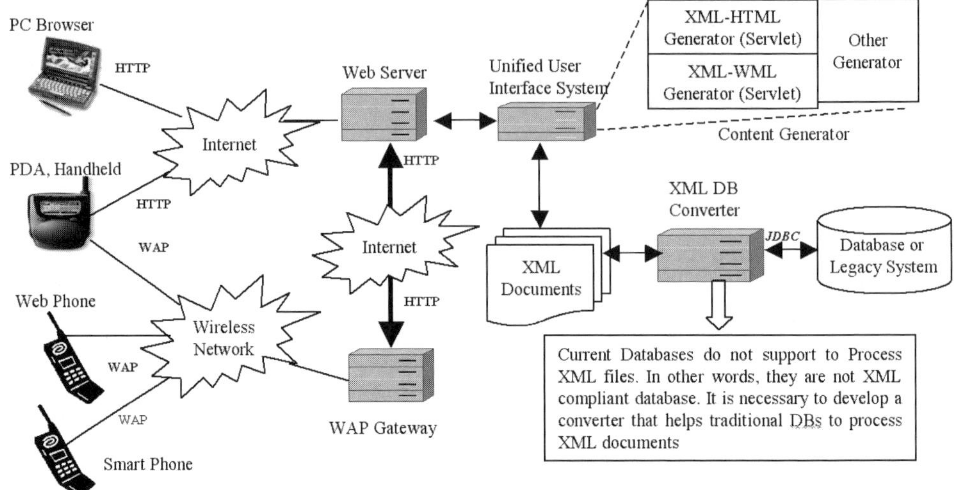

Figure 13.6   An XML-based unified mobile user interface system.

values: User-Agent and Accept. Different XSLT style sheets are then chosen for the XSLT engine, enabling it to produce WML documents for diverse mobile devices. The application developer must manually construct a large number of XSLT style sheets for only one wireless application. This task could be very lengthy and cumbersome, depending on the number of device types that the system intends to support. It is necessary to find a systematic solution for easily generating XSLT style sheets to support different devices.

Kwok et al. [13] present an approach to automatically generate XSLT style sheets based on presentation rules and content rules. Figure 13.7(a) shows a systematic process to create presentation contents and define the content rules in a setup process for wireless-based application systems. These presentation rules must be defined based on the configuration and profiles of mobile devices. Figure 13.7(b) presents the detailed workflow to generate XSLT style sheets automatically for runtime use. The device information defined from the WAP header is passed to the XSLT style sheet selector during runtime, as shown in Figure 13.7(c).

The selector chooses a style sheet with respect to the mobile device information and the DTD document (or XML schema) from the XSLT style sheet pool. An XSLT engine generates a WML document with an appropriate content and presentation style, based on the input of the chosen style sheet and XML document. This solution also can be used handle the content pages in other formats, such as HDML, cHTML, and XML.

13.4.3   Adding Mobile Links to Online Systems

Most businesses and enterprises have established online Web sites for customers and users. Content providers, publishers, and online portals have also provided a great number of accessible HTML pages on the Internet. These organizations face the urgent need to provide mobile online sites to mobile users, due to the rapid increase in the number of mobile device users. They have two choices for achieving this. The first is to establish a new mobile site for mobile users to access newly generated

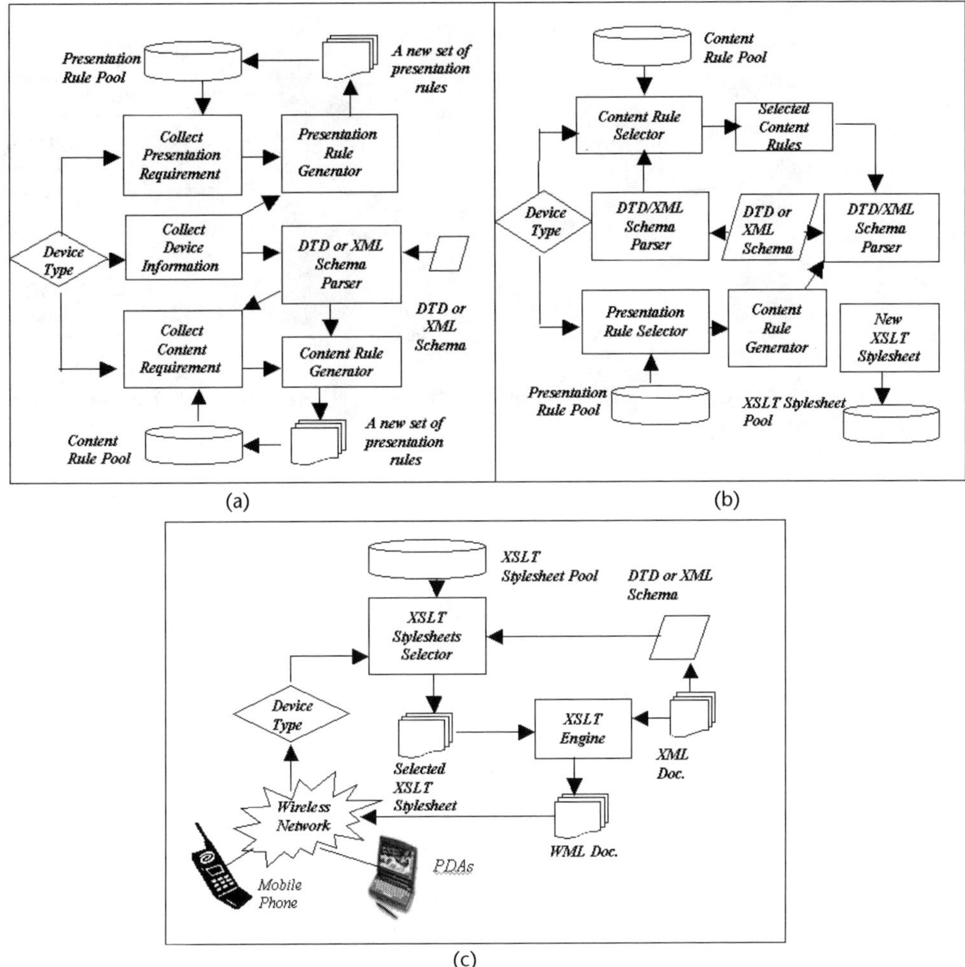

Figure 13.7   Systematic approach to generate XML documents and XSLT style sheets: (a) creation of content and presentation rules pool, (b) generation of XSLT stylesheets, and (c) production of XML documents.

mobile contents (e.g., WML-based content pages). The other alternative is to convert the existing Web site to WML pages (or cHTML/XHTML pages).

Converting HTML Pages to WML Pages

Kaasinen et al. [14] present a systematic approach for converting HTML pages to WML pages. Their system includes separate specialized proxy servers that perform three tasks in content conversion, caching, and content adaptation, as shown in Figure 13.8.

- Proxies support the adaptation of Web application content to different terminal and network environments.
  - A proxy server can try to minimize the information flow over low-speed or medium-speed wireless links.

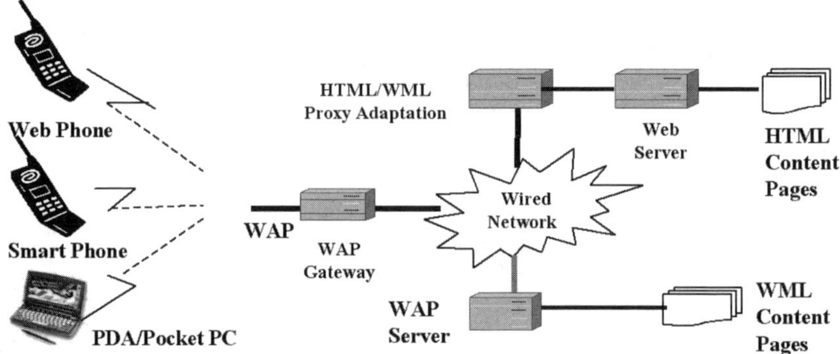

Figure 13.8 Two approaches to adding mobile links. (*From:* [14]. © 2000 Elsevier Science B.V. Reprinted with permission.)

- The proxy server may filter out some content types of HTTP streams (e.g., images, Java script, or Java Applets).
- The proxy server can modify the content (e.g., image depth and size), based on the user's preferences and channel throughput.
- Content converters perform two basic tasks:
  - They parse HTML documents, extract logical elements (e.g., attributes and texts), and correct errors based on the given DTDs.
  - They convert HTML content to WML content, based on a given set of conversion rules. Some data inevitably may be lost during the conversion process.
  - A tree structure is created to represent the resulting WML decks. The tree is then manipulated according to adaptation rules.
- The conversion rules are generated, based on configuration parameters, user's preferences, and capabilities of the requesting user agents.

A set of conversion rules is given here.

- Each HTML frame of a frame set is converted to one or more WML decks.
- Frame sets are converted into decks that provide indices to WML decks, which correspond with individual frames.
- The elements of user inputs on a form are converted to a single card, where they appear in the same order as the original HTML code.
- Graphic layouts usually are lost in the conversion.
- Groups of user inputs can be converted into WML cards, and each group can be converted into a selective input list.
- An HTML table can be converted into one of the following, based on the size of the table and user agent capabilities.
  - A WML table, which is usually not nested, contains one or two columns, due to the small size of display screen on mobile devices. This conversion is only useful when the HTML table is simple.

- An indexed subtree is a two-level tree, in which the root card is an index of links to the leaf cards. Each leaf card presents the content of a single table cell. This method works well when converting layout tables.
- A list is the simplest way to convert HTML tables. It concatenates the contents of the table cells to form a list that produces compact and clear output, but loses the table structure.

Implementation issues during page conversions are listed as follows, according to [14].

- An HTML page that contains too much information may need intelligent support to identify and extract the important information elements.
- One problem in creating indices is to find informative labels to the links. HTML 4.0 provides many ways to include the metadata, which is not presented in many cases.
- Online tables generate layouts or organize information. It is difficult to define the purpose of the table using automatic conversion. Table sizing and user agent type may be useful.
- Image buttons often cause problems when filling in and submitting forms.
- Preformatted text also causes a conversion problem because the formatting does not function on a small screen.

13.5 Application Examples of Mobile Client Software Design

We present in this section a mobile client design and implementation example, based on a research project and its prototyping system. The system is a mobile messaging system, known as Mobile Messenger. Its high-level function requirements are given in Chapter 9. Figure 13.9 shows its function blocks in its J2ME-based mobile client software.

Mobile Messenger is a wireless-based multimedia messaging system that helps mobile device users manage and maintain their personal multimedia messages using mobile devices. It supports mobile devices, such as mobile phones with 3G network

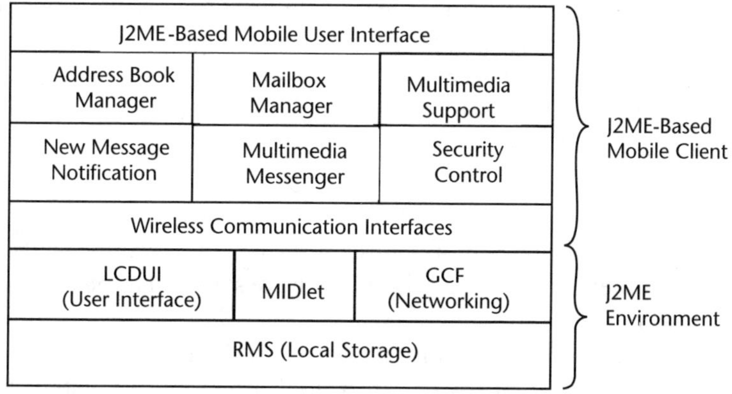

Figure 13.9 Functional blocks of mobile client software for Mobile Messenger.

connectivity and WAP technology. J2ME is used to develop mobile client software to support system-and-user interactions, text-based inputs and outputs, and multimedia accesses (e.g., digital audio messages, images, or short video clips). The mobile client software can be grouped into two parts, as shown in Figure 13.9. The first part consists of J2ME components in a J2ME platform, which has been covered in the chapters of Part II. The other part consists of the following components:

- User interface, which supports system-and-user interactions;
- Address book manager, which supports the address book function, including adding, deleting, updating, and viewing the address book and its entries;
- Mailbox manager, which manages and maintains a multimedia mailbox for a mobile user on a client site, such as adding, deleting, viewing (or listening to) a message;
- Multimedia support, which supports multimedia interactions, such as inputting or outputting a wireless message;
- Multimedia messenger, which supports the MMS communication protocol, including packing and unpacking MMS-based messages;
- New message notification, which provides the MMS-based notification for newly arrived messages in a mailbox;
- Security control, which performs necessary system security and access control for mobile users;
- Wireless communication interface, which supports the wireless network connectivity.

Figure 13.10 shows the result of dynamic analysis and design for the mobile client. It presents its dynamic behaviors in terms of states and transitions in a UML-based state diagram. This diagram does not present some abnormal states and their handling (e.g., network connectivity failure and incorrect messages).

We have used a model, known as a navigation diagram, to present the system-and-user interactions of a mobile interface. Figure 13.11 shows an example. Each text box in this model represents a mobile user interface entity, such as a screen page. Each link represents a mobile user interaction event, such as a function button click or an option selection. A number of research projects and implementations have found this model to be very useful in designing a user-friendly mobile interaction. Designers can easily find the missing links (incomplete functions), unreachable pages, and other issues in operation flows. Designers can easily verify the usability and operation ability of a mobile client interface.

Figure 13.12 shows an example of a multimedia message using this mobile client, which allows a mobile user to listen to an audio message with a short text message.

Figure 13.13 presents the mobile user interface and its layout screens. The operation sequence shows a mobile user preparing and sending out a multimedia message to another user. The details about the design and implementation of this mobile client can be found in [15].

Figure 13.14 shows the mobile client interface and layout screens for supporting a mobile user editing a roster list in the Mobile Jabber-Instant Messaging System,

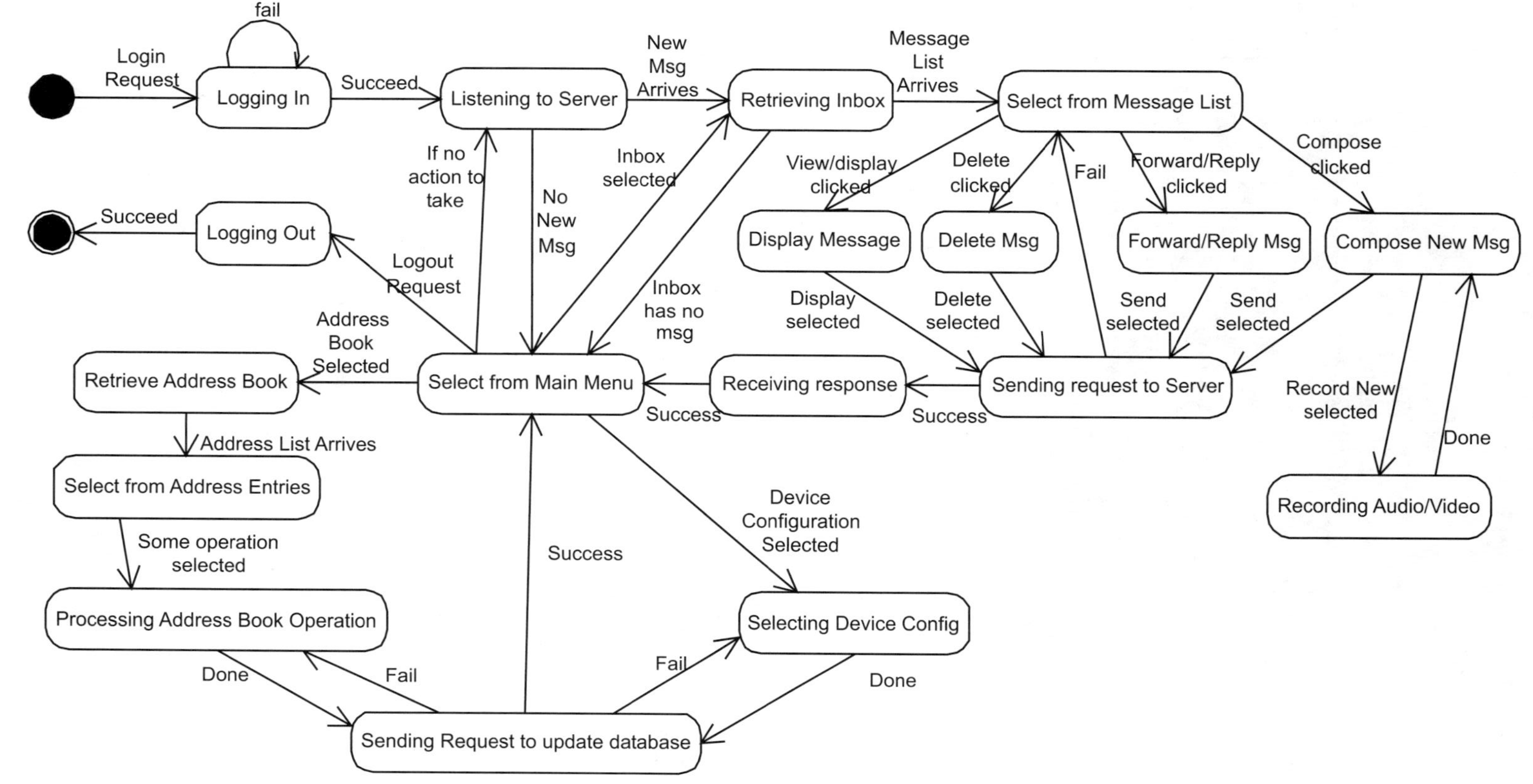

Figure 13.10   The UML-based state diagram for mobile client of Mobile Messenger.

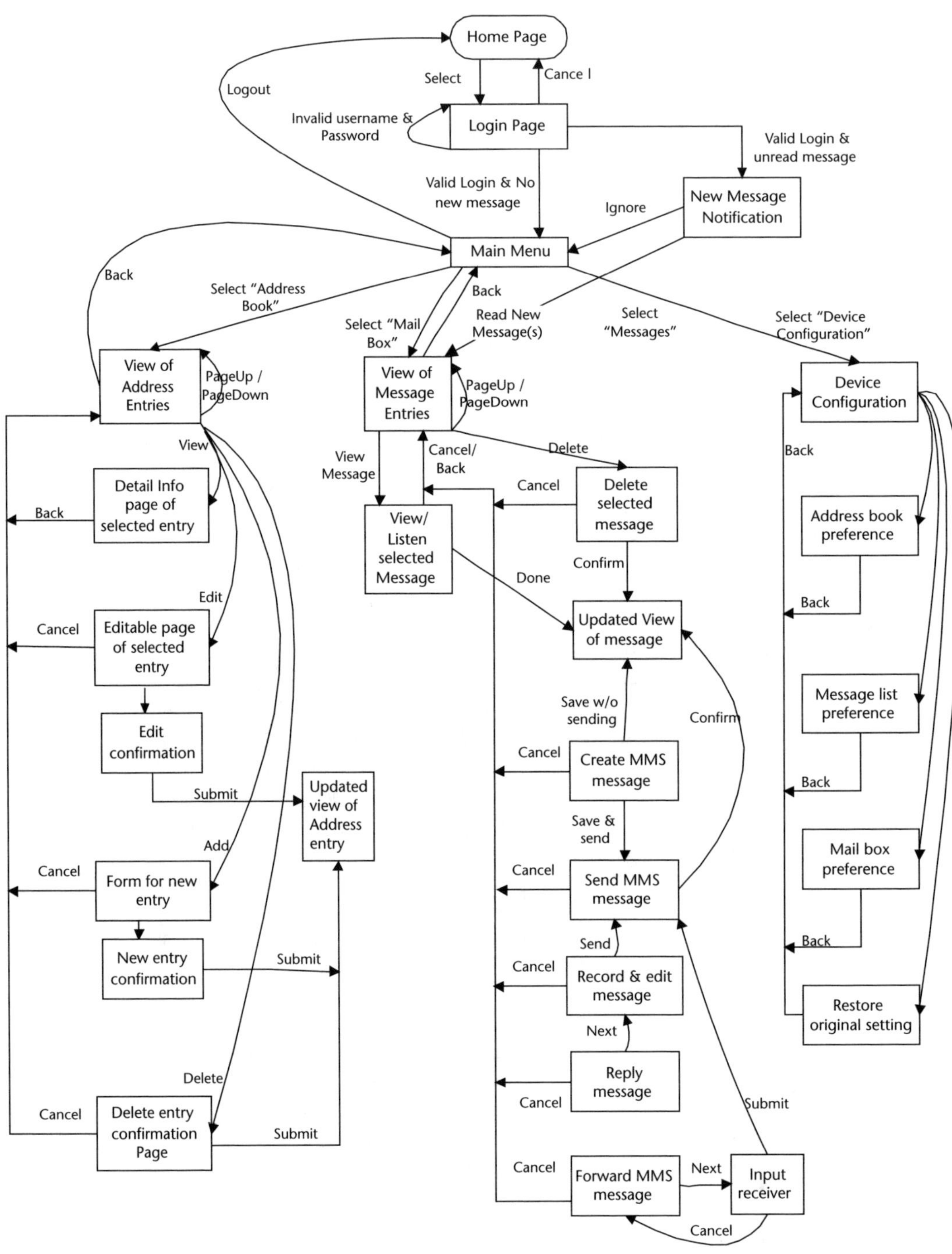

Figure 13.11   The navigation diagram for mobile client of Mobile Messenger.

Figure 13.12   Screen shots for displaying and viewing an MMS message.

known as Jabber-IM. This is a text chatting system, developed at San Jose State University in 2003. Figure 13.14 shows the detailed operation sequence to edit the roster list, and the details about this system can be found in [16].

13.6   Summary

The development of mobile client software for wireless-based application systems has some new design issues and challenges, compared to the development of conventional user interfaces. Engineers need to do the following before beginning wireless programming.

- Select the right type of mobile clients (e.g., thin, smart, and thick), and define its architectural model and interaction modes, based on the given system requirements in network connectivity, functional features, and services;
- Select and determine proper mobile platforms and technologies for developing and deploying mobile client software;
- Partition mobile client software into function blocks, and define the organization and interactions between them;
- Perform logic analysis and design of thick or smart mobile clients, by focusing on classes and their relationships;
- Conduct dynamic modeling for mobile client software to focus on its state-based processing and communications;
- Identify and determine the layout and operation sequences of mobile client software.

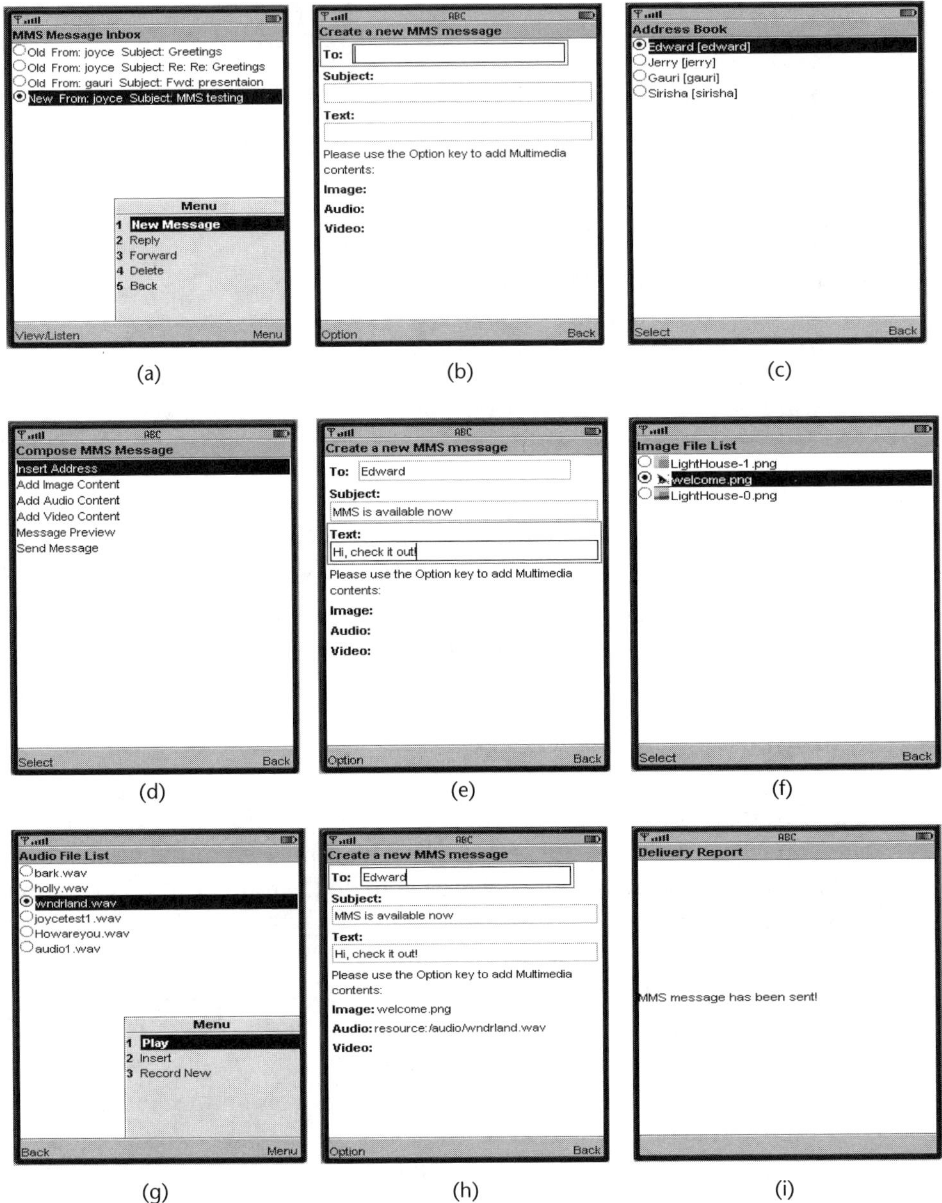

Figure 13.13 Screen shots for composing and sending a new MMS message: (a) make a selection, (b) create a new message, (c) select a receiver, (d) select message media, (e) edit a text message, (f) add an image to the message, (g) select the format, (h) complete the message, and (i) send out the message.

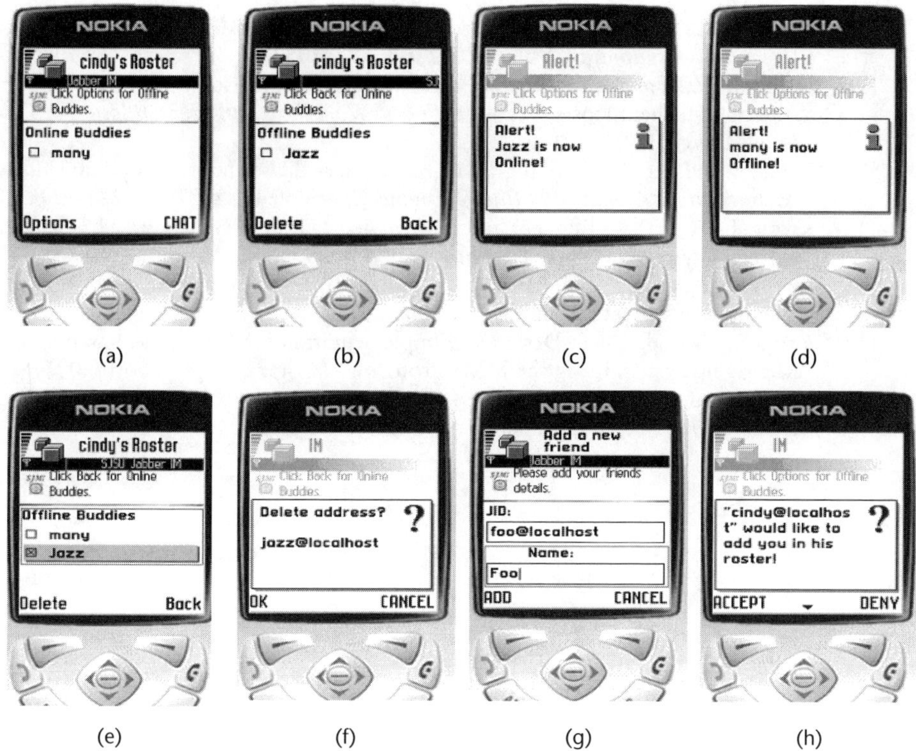

Figure 13.14   Screen shots for editing a mobile user's roster list: (a) online buddy list, (b) off-line buddy list, (c) friend comes online, (d) friend goes off-line, (e) add a friend UI, (f) accept request for add, (g) delete a friend, and (h) confirmational alert.

References

[1]   Albers, M. J., and L. Kim, "Information Design for the Small Screen Interface: An Overview of Web Design Issues for Personal Digital Assistants," *Technical Communication*, February 2002, Vol. 49, No. 1, pp. 45–60.

[2]   Satyanarayanan, M., "Fundamental Challenges in Mobile Computing," *Proc. 15th Annual ACM Symposium on Principles of Distributed Computing*, Philadelphia, PA, May 1996, pp. 1–7.

[3]   Tarasewich, P., "Designing Mobile Commerce Applications," *Communications of the ACM*, Vol. 46, No. 12, 2003, pp. 57–60.

[4]   Jing, J., A. Helal, and A. Elmagarmind, "Client-Server Computing in Mobile Environment," *ACM Computing Surveys*, Vol. 31, No. 2, June 1999, pp. 117–157.

[5]   Grundy, J., X. Wang, and J. Hosking, "Building Multi-Device, Component-Based, Thin-Client Groupware: Issues and Experience," *Proc. 3rd Australasian Conference on User Interface*, Melbourne, Australia, 2002, pp. 71–80.

[6]   Heins, J. E., and D. S. Steiner, "Taking Your Information Into the Wireless World: Developing Information for Delivery to Mobile Devices," *Proc. IEEE Int. Professional Communication Conference (IPCC 2001)*, Sante Fe, NM, October 2001, pp. 237–244.

[7]   Gong, J., and P. Tarsewich, "Guidelines for Handheld Mobile Device Interface Design," *Proc. DSI 2004 Annual Meeting*, Boston, MA, November 2004.

[8]   Yoshimi, B., et al., "Lessons Learned in Deploying a Wireless, Internet Application on Mobile Devices," *Proc. 4th IEEE Workshop on Mobile Computing Systems and Applications (WMCSA'02)*, IEEE Computer Society, Callicoon, NY, June 2002.

[9]   Chan, S., et al., "Usability For Mobile Commerce Across Multiple Form Factors," *J. of Electronic Commerce Research*, Vol. 3, No. 3, 2002.

[10] Brewster, S., "Overcoming the Lack of Screen Spaces on Mobile Computers," *Personal and Ubiquitous Computing*, Vol. 6, 2002, pp. 188–205.

[11] Buchholz, S., et al., "Transaction Processing in a Mobile Computing Environment with Alternating Client Hosts," *Proc. 10th Int. Workshop on Research Issues in Data Engineering*, 2000, pp. 1–8.

[12] Yang, S. J. H., et al., "Building XML-Based Unified User Interface System Under J2EE Architecture," *Annals of Software Engineering*, Vol. 12, 2001, pp. 241–256.

[13] Kwok, T., et al., "An Efficient and Systematic Method to Generate XSLT Stylesheets for Different Wireless Pervasive Devices," *WWW 2004*, New York, May 17–22, 2004.

[14] Kaasinen, E., et al., "Two Approaches to Bring Internet Service to WAP Devices," *Computer Networks*, Vol. 33, 2000, pp. 231–246.

[15] Yang, J., "Mobile Client Design and Implementation for A Wireless-Based Multimedia Messaging System Using the MMS Protocol," Master Project Report, 2005.

[16] Gao, J., et al., "Mobile Jabber IM: A Wireless-Based Text Chatting System," *Proc. IEEE E-Commerce Technology*, San Diego, CA, July 6–9, 2004.

Mobile Commerce Systems

The wide deployment of wireless networks and the explosive growth in the number of mobile users have created a very strong demand on mobile commerce applications and a variety of mobile services that deliver contents. According to Nokia, there are 1.4 billion mobile phone users in the world, and the number is rapidly increasing. In Europe and Asia, transactions of mobile commerce and services are reaching billions of dollars per year. Mobile commerce and services are heralded as the mobile revolution. They are providing individuals and companies with unprecedented new business opportunities.

The following chapters examine the engineering aspects of mobile commerce in detail. Chapter 14 introduces the design and implementation of mobile commerce systems. Chapter 15 examines multimedia messaging services. Chapter 16 discusses wireless advertising and marketing systems. Chapter 17 presents various mobile payment systems.

CHAPTER 14

Introduction to Mobile Commerce Systems

The mobile Internet has brought dramatic changes to individuals and businesses alike. Individuals and businesses are communicating through e-mail and instant messages, and engaging in mobile commerce (m-commerce). M-commerce allows goods and services to be bought and sold electronically through mobile devices, and provides exchange of electronic information on transactions. The information from these transactions further facilitates the study of focused groups of consumers. Merchants can use the information to more specifically address the needs of the target customers, according to their profiles and behaviors. Broadband connections to mobile phones, and the convergence of video, voice, and data is an uptrend in the industry. Few things can compare to the success of mobile handheld devices in today's post-PC consumer electronics market. Mobile personal devices, such as Palm Pilot devices, smart cellular phones, and personal digital assistants (PDAs), have been popular with both consumers and businesspeople. The decrease in price of the mobile devices due to market competitiveness and the established open standard for wireless application development are fueling this hypergrowth of wireless e-commerce applications or mobile commerce.

M-commerce refers to the ability to conduct wireless commerce applications using mobile devices. M-commerce applications can range from a simple program to synchronize an address book to a software package, to something as complicated as credit card transaction management and wireless location-based solutions. M-commerce is dramatically growing, and supporting complex commerce transactions and services. The well-defined wireless application protocol (WAP), which facilitates the development of wireless applications, may not be sufficient to handle complex mobile transactions that require the cooperation of different service components. An intelligent, robust, and scalable framework that provides diverse m-commerce services is required to efficiently handle these increasingly complex mobile commerce applications.

Surveys by marketing research firms show that mobile commerce transactions alone will amount to a few hundred billion dollars in the next couple of years [1]. These numbers might sound too optimistic to some industry critics, due to the fact that the use of mobile applications in the United States has been lagging behind that in Europe and Asia. There is no significant evidence that the U.S. mobile applications market will catch up with the market in those parts of the world in the near future. One common criticism is that the WAP [2] applications basically have not been proven to be user-friendly. Few useful WAP applications are available in the market. Despite some discouraging industry news due to recent economic conditions, the mobile device industry in general, and mobile application developers in particular, are still cautiously optimistic about the future of mobile commerce, in terms of the development and manufacture of new and better mobile devices that offer additional functionalities.

Factors such as user interface infrastructure design, underlying wireless networks, bandwidth requirements, and wireless application standards are critical in determining the success of mobile commerce. We believe that the ability to successfully manage the interactivity and connectivity of different mobile devices and wireless services is one of the most important aspects to the success of mobile commerce.

14.1   M-Commerce Versus E-Commerce

M-commerce differs from e-commerce in two core dimensions: "mobility" and "locatability." The mobile devices offer the mobility. The mobile networks, sometimes in conjunction with mobile devices, offer the locatability (or traceability).

The following is the comparison between m-commerce and e-commerce:

- *Mobility:* Mobility is limited with e-commerce. The user performs the e-commerce transactions using a desktop computer. Mobility is ubiquitous with the m-commerce. The user can carry out m-commerce transactions any time, anywhere, as long as he is in the mobile coverage area.

- *Locatability:* It is difficult to find a physical location of the user in the e-commerce domain, unless the user provides it. The user location with m-commerce can be identified with the help of the mobile network, as well as sometimes in conjunction with the mobile devices. This locatability attribute of m-commerce promotes location-based, as well as time-sensitive, services to the users. This feature provides a competitive edge to the m-commerce provider.

- *User identity:* A user in the e-commerce domain can be identified based on the IP address of the desktop machine through which he is doing the transactions, which is not always the same (due to DHCP/NAT and other protocols). The only way to identify the user is through user log-in. A user in the m-commerce domain can be easily identified through the mobile device, which is mostly the same as for a user. This helps to create a user-profile management and provide a personalized, customized service.

- *Personalization:* The e-commerce user needs to log in for a personalized user interface and profile management. M-commerce can provide this information with only the identity of the mobile device that the user generally uses. Losing a mobile device could turn disastrous to a user, but with the latest security and forensic advancements, this issue can be addressed.

- *PUSH marketing:* PUSH marketing refers to the pushing of an item of interest to a customer without his explicit request. PUSH marketing is difficult with e-commerce. Marketing personnel can push time-sensitive items of interest to a mobile consumer with m-commerce, due to its profile management and locatability feature.

- *Key customer concern:* E-commerce customers mostly use the free Internet, which makes them cautious about money or security. The m-commerce customers are already paying for the charges of the mobile services. These mobile customers are mostly time-cautious due to mobile usage restrictions. They want to perform the transactions faster than do the e-commerce customers, in order to conserve usage charges.

- *Customers service area:* E-commerce customers are served only at the location where desktops are installed. This puts the restrictions on the service area for e-commerce. For instance, one cannot do the e-commerce transactions while driving. There are no such restrictions for the m-commerce service area. The m-commerce customer will be served as long he is in the mobile coverage area, which expands the m-commerce service area.

- *Customer expectations:* M-commerce customers have the same expectations of services from their wireless provider as e-commerce customers have from their wired Internet provider.

14.2 Wireless Device Constraints, Application Usability, and Interface Issues

A user interface is a point-of-interaction with the user. A well-designed, user-friendly, adaptable, and usable user interface is critical to the success of any application or technology. M-commerce applications are no exception to this rule, especially considering unique factors, such as the limited size of handheld devices' screens, limited battery life, mobile nature of the applications, and so forth.

The three entities involved in information access in mobile environment are mobile device, mobile applications, and mobile users, as shown in Figure 14.1.

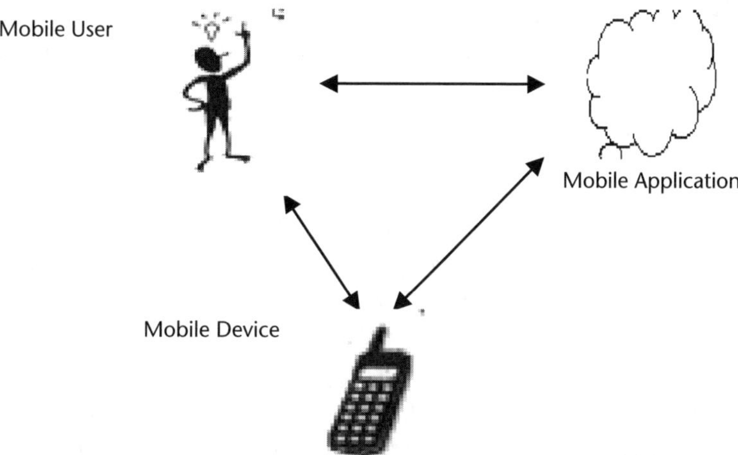

Figure 14.1 Mobile environment entities.

14.2.1 Device Constraints

Human interactions and user requirements must be considered when presenting the contents on the handheld device. The unique characteristics and the physical limitations of the device must be considered for content presentation, as well as for content management.

Following are some of the well-known limitations of the mobile devices that affect the user interaction with the mobile applications, as compared to traditional e-commerce applications [3].

- *Small screen size and color:* The small screen size of handheld devices reduces the amount of information that can be displayed. A Web page designed for the traditional e-commerce (with typical resolution of 640 × 480) cannot be displayed on the mobile device's screen. The screen resolution (typically 160 × 160), as well as the colors supported on the mobile device, affect the look of the Web page. The screen size will be small due to the portability of mobile devices, although the screen resolution and color of new products are improving. The screen size issue can be resolved when new technologies, such as electronic paper (in which the screen can be folded like paper), become commercially available. New techniques address the screen size issue, such as device-specific authoring (designing a Web page for a specific device, as is done in NTT i-mode); multiple-device authoring (designing a Web page for multiple devices); client-side navigation (allowing users to modify, zoom, or collapse a Web page content when it is displayed); and automatic-reauthoring (converting a traditional Web page to a device-specific mobile page using software).

- *Limited memory:* Mobile devices have limited memory storage. This issue will have be resolved in order for the use of m-commerce to increase.

- *Limited processing power:* Many of the Internet services, such as music, gaming, or video, require faster CPUs and higher memory levels. The mobile device CPU is less powerful than that of a desktop. This issue is also being addressed.

- *Battery life:* Batteries affect the portability (weight) and the processing power of handheld devices. Battery management technology is developing to provide light, efficient sources of power.

- *Slow download speed/limited bandwidth:* Mobile devices suffer from slow download speeds, due to the limitations in CPU, memory, and power consumption, as well as to the increased number of mobile network users. This problem will be addressed by third generation (3G) wireless networks. For example, GPRS achieves speeds equal to that of traditional analog modems.

- *High latency:* Due to mobile network behavior, the mobile devices suffer from high latency.

- *Small keyboards and input method:* Mobile devices have limited means for input. They have small keypads and a small numbers of cursor keys, which adversely affects user interactions. A fair amount of learning is involved with Graffiti-style keystroke inputs. Voice activated inputs are also being supported.

- *Limited output method:* The mobile device's typical output is a screen. Other output modes are already on the market or under development.

- *Different technologies and standards:* Standards are now under development for mobile devices and for m-commerce, since there are presently no clear-cut standards.
- *Security issues:* Complex security algorithms are difficult to implement on mobile devices, due to their physical limitations in memory and CPU power.
- *User interface:* Users need to reach the information destination in as few keystrokes as possible. Developing a good user interface for a mobile device is a challenge. The user interface needs to be attractive and user-friendly.
- *Dropouts:* Connections may drop out in the middle of important transactions, due to the mobility (or cell handovers) of the devices.

14.2.2 Mobile Application Usability

Usability describes how well an application is designed or presented, so that the user can easily and effectively perform the desired operations or tasks. Usability gauges the quality of the user's experience in interfacing with a product, system, or service. The learning curve for the user needs to be small. Usability issues involve user-interface design, as well as methods for development, testing, and deployment. The current wireless technology imposes several constraints on efficient interface design, such as inconsistent connectivity, limited bandwidth, absence of standards, and so forth. The designer needs to carefully consider the interaction design of user operations, form factors, and purpose of applications during interface development. The context of the environment and the device constraints also need to be considered.

Generic usability principles include [4]:

- *Ease of learning:* This includes the length of the learning curve for novice users to use the system or applications to perform some basic tasks.
- *Efficiency of use:* This includes the speed with which the user can perform the task on the system or application.
- *Memorability:* This includes the user's efficient recollection of the input operations required to perform the tasks in an efficient manner.
- *Error frequency, severity and recovery:* This includes the number of times that the user commits errors during usage, the severity of the errors, and the recovery from the errors.
- *Subjective satisfaction:* This includes the user's satisfaction in using the application.

The following guidelines are suggested by researchers to improve the application usability [4].

- *Web page look:* The Web page should not be cluttered with the text and/or graphics. It should look simple, and should be easy to use or navigate. Horizontal scrolling should be avoided.
- *Navigation:* The navigation links and buttons should be self-explanatory.
- *Categorization:* The products and their information need to be properly and adequately categorized, with levels of categorization.

- *Information:* The information needs to be clearly organized and presented. The complete usage, description, pricing, and pictures for every product should be provided.

- *Logging support:* The customers should be able to browse the site without logging.

- *Shopping cart support:* The user should be able to add the items to a shopping cart and continue shopping.

- *Customer support:* A proper link to the customer support needs to be provided.

14.2.3   User Interface Issues for M-Commerce

The five major issues concerning the user interface for m-commerce include: technology, user goals and tasks, content presentation, application development, and the relation between m-commerce and e-commerce [4].

Technology Issues

The bandwidth limitations and other factors impose restrictions on the user interface design.

- *Bandwidth/connection limitations:* Mobile communication data rates are usually less than the data rates of regular modems. Most of the mobile standards support data rates less than 28.8 Kbps. The connection to the base station is not consistent. The amount of information downloaded from the base station needs to be limited. Graphics and lengthy messages should be avoided. The status of the download and the connection should be indicated to the user, since the download takes more time and the process might get terminated due to an unstable connection. Caching can be a good solution to cope with the disconnection problems. A friendly recovery from the broken connection (such as session resumption from the termination point of the original session) would be a desirable feature.

- *Form factors:* Different mobile devices have different operating systems, features, and functionalities, which affects the interface designs. The unique characteristics of the device form factor must be considered at the time of developing the interface.

User Goals and Tasks

For e-commerce customers, information is more important than time. For m-commerce customers, time is more important, due to the limited time available while moving, and due to other mobile device constraints. M-commerce customers use mobile devices for performing mobile value services, time-critical applications, and spontaneous need services, for which time is important. The following issues must be considered from a user's perspective when designing m-commerce applications.

- The tasks that are essential to a mobile user, as well as the tasks that are compatible with the mobile environment, must be determined. The user must perform tasks in a short time, due to device constraints and limited connectivity. This affects the selection as well as the design of applications for the mobile platform.

- A mobile device application interface should support the expert (i.e., existing) customers as well as the novice (i.e., new) customers. The existing users moving from a desktop-based Web page to the mobile-based page should not find much difference, since the mobile page would be a miniature version of the Web page. However, it will be a totally new domain for the new user. These criteria must be considered in the design phase.

- Location-based services need to be invoked for the applications that have geographical differentiations and the applications that do not trigger an extensive search.

Content Presentation

The content of the e-commerce Web page and the corresponding m-commerce Web site is bound to be different, due to the device constraints and the bandwidth limitations. The content presentation is simplified in the wireless Web site. Following are the guidelines for the content presentation.

- *Amount of information:* Large amounts of information cannot be downloaded to the wireless user, due to the mobile device constraints and bandwidth limitations. The amount of information to be displayed on a mobile device must be filtered. Adequate information to perform tasks must be provided to the user.

- *Navigation:* Some devices have function keys for navigation, while other devices have display buttons for navigation.

- *Information organization:* The organization of information on the mobile Web page is important, due to the user's limited browsing time. The organization should support more options, and the information should be displayed in as few clicks as possible.

- *Content display:* Graphics is generally better than text for information display. In case of the mobile environment, graphics can affect information display and download speed, due to limitations in screen size, resolution, and color.. Small graphics can work fine. These issues, as well as the form factor of the target device, must be considered when deciding between text or graphics display.

Application Development

In traditional e-commerce, application development and testing are performed in a fixed context of use (i.e., single computer), while for m-commerce, this must be done in a real life context. This includes the context of the user, based on requirements, mobility, form factors of devices, and so forth. The ethnographic approach (i.e., observing user activities in a real-time context) must be used. The interactions between the user, the mobile device, and the mobile network affect the user interface. These must be considered during development and testing. More complex jobs must be left for the desktop-based applications.

Relationship Between E-Commerce and M-Commerce

Many e-commerce Web sites have extended their support to mobile devices. The functionality offered on this new version of the Web site is limited, and somewhat different than the traditional Web site. A careful mapping of the services and the way it is offered must be done to prevent user frustration with m-commerce Web sites. M-commerce needs to be considered as an extension of e-commerce, instead of as an independent channel, due to the current state of wireless technology, mobile device limitations, and poor usability of most wireless sites.

M-commerce can be used to strengthen the relationships with the customers. It can personalize contents and services for a user, track the user's location to provide timely location-based services, and initiate a two-way communication with users. These features of m-commerce make it an ideal channel to implement customer relationship management (CRM).

References

[1]   Arehart, C., and N. Chidambaram, *Professional WAP*, Chicago, IL: Wrox Press, 2000.

[2]   WAP Forum, WAP 1.1 Specification, http://www.wapforum.org.

[3]   Cyberatlas.internet.com, "Mobile Commerce Frustrating Many Early Users," http://cyberatlas.internet.com/markets/wireless/article/0,,10094_513431,00.html, November 16, 2000.

[4]   Gray, J., and A. Reuter, *Transaction Processing: Concepts and Techniques*, San Mateo, CA: Morgan Kaufmann, 1993.

Selected Bibliography

Brady, M., "Out on a Limb with M-Commerce," *E-Commerce Times*, September 2000, http://www.ecommercetimes.com/.

ClickService.com, "ClickService.com Unleashes a New Web Portal for Wireless Internet," Wow-com, February 25, 2000.

McDonough, B., "Cingular Gives M-Commerce a Kick," *Wireless News Factor*, June 2001.

Muller-Veerse, F., *Mobile Commerce Report*, Durlacher Corporation, London, http://www.durhacher.com/downloads/mcomreport.pdf.

Varshney, U., R. Vetter, and R. Kalakota, "Mobile Commerce: A New Frontier," *IEEE Computer*, October 2000.

Varshney, U., and R. Vetter, "Emerging Wireless and Mobile Networks," *Communications of the Association of Computing Machinery (ACM)*, June 2000.

Varshney, U., and R. Vetter, "A Framework for the Emerging Mobile Commerce Application," *Proc. of 34th Annual Hawaii International Conference on System Sciences*, 2001.

CHAPTER 15

Multimedia Messaging Service

Mobile commerce consists of different types of services, such as headline news, online stock trading, e-mails, entertainment services, and many others. Mobile devices are ubiquitous, so these services can be accessed anywhere. Mobile phones that use third generation (3G) high-speed networks will soon advance from the current voice market to mobile data communication. Mobile phones will soon reach the data transmission rates of personal computers. Multimedia messaging services (MMS) is the next important business opportunity for mobile operators. MMS is sends text, voice, audio, images, and video over mobile networks. The multimedia messaging service center (MMSC) sends these MMS messages to mobile user agents. The MMSC provides a variety of services, including interfaces to user databases, legacy systems, and value added service providers (VASP). No current standard mechanisms exist for deployment of most of these mobile services. Extensive studies are needed to standardize the business model and to determine the technical requirements for a successful mobile service architecture. The protection of data exchange is another major issue that must be addressed.

One service that is unique in mobile commerce is short messaging service (SMS), which is limited to text-based messaging up to 160 characters, and uses mobile connections with limited capacity [1]. The new generation of mobile phones has multimedia capability to enable capturing photos, voices, and videos. SMS is insufficient for providing messaging services for multimedia phones. MMS allows a user to create, send, and receive multimedia messages among mobile phone users. A typical message contains text, but it may also include images, audio, and video. This additional capability of adding multimedia content should be easy. The multimedia messages can be exchanged between mobile phones and email addresses.

MMS is the first mobile messaging service to embrace Open Internet standards for messaging [2]. It supports standard image formats, such as GIF and JPEG; video formats, such as MPEG-4; and audio formats, such as MP3 and MIDI. The high transmission speeds required for these formats can be provided by 2.5G (e.g., GPRS and EDGE) and 3G networks [3]. The higher bandwidth available for 3G opens up new services for mobile receivers.

15.1   MMS Overview

MMS is a rapidly developing new mobile communications technology [3]. MMS can be seen as a step toward the services that will be offered in the third generation mobile networks. The telecommunication industry needs services in order to make use of the high-capacity networks, and the services are essential for giving concrete examples of new innovations to the customers. MMS is the first example of such a service that offers rich mobile content.

Media data could be obtained from a local file, a camera phone, a video recorder, or a microphone. Media files often contain multiple channels of data, called tracks. A track's type identifies the kind of data it contains, such as audio and video. The format of a track defines the arrangement of the data for the track. Some of the common media formats are AVI, MIDI, MPEG, WAV, AU, and QuickTime. These formats are designed with particular applications and requirements in mind. Quality depends on compression ratio, client processing capability, and the network bandwidth.

MMS is a nonreal-time messaging service that provides a new extension to mobile communication. It is an extension to the popular SMS that is prevalent today. SMS sends and receives short text messages, while MMS sends multimedia contents in a single "rich" message [4].

Multimedia messaging allows for a service provider to leverage the business model of the SMS, since higher bandwidths are available for 2.5G and 3G messaging [3]. According to Mobile Streams [1], "The transition from Short Message Service (SMS) to Multimedia Messaging Service (MMS) is as important on mobile phones as the transition from DOS to Windows was for the PC. It represents a revolution."

SMS also enhances its service by providing multimedia features, such as transmitting and receiving simple pictures, sounds, and animations. However, this enhanced service uses the existing SMS infrastructure of the circuit-switched GSM networks with transmission rates of 9.6 Kbps. MMS will depend on 2.5G (e.g., GPRS and EDGE) and 3G (e.g., UMTS) networks [3]. MMS messages use the store and forward model, similar to SMS messages.

15.1.1 MMS Protocol

One of the important features of MMS is the ability to support interoperability with other messaging systems. This is shown in Figure 15.1, which gives an abstract view of an MMS network diagram. It is expected that specific MMS networks may have more than one legacy messaging connection, and include other services, such as fax and voice mail systems. The MMS architecture elements are shown in Figure 15.1. All MMS network elements (MMSE) are directly or indirectly connected to the MMSC. The MMSC is a logical unit made up of both the MMS proxy and the MMS server [3]. The MMS proxy is the primary routing element passing messages and notifications between other entities. The MMS server stores messages for later retrieval, because the MMSC uses the store-and-forward paradigm to pass messages, which means that the MMS server stores messages until the MMS proxy has successfully delivered them [2]. The MMSE network provides functions that implement off-line and online charging mechanisms on the bearer (e.g., GPRS), subsystem (e.g., IMS), and service (e.g., MMS) levels. An MMS client interacts through the MMS proxy with users, e-mail servers, and other servers listed in the Figure 15.1. The servers may include legacy systems, such as messaging servers. The MMS proxy, while interacting with other servers, uses well-defined interfaces, such as e-mail (E-M API) and legacy (L-M API) systems, which utilize SMTP and proprietary protocols, respectively.

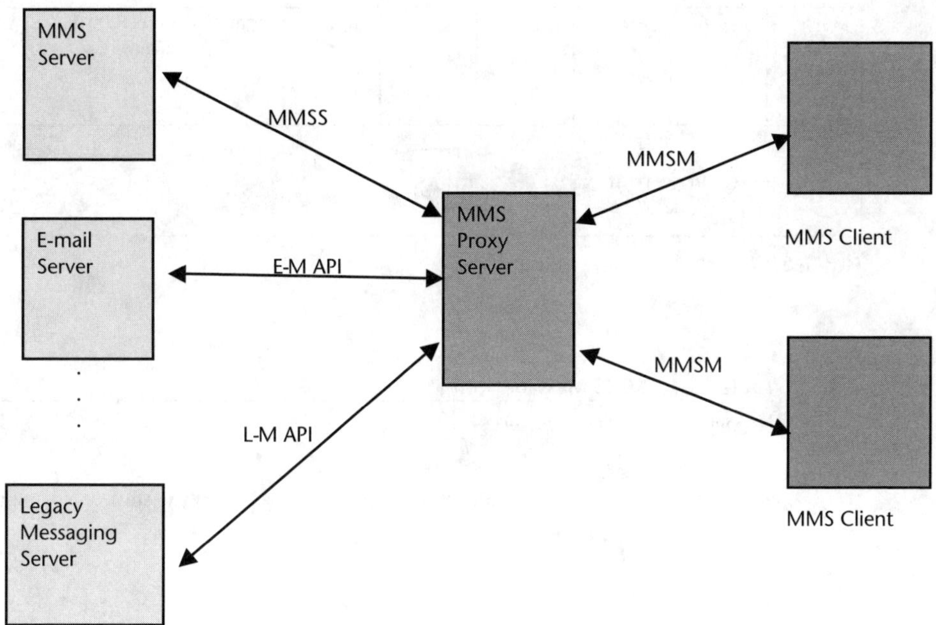

Figure 15.1 MMS architecture.

15.1.2 Message Types

There are seven different types of messages that are passed between a client and a server. The client and the server may:

- Send a multimedia message to an MMS server;
- Fetch a message from an MMS server;
- Request a list of messages;
- Fetch a list of messages from the server;
- Request a particular message;
- Confirm a sent message;
- Notify MMS about a new message.

15.1.3 Message Format

A message consists of a header and a multipart body [5]. Figure 15.2 shows the format of the message and Table 15.1 describes the format.

The MMS headers contain MMS-specific information, such as information about the type of message and the details of the recipient. The message body is used when the multimedia message is sent or retrieved. All other messages contain only the MMS headers. The message body consists of multimedia objects.

Each of the components may be in a separate part, or with the root or parent part. An optional part called a presentation part also exists. The presentation part contains instructions regarding the rendering of the multimedia content contained in different parts to the mobile receiver [5]. Multiple presentation parts may be contained in the message body. In the absence of the presentation part, the multimedia

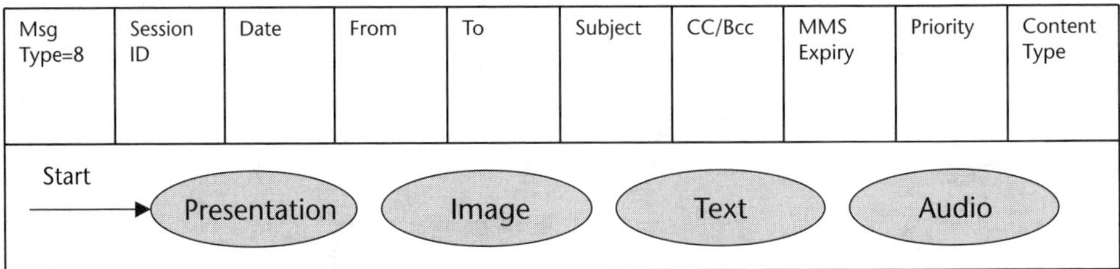

Figure 15.2 MMS message format.

Table 15.1   MMS Message Format

Header Name	Description
Msg Type	Type of the message
Session ID	Session identification assigned by a server at the start of a session
Date	Message creation date
From	Address of the message sender
To	Address of the message receiver
Subject	Subject of the message
MMS Expiry	Expiration date of the message

content is presented as allowed by the terminal specifications. The presentation part is written using specialized languages, such as synchronized multimedia integration language (SMIL) or wireless markup language (WML).

15.1.4 Addressing Model

The MMS addressing model contains two addressing modes: e-mail addressing and MSISDN addressing.

With e-mail addressing, a subscriber address is expressed in the form of an e-mail address, using the following format: user@system. With MSISDN addressing, users are identified with MSISDN numbers. E-mail addressing is used for the exchange of messages between MMS user agents attached to different MMS environments. Each MMSE has a unique domain name in the form of mms.serviceprovider.net or mms.network.net.

Address hiding refers to the possibility of a message originator to hide its address from message recipients. The anonymous multimedia message is delivered without providing the originator details to the message receivers. The originator MMSC knows the originator address, even if the message originator requests address hiding.

15.2 Multimedia Presentation in MMS

The concept of MMS presentation means that the ordering, layout, sequencing, and timing of multimedia objects are specified. The sender of the multimedia message must organize multimedia contents in a meaningful order and show how the multimedia objects are rendered to the receiver. MMS presentation is optional, because some receivers have limited presentation capabilities. A receiver may be able to present a particular content if the receiver supports that type of media, even if the receiver does not support the sequencing and timing information.

There are various languages available for presentation of content on MMS-enabled devices. The Third Generation Partnership Program (3GPP) recommends three languages for this presentation: WML, SMIL, and XHTML. We will examine these three standards in the following sections.

15.2.1   WML

WML is an extensible markup language (XML) language that is used to specify the content and user interface for WAP devices. Almost every mobile phone browser in the world supports WML. It allows the text portions of Web pages to be presented on mobile phones and PDAs through a wireless connection. WML pages are requested and served in the same way as HDML pages. WML presentation for multimedia messaging offers the same sequencing and layout capabilities as browsing.

15.2.2   SMIL

SMIL became an official recommendation by the W3C in mid-1998 [6]. SMIL documents are XML documents. SMIL has its own document type definition (DTD) in the XML syntax. SMIL uses DTD to define unique tags to be used in an SMIL file. Designers make use of SMIL-specific tags to specify the synchronization between audio, video, image, and text components in a multimedia presentation.

An SMIL element contains a head element and a body element. The body element contains information that is related to the temporal and linking behavior of the document. The head element contains layout elements that define regions where elements contained in the body element are placed. Presentation information is described in the body element. A media object element in the body element describes a media object, which can be audio, video, image, text, text stream, animation, or reference. Media object elements may be arranged using parallel or sequential elements to prescribe presentation sequence in time. Startup delay can be specified either in time or by events before a media object can be displayed. The end of a media object or a timer expiration event may trigger the display of other media object(s). The following is a simple example of an SMIL file.

```
<smil>
   <head>
      <layout type="smil-basic">
      <root-layout height="400" width="500" back-
ground-color="192,192,192"/>
      <region id="left-video" left="20" top="0" width="180"
height="150" z-index="1"/>
      <region id="right-video" left="280" top="30" width="150"
height="150" z-index="2"/>
      <region id="left-image" left="100" top="200" width="180"
height="180" z-index="1"/>
      <region id="right-text" left="300" top="300" width="180"
height="30" />
      </layout>
```

```
        </head>
        <body>
            <par>
                <video src="foo.mov" region="left-video"/>
                <seq>
                    <video src="bar.mpg" region="right-video"/>
                    <img src="image.jpg" region="left-image" dur="8s"/>
                    <text src="text.txt" region="right-text" dur="8s"/>
                </seq>
                <audio src="bar.wav"/>
            </par>
        </body>
    </smil>
```

The example shows an SMIL file with two videos, one image, one text, and one audio object. An SMIL file includes its own default tags, which are specified in the "smil-basic" file. The "smil-basic" file is a DTD file that includes the syntax and definition of default tags. An XML parser will check the syntax of an SMIL file using this DTD file. Each region in the head element is defined by specifying the coordinate position in pixel units or percentage values from the upper left corner of the viewing canvas. Each region has its own identifier that is used in the body element. The z-index is used to specify the depth of overlaying regions. In the body element of the example, the video "foo.mov" and the audio "bar.wav" are played in parallel (enclosed by par tag); and "bar.mpg," "image.jpg," and "text.txt" are displayed in sequence (enclosed by seq tag).

The XML parser generates an object tree, since SMIL is a declarative language and conforms to XML. The object tree structure contains all the media objects and presentation timing information. An SMIL player use the object tree to render the display for the media objects described in the given SMIL document. Playback events are scheduled on a time line, and media objects are displayed either in parallel or in sequence.

15.2.3   XHTML

XHTML is a language that can also be used for representing message scene descriptions. In particular, XHTML Mobile Profile (XHTML MP) extends the HTML Basic Profile published by the W3C. XHTML MP has been specifically tailored for resource-constrained devices. However, HTML MP remains a suitable language for the definition of MMS scene descriptions.

15.3   MMS Client

The user selects the MMS application upon launching the software. Figure 15.3 represents the architectural diagram of the mobile client software. The complete functionality of the mobile client is divided into these modules.

The MMS client application contains the address book management module, which allows the user to add, delete, edit/update, and list available entries. The mailbox management module maintains the list of incoming and outgoing messages. The

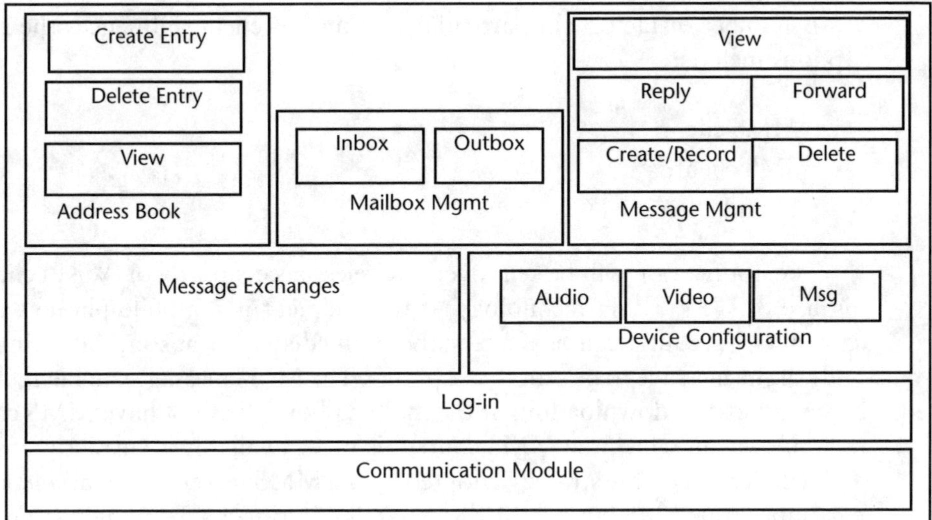

Figure 15.3 MMS client architecture.

message management module is responsible for creation, deletion, forwarding, replying, and viewing of multimedia messages. The device configuration module is responsible for sending mobile terminal settings to the MMS server after a successful login to the application. The client notifications module collects any new message notifications from the server and displays them to the mobile user.

Table 15.2 outlines the current set of media formats supported by the MMS standard. The SMIL and XHTML standards define the layout and behavior of the various media components within an MMS message.

15.4 MMS Content Delivery

MMS can be thought of as SMS with the additional capability of sending a combination of multimedia objects and text messages. These messages are forwarded to an MMS server, which stores them until they can be delivered to the intended receiver. Notification of an impending MMS message is sent at the earliest opportunity to the recipient. The scenario then diverges from the SMS scenario.

Table 15.2 MMS Supported Media Formats

Media	Format
Text	Plain text, Unicode characters only
Speech	AMR & AMR-WB
Audio	MPEG-4 AAC LC (LTP Optional)
Synthetic audio	SP-MIDI
Still images	JPEG, JFIF
Bitmap images	GIF 87a, GIF 89a, PNG
Vector graphics	SVG-Tiny

A number of factors, in particular the kind of client, influences the next step. Options include:

- MMS client;
- SMS client;
- E-mail client.

The notification will be sent over a wireless access protocol (WAP) client, making use of WAP PUSH technology, if the recipient is a mobile phone with MMS capability. A request can be made, either immediately or at some later time, for the body of the message to be sent to the phone. The MMS message can be freely viewed at any time after downloading. If the mobile phone does not have MMS capability, then the way in which the MMS content is made available is currently unspecified, and will depend on the strategy adopted by the MMS server. For example, for terminals supporting SMS but not MMS, Nokia's Multimedia Terminal Gateway stores the MMS content in its internal storage and sends an SMS message to the intended recipient, notifying them of the URL of a Web page that they can visit to view the MMS message using a suitable browser.

It is also possible to send an MMS message to an e-mail address, in which case it will be delivered directly to the recipient's mailbox, without any prior notification being necessary. It can then be downloaded and viewed in the standard way through a suitable e-mail client. The client's handling of the data depends on the capability of the e-mail client program. If the e-mail program supports MIME, then it should be capable of presenting the separate multimedia objects and text within the MMS message. Otherwise, the e-mail client may make them available as detachable files. If it also has MIME support for the markup language that is used, for example "application/smil," then it should be able to display the content as intended by the sender.

15.5 Client State Diagram

The client state diagram represents the various states of the client application, based on all the requests that are allowed by the application. See Figure 15.4.

The mobile user starts by logging into the MMS application and entering the *Connecting to Server* state. The mobile client retrieves the mailbox and displays its contents, as soon as the server responds with successful log-in by validating the user. If the log-in was unsuccessful, then the client enters into the *LoginFailed* state. If the client is not able to connect to the server, then it enters into the *Timeout* state. If an erroneous message is retrieved, then a screen message is displayed. If the client chooses to view the multimedia message, then it connects to the server to receive the message, and on receiving the correct message, enters into the *Display Message* state. If the client chooses to create a new message, then it enters into the *Compose Message* state. The client can switch between recording audio or video states at the time of message composition. On selecting the log-out option from main menu, the client returns to the *Prompting username and password* state.

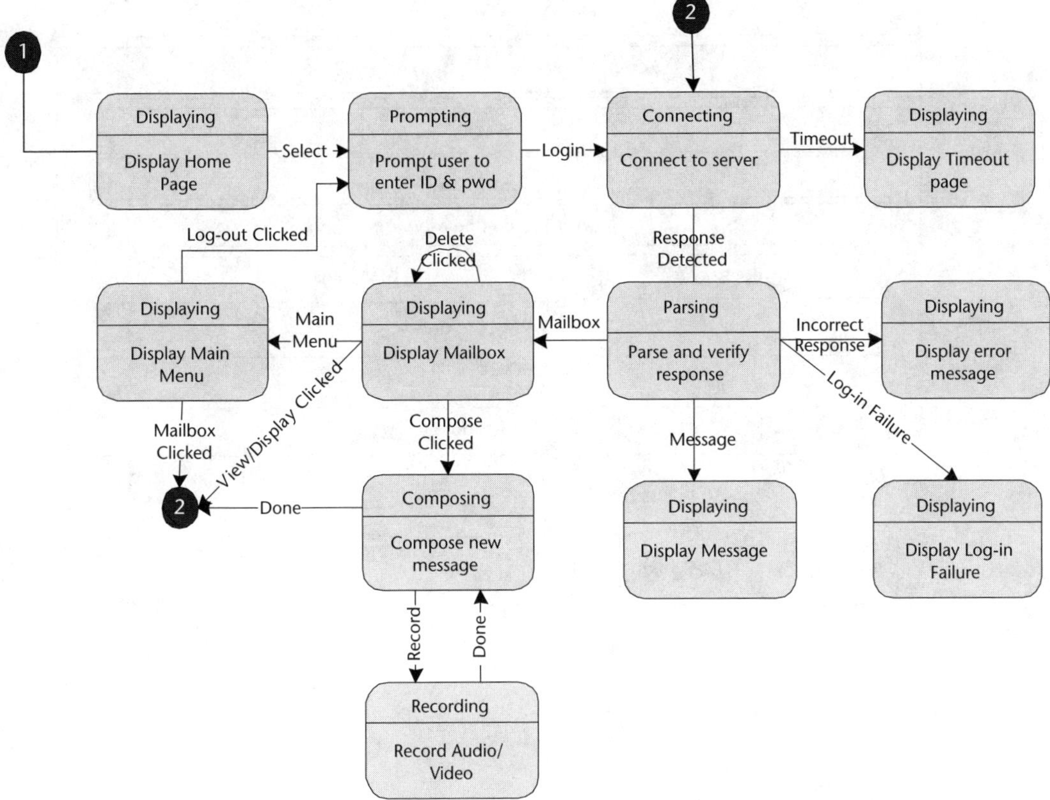

Figure 15.4   Client state diagram.

15.6   MMS Server

The MMS server is responsible for listening to the mobile client requests and responding to them. See Figure 15.5 for the architecture of the MMS server. The wireless communication module handles the concurrency and multiuser environment. The access control module performs the user authentication when the mobile user logs into the client application. A secured session is then maintained for that user until he logs out. The session also keeps track of the device settings for the particular user that were sent after successful log-in to the application. These settings will be applied to the multimedia messages that are sent to that user, and will handle the user device capabilities. The security module delivers secured messages to the user. The caching module buffers messages, message lists, or address book entries for that user, depending on the user request it is currently handling. This improves performance over the wireless network, by dealing with device limitations while playing the larger messages, or by dealing with user preferences while rendering message lists or address book entries. The MMS engine takes care of all the MMS protocol-related functionalities, such as combining the different parts of a message using SMIL. MyMMS retrieves and applies the user preferences that are submitted using the Web client. The message, mailbox, and address book modules are responsible for storing and retrieving from the database messages, mailbox entries, and

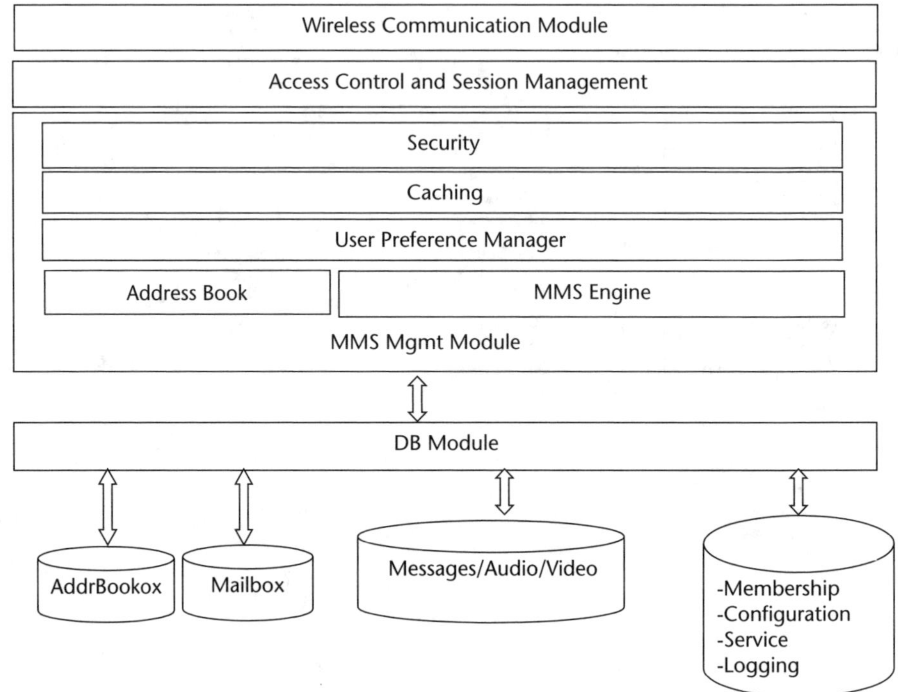

Figure 15.5   MMS server architecture.

address book entries, respectively. The message table will be indexed in message-id for quicker retrieval of messages.

15.7   Server State Diagram

The server state diagram represents the various server states, based on different user requests allowed by the application, and is shown in Figure 15.6.

The server starts up and keeps listening to the requests from clients. When the server receives a log-in request, it creates a user session, retrieves the mailbox, creates a multimedia message (MM), sends it to the client, and continues to listen. When the server receives an MM request, it parses the request to determine its type, and processes it based on its type. New requests are saved to the database, based on the type, and the output is buffered for faster retrieval. Old requests are retrieved from the buffer. The server creates a new MM and sends it back to the client and continues to listen. When the server receives a log-out request, it saves the user session.

15.8   Case Study: Nokia

Nokia's MMS comprises a complete end-to-end solution for person-to-person mobile messaging, from terminal to terminal, from terminal to e-mail, or from e-mail to terminal. It allows full content versatility, including images, audio, video,

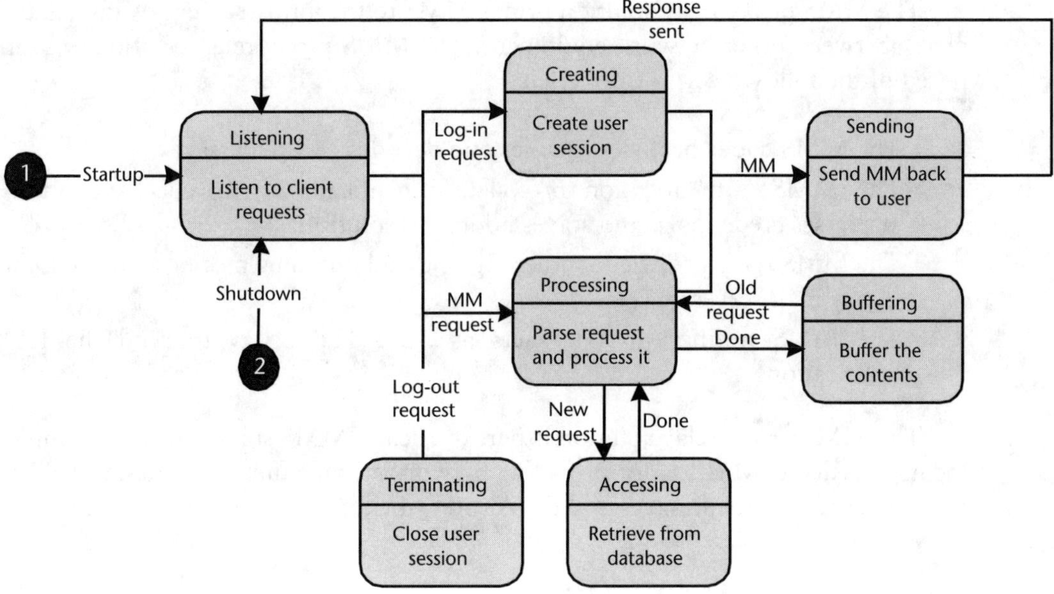

Figure 15.6   Server state diagram.

data, and text, in any combination. MMS delivers a location-independent, total communication experience.

Nokia is also actively participating in the development of WAP to support MMS, and in 3GPP, a bearer service that includes optimal support for multimedia messaging. WAP, as a global, bearer-independent solution, allows multimedia messaging in all product categories.

MMS content can include one or several content types (e.g., picture, data, text, audio, and video), with minimal restrictions to message size or format. The MMSC is needed from the network perspective, to perform the required store-and-forward operations of multimedia messages.

15.9   Case Study: Alcatel and Intel

The MMS proxy relay allows mobile operators to deploy multimedia messaging services seamlessly by opening access to existing messaging servers. The proxy relay also allows end users registered in the same mobile network (or in a foreign MMS-enabled mobile network) to exchange multimedia messages. The MMS proxy relay is the only component that is required to make a mobile network MMS-enabled; it handles high loads of MMS with complete reliability. The MMS proxy relay architecture uses carrier-grade servers powered by Intel to provide scalability, high performance, reliability, and availability, while also offering opportunities for growth and flexibility to manage future network evolution and optimization. It complies with the 3GPP architecture and the WAP Forum specifications.

The MMS proxy relay solution brings MMS to the infrastructure by inserting a proxy between other network equipment. The MMS proxy relay solution is composed of the following set of proxylets.

- The MMS codec proxylet encodes and decodes MMS messages.
- The MMS authentication proxylet authenticates users, checks emitters, addresses credentials, and solves address resolution.
- The MMS relay proxylet handles ingoing and outgoing messages through the legacy mail server.
- The MMS notifier proxylet acts as an SMTP proxy to send the PAP notification.

The MMS proxy relay solution is part of Alcatel MMS suite, which also implements the Alcatel MMBox for providing network storage functionalities, as well as unified access to the messages and the Alcatel MMS Applications-in-a-Box.

15.10 Comparison

The external application interface (EAIF) provided by Nokia is not a complete MMS, since it only enables multimedia messaging services to be developed in the Nokia MMSC. Similarly, the MMS proxy relay solution provided by Intel and Alcatel claims to be the only required component to MMS-enable a mobile network. It also claims to rely on the Alcatel proxy platform, and run on the Intel Xeon processor family-based carrier-grade servers under the Linux operating system.

None of the solutions described above provides a complete solution to MMS, as the comparison in Table 15.3 shows.

Table 15.3   Comparison Between Nokia and Alcatel Solutions

	Nokia	*Alcatel*
Protocol	EAIF, MM7	WAP, WSP
Security	RSA Key Exchange	Password Authentication Protocol (PAP)
Image formats	Baseline JPEG with JFIF as the exchange format: GIF 87a, GIF 89a, WBMP	BMP, WBMP, GIF, aGIF, JPG, PNG
Text	US-ASCII, Utf-8, and Utf-16 with explicit byte order mark	—
Audio	AMR	MIDI, SP-MIDI, AMR IETF, iMelody
Message size	Minimum 30 KB	Maximum 50 KB
Gateway	WAP gateway, push proxy gateway	WAP gateway, push proxy gateway
Networks	GPRS, GSM, 3G	GPRS, GSM, 3G, UMTS
Hardware	Any Pentium processor	Intel Xeon processor family-based carrier-grade servers
Operating system	Windows 2000, Windows XP	Linux

15.11 Conclusion

The MMS architecture shows a multimedia messaging service based on the standards. It supports various audio, video, and image standards that overcome the limitations of the mobile terminals. The MMS solution must work with the existing wireless networks. This is considered to be a basic feature, while the wireless networks are being standardized. The support for large message size is also essential with MMS services.

References

[1]   Ajit, J., MMS FAQ, 1999–2002, http://www.mobilemms.com/mmsfaq.asp#3.5.

[2]   Mobile Streams, 1999–2002, MMS Overview, http://www.mobilemms.com/mmsoverview.asp.

[3]   Logica. The Essential Guide to Multimedia Messaging, http://www.logicacmg.com/pdf/telecom/Mmsguide.pdf.

[4]   GSM Association, MMS, http://www.gsmworld.com/technology/mms/index.shtml.

[5]   WAP Forum, Wireless Application Protocol, MMS Encapsulation Protocol, June 2001, http://www.wmlclub.com/docs/especwap2.0/WAP-205-MMSArchOverview-20010425-a.pdf.

[6]   Synchronized Multimedia, http://www.w3.org/AudioVideo/.

Selected Bibliography

3GPP, Technical Specification Group Terminals; Multimedia Messaging Service (MMS); Functional description; Stage 2, October 1999, http://www.3gpp.org/ftp/tsg_t/WG2_Capability/SWG3/SWG3_01_Ascot/Docs/MMSstage/Tdocs/T2M99105(23.140 V 0.1.0 MMS Stage2).doc.

3GPP2, Multimedia Messaging Services, http://www.3gpp2.org/Public_html/specs/S.R0064-0_v1.0.pdf.

Alcatel, http://www.intel.com/business/bss/solutions/blueprints/pdf/alcatel_sb.pdf.

MMS Whitepaper, http://www.csgsystems.com/uploadedDoc/MMS_Whitepaper.pdf.

Nokia, http://www.nokia.com/downloads/aboutnokia/press/pdf/mmsA4.pdf.

WAP MMS Architecture Overview, http://www.openmobilealliance.org/tech/affiliates/wap/wap-205-mmsarchoverview-20010425-a.pdf.

WAP Architecture, http://www.openmobilealliance.org/tech/affiliates/wap/wap-210-waparch-20010712-a.pdf.

Wireless Application Protocol, http://www.cellular.co.za/wap.htm.

CHAPTER 16

Wireless Advertising and Marketing Systems

The fast development of wireless networking technology and the significant increase in mobile device users has made wireless advertising and marketing a hot topic. According to Cyberatas [1] and WindWire [2], the number of mobile users exceeded 468 million in 2000—a much higher number than the 365 million people using the Internet that year. Jupiter Media Metrix [3] predicts that the wireless advertising market will reach $700 million annually in the U.S. market over the next 4 years. Studies by wireless media research companies, such as WindWire [2] and SkyGo [4], indicated that delivering permission-based alerts to wireless phones captures consumers' attention, drives response actions, and builds brand awareness. Microsoft, Yahoo, AOL, and other large companies have created subsidiaries to focus on this wireless advertising market.

As discussed in the previous chapters, wireless devices have three main advantages over PCs and other conventional platforms [5]. They are:

- *Accessible:* Wireless devices are rapidly becoming personal devices because they are handy, portable, and available at all times.
- *Personal:* The typical wireless device belongs to a single person, thus becoming uniquely identified with that individual.
- *Location-aware:* If a wireless device is turned on and connected, then it can be used to track a user's physical location. This is desirable in electronic advertising.

These advantages allow marketing and advertising vendors to post highly targeted, flexible, and dynamic ads to mobile users, in spite of the limited display screen and battery life. The fast growth of the wireless users creates new business opportunities, and brings new demands and challenges in developing new solutions to deliver wireless ads to mobile users [6].

This chapter is dedicated to wireless advertising and marketing by exploring the engineering issues, challenges, and solutions in delivering wireless ads to mobile device users. The chapter is structured as follows. Section 16.1 provides readers with the basic concepts of wireless advertising, including business perspectives, technical challenges, wireless ad types, and standards. Section 16.2 focuses on the engineering perspectives of developing wireless advertising solutions. Section 16.3 presents a summary of major players and their solutions. Section 16.4 summarizes the chapter.

16.1   Understanding Wireless Advertising and Marketing

Wireless advertising refers to a process in which various advertising and marketing activities are performed using mobile advertising solutions to deliver advertisements to mobile devices. Wireless advertising creates a new marketing channel between advertisers and mobile users. It allows advertisers to use systematic approaches to deliver location-based, individually targeted personal ads to mobile users over a wireless network or a wireless Internet. Mobile devices, including mobile phones, personal digital assistants (PDA) and pocket PCs, are the major platforms of wireless advertising. Similar to conventional advertising on traditional media (e.g., TV, radio, and newspapers) and online advertising, wireless advertising can promote the sales of goods and services, and carry on business promotions and brand-name building. Table 16.1 presents the major differences among these forms of advertising.

16.1.1   Wireless Ads

Various kinds of wireless ads can be delivered to different mobile devices, such as mobile phones, PDAs, or pocket PCs. The Wireless Advertising Association (WAA) groups them into SMS ads, PDA ads, voice/speech ads, and location-based ads. Wireless ads, unlike online ads, must work within the constraints of the small display screen on mobile devices. We classify various wireless ads into the following types.

- *Location-based ads:* These ads are delivered to mobile users based on their mobile locations. For example, users may register to receive alerts (or offers) on their mobile phones from the stores located in a specific location. Advertis-

Table 16.1   Comparing Wireless Advertising with Other Advertising Approaches

Advertising Features	Wireless Advertising	Online Advertising	Traditional Advertising
Communication style	Two-way wireless	Two-way Internet	One-way delivery
Ad delivery	Electronic and dynamic	Electronic and dynamic	Ad hoc delivery
Reachable consumers	Global mobile users	Global online users	Regional readers/viewers
Interacting with users	Supported	Supported	Not supported
Ad platforms	PDAs, mobile phones, and Pocket PCs	Web browsers on laptops and desktops	Traditional media platforms: TV, radio, and newspapers
Linkage to product catalogs	Supported	Supported	Not supported
Enabling with trading transactions	Supported	Supported	Not supported
Advertising management	Systematic	Systematic	Ad hoc
Ad targeting	Personal, content-based, and location-based	Content-based and IP-based	Ad hoc; regional subscribers or receivers
Ad tracking	Systematic	Systematic	Very limited
Ad performance measurement	Systematic	Systematic	Not systematic

ing service agencies may send wireless ads to the registered mobile users. This type of advertising relies on location-based technology. Figure 16.1 shows Vindigo's applications for finding the best places for activities, such as dining, shopping, and so forth, with the respect to the user's location, as determined from a mobile device. Zip code–based advertisements can be delivered to a mobile user whose current mobile location is within the area of a given Zip code.

- *Brand-building ads:* These ads are designed and delivered to mobile users to increase the awareness of a specific business or a product. Figure 16.2 shows two of SkyGo's wireless ad samples for brand-building.

- *Home page ads:* These ads refer to the advertisements that are placed on the home pages of wireless service vendors (e.g., AvantGo), with premier placement header and anchor ads for maximum visibility. When a mobile user accesses AvantGo's wireless service, the first page seen is the home page, thereby increasing brand awareness from the ad posted. Figure 16.3 shows a wireless home page sample ad.

- *Category ads:* These ads provide customers with easily traceable and well-classified advertisements. AvantGo's wireless category pages, as shown in Figure 16.4, are typical examples. They target specific demographics with messages focused on the consumers who browse the marketing channels within a specific interest category. Wireless category ads can help advertisers reach the consumers registered in a particular geographical location.

- *Interstitial ads:* These ads refer to the advertisements that appear on mobile devices during the mobile users' interstitial time; for example, page loading,

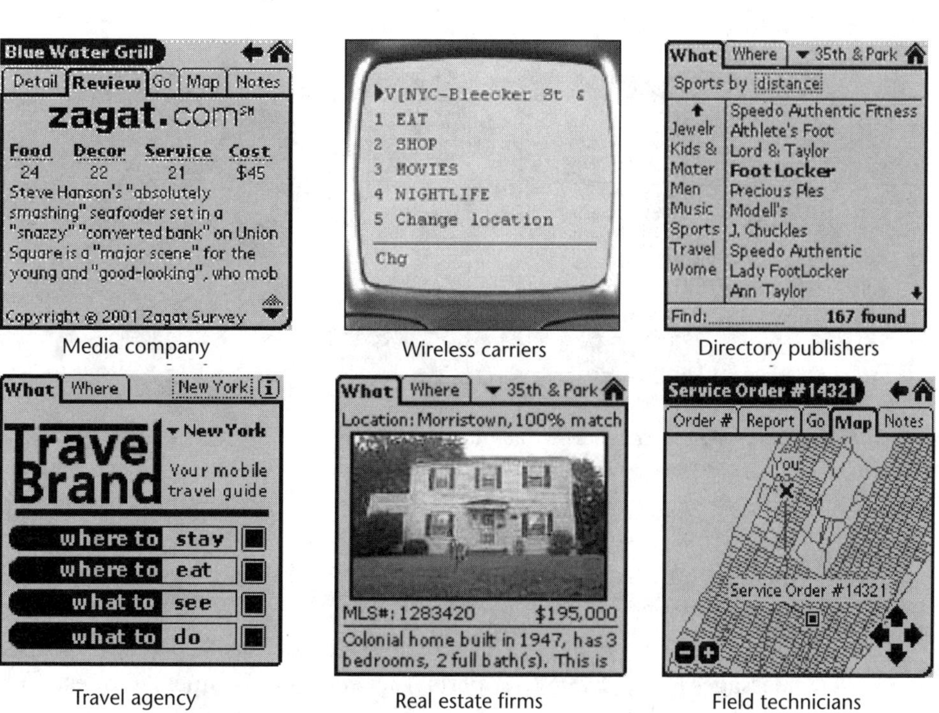

Figure 16.1 Vindigo's wireless advertisements.

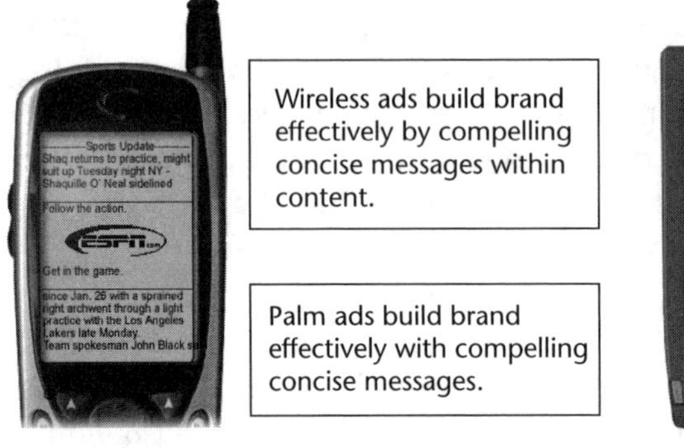

Figure 16.2   SkyGo's wireless ad samples for brand-building.

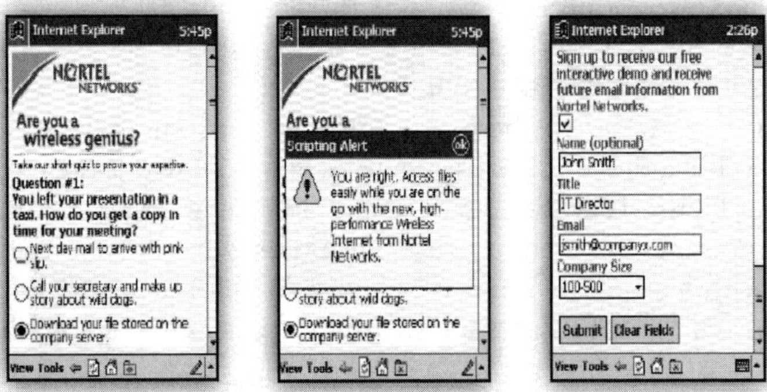

Figure 16.3   AvantGo's wireless home page sample ads.

Figure 16.4   AvantGo's wireless category ad samples.

wireless Web surfing, or waiting for wireless connections. Interstitial ads can increase interactivity by click-through links and buttons, and can provide audio features.

- *Short message ads:* These ads are delivered to mobile device users as short alert messages (e.g., SMS/WAP alerts) to meet the consumers' requests. These alerts mostly use the permission-based alert model to deliver ad messages to mobile users, based on the users' selections, interests, and current mobile location. Samples of SMS ad alerts can be found in [7].

- *Game-based ads:* These ads let the users play games using mobile devices, while the ads pop up between games. AvantGo's game-based ad samples on PDAs can be found in [7].

- *Mobile coupons:* These ads are delivered to mobile users to give them permission to receive discount coupons. Electronic coupons have been used in online advertising, and have proven to be an effective marketing method.

16.1.2   Business Models

A business model is a blueprint that describes the business strategy for making revenue, by specifying markets, products, customers, and positions in the value chain. Five different wireless advertising business models are discussed and compared in Table 16.2 [6]. They are illustrated here.

- *Voice-subsidized model:* This model seeks to provide competitive offerings, with no intention of generating additional revenue. The wireless service providers must support wireless Internet services and contents, using only subscriber fees and wireless service provider (WSP) resources.

- *Safe income, no upside revenue model:* This model seeks to subsidize acquisition of premier content to elevate the quality of wireless Internet business offerings. WSPs derive revenue from subscriber fees and placement fees in this model. The business charges these fees to content providers who want prominent placement or general distribution through the WSP's network. This strategy also refrains from taking a share of any revenue generated by the content providers as a result of ad distribution over the WSP network. This model, similar to the voice-subsidized business model, requires no wireless ad technology.

Table 16.2   Business Models for Wireless Advertising

Business Model	Subscriber Fees	M-Commerce Fees	Content Placement Fees	Advertising Revenue	Partner Technology Requirements
Voice-subsidized	Majority	None	None	None	None
Safe income, no upside revenue	Majority	None	Majority	None	None
Diversified revenue	Share	Share	Share	Share	WSP-sanctioned publisher, ad serving solution, outsource media sales, measurement capability
Media-dependent, outsource media sales revenue	None	Share	None	Share	WSP-sanctioned publisher, ad serving solution, outsource media sales, measurement capability
Media-dependent, in-house media sales revenue	None	Majority	None	Majority	Carrier ad serving solution, extensive in-house media sales, campaign management interface

Source: [6].

- *Diversified revenue model:* This model allows WSPs to generate revenue either through placement fees or by receiving a share of advertising revenues and m-commerce fees. The content and service providers receive these fees directly, using the WSP's distribution channel, in addition to subsidizing subscriptions or content fee acquisition. The WSP can either subsidize the cost of acquiring content, or discount the consumer service fees, using the placement fees and ad revenue generated. The model also has the benefit of increasing profitability, with minimal effort by the WSP. This model requires the WSP to have its own wireless ad server or a sanctioned ad server for publisher usage.

- *Media-dependent, outsource media sales revenue model:* This model seeks to grow and retain the subscriber base, while generating significant revenue. This model generates all of its revenue from a share of all advertising and m-commerce revenue generated over the WSP distribution channel. Thus, it requires a greater commitment to wireless advertising from WSPs. Media sales for a publisher's content is an outsource task that is largely the publisher's responsibility. WSPs must track the ads served over the network to ensure proper revenue sharing, and may opt to control the number and frequency of ads served to wireless consumers based on subscriber profiles or restrictions.

- *Media-dependent, in-house media sales revenue model:* This model seeks to expand the carrier's business function to include media sales, thereby controlling and drawing revenue from wireless Internet property. This model depends fully on wireless advertising and m-commerce fees for wireless content acquisition and subscriber access. The model differs from the previous model in that it requires a WSP to create an in-house team that is devoted to media creation and sales, as well as processing and maintaining wireless ads. The WSP provides a complete solution to support wireless advertising in ad creation, management, and service.

16.1.3 Business Issues and Technical Challenges

The wireless advertising industry is still in its early stages. A wide range of business issues and technical challenges must be studied and solved before wireless advertising gains wider use and acceptance. There are two groups of issues in wireless advertising: business issues and technical issues, according to [6].

Business Issues

There are a number of business issues relating to wireless advertising. Understanding and resolving these issues is essential for wireless advertising businesses and service agencies in business planning, solution delivery, and market positioning.

- *Cost:* Wireless customers, advertisers, marketing vendors, and advertising service agencies first must face the cost issue. The current costs of wireless connection services and mobile devices are still too high for most consumers. As pointed out by IDC analyst Kevin Burden in [8], the current wireless access for handhelds is "too expensive for what they are getting back."

- *High business risk:* Since wireless advertising is a new market, there is much uncertainty, which means high business risk and low advertising revenue. This will result in a lack of infrastructure development and investment research, forcing pioneers of this market to navigate uncharted territory with little support and few resources. Early entrants will be faced with a tough choice in selecting from different business and marketing models, due to the scarcity of market research.

- *Ad reception:* This is an important business issue for wireless advertisers, advertising businesses, and marketing service agencies. These groups need to determine whether or not consumers will accept ads on mobile devices before they launch their businesses. Traditional TV and radio ads have become socially accepted forms of intrusion for consumers. Delivering wireless ads to personal mobile devices is still considered as a form of intrusion to most consumers. According to a recent study by Jupiter Media Metrix, 46% of mobile users want no advertising at all, even if it pays for all or a portion of their services or devices. In Sweden, Ericsson found in a large trial that 60% of its mobile device users liked to receive SMS advertising messages, provided that they were targeted to their profiles and interests. According to WindWire's survey of 260 mobile users, 86% indicated that they preferred free or advertising- subsidized wireless content over free-based content, and approximately 50% of consumers will accept wireless advertising if they get something in return.

- *Privacy:* The issue of privacy must be addressed when discussing consumer profiling. Laws must be passed to govern what information that ad agencies are allowed to request and share. Data presented by the Yankee Group [9] during the WAA meeting indicated that more than 50% of consumers have privacy concerns about the carrier's use of their personal information.

- *Ad pitch:* This is one method used by advertising agencies to attract consumers. It has multiple forms, ranging from paper coupons to online ad coupons. Vendors can target ad pitches based on the consumer's current mobile location, using wireless advertising.

Technical Issues

Wireless advertising pioneers must overcome several technical challenges, as listed here.

- *Smaller screen size:* The mobile user interface is quite limited, since the compact size of most mobile devices cannot display information-rich content well. Advertising companies must create effective ads that display on a limited screen size. The inconsistency of screen sizes on mobile devices presents another problem. For example, the screen on a mobile phone screen is much smaller than on a PDA. Mobile device manufactures must adopt and enforce some standards in the wireless advertising community to solve this problem. Advertising vendors must create new solutions to generate and deliver wireless ads based on those standards. Tables 16.3 to 16.5 list three sets of stan-

Table 16.3 GSM and Non-GSM SMS Standards for Message Alerts

Features	GSM SMS Standards	Non-GSM SMS standards
Locations	Applied to messages carried over GSM mobile networks, predominantly in Europe	Applied to all SMS messages carried by TDMA and CDMA mobile networks, predominantly in U.S. and Asia
Sponsorship	Message up to 34 characters in length (two lines of text on most phones), designed for sponsoring content that precedes or follows the ad unit	Message up to 34 characters in length (two lines of text on most phones), designed for sponsoring content that precedes or follows the ad unit
Full message	Message up to 150 characters in length, typically delivered standalone	Message up to 100 characters in length, typically delivered standalone

Table 16.4 WAP Phone Standard for Display

	Text Ads	Graphic-Only Ads	Graphic and Text Ads	Comments
Standard	Fixed text size with 15 characters Two-line fixed text with 30 characters One-line marquee text with 34 characters	80 × 8 pixels 80 × 15 pixels 80 × 20 pixels	80 × 8 pixels and l-line text 80 × 15 pixels and l-line text	Content-friendly Can be run as interstitial
Four-line high-display	—	80 × 31 pixels	80 × 15 pixels and 2-line text	Can be run as interstitial
Screen devices	—	—	80 × 20 pixels and 1/2-line text	Can be run as interstitial

Table 16.5 PDA Standard for Screen Display

	Palm OS	Pocket PC OS	Comments
Recommended	150 × 24 pixels	215 × 34 pixels	2 lines of text
	150 × 32 pixels	215 × 46 pixels	—
Supported	150 × 18 pixels	215 × 26 pixels	1 line of text
	150 × 40 pixels	215 × 58 pixels	3 lines of text

dards for GSM and non-GSM messages, WAP-based mobile phones, and PDAs, respectively [10].

- *Inconsistent formatting:* When deciding upon the presentation formats of wireless ads, vendors must consider whether or not these formats can be easily accepted by mobile devices. For example, special fonts and colors might not display well on some mobile devices. Advertising solutions must provide good customization capability and strong interoperability in order to generate and deliver ads that fit different mobile devices.

- *Slow download speeds:* Wireless Internet is facing the slow download speeds, similar to that faced by the wired Internet several years ago, due to the limitation of current wireless system. This issue affects the users' acceptance of wireless ads, and reduces advertising performance. Several technologies, such as 3G and 4G, are trying to address and resolve this issue, but these technologies have not yet been widely adopted.

- *Broad spectrum of technologies:* Powerful players, such as Microsoft, Nokia, and Phone.com, are pushing various wireless technologies and solutions. Since wireless advertising solutions involve a broad spectrum of technologies, advertising agents must decide which technology to use. The industry is trying to develop the means for pushing ads to mobile devices, and to create organizations that regulate this practice.
- *Standardization:* The wireless industry will become too competitive to sustain itself without standardization. Wireless advertising currently lacks well-defined standards and regulations on wireless ad formats, delivery, payment, and measurement.

16.2 Engineering Wireless Advertising Solutions

As discussed in the previous section, vendors need cost-effective wireless technologies and new advertising solutions to support various wireless advertising activities. Engineers must first analyze, understand, and select (or define) a wireless advertising business model by considering its customers, vendors, business value chains, business rules and constraints, and operation workflows.

Engineering a wireless advertising solution consists of four tasks. The first task is to analyze and define domain-specific enterprise processes and workflows for the different parties in wireless advertising, including advertisers, WSPs, content providers, mobile portals, and service vendors. The second task is to understand and specify functional components in a wireless advertising solution. The third task is to specify the required wireless advertising techniques and metrics, based on the underlying business model and advertising standards. The fourth task is to develop, implement, and deploy the required wireless advertising solutions. Since we have covered the details about the application of software engineering methods in developing wireless-based software systems in Part IV, we will now provide readers with basic understanding and insights only about the first three tasks.

16.2.1 Enterprise Processes and Workflows in Wireless Advertising

Understanding wireless advertising processes and service operation workflows is very important for advertisers, publishers, and advertising service vendors, in order to establish their systematic advertising systems and cost-effective solutions. There are specific types of wireless advertising processes for each of these three groups, as examined next [11].

An Enterprise Process for Advertisers

Figure 16.5 shows the basic workflows for advertisers to support electronic advertising. The process consists of the following five phases.

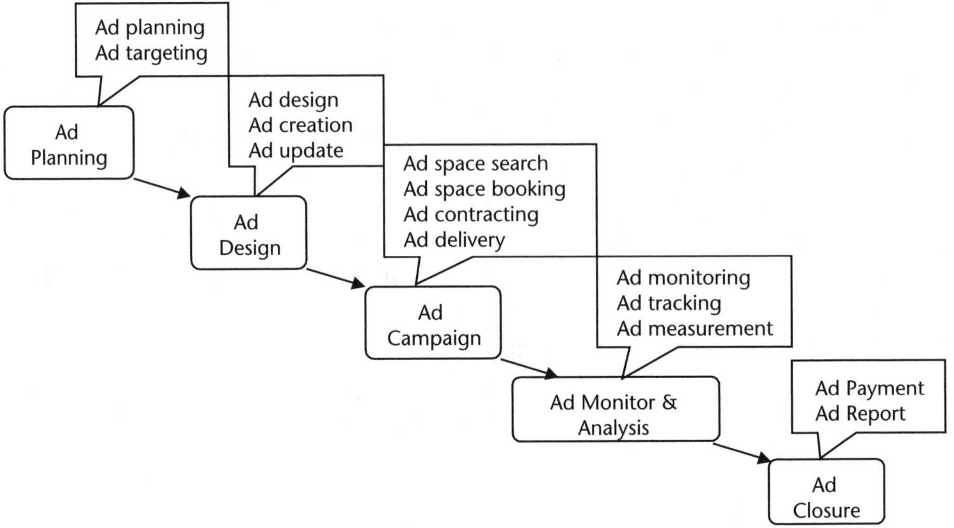

Figure 16.5 An engineering process for advertisers.

- *Advertisement planning (ad planning):* Marketing people first create a plan for wireless advertising based on a well-defined marketing strategy. They must conduct product and customer targeting analyses. Based on the results, they develop a plan that specifies the decisions on the media and publisher selection, presentation approach, targeted audience, posting schedule, and ad content.

- *Advertisement design (ad design):* This phase allows ad designers to design, create, and update advertisements.

- *Advertisement campaign (ad campaign):* Advertisers conduct the ad campaign by interacting with the selected publishers to find the desirable ad spaces and available schedules. They then negotiate with them to reach a business deal, and an advertising contract is then generated for each scheduled ad space. The contract specifies the posting spaces, schedules, payment methods, and costs of the ads. Finally, the ad is delivered to publishers for posting.

- *Advertisement monitor and analysis:* The advertisers track and monitor ad posting (or delivery) status, and collect ad performance data from their publishers. They evaluate the performance of each posted (or delivered) advertisement, based on this performance data.

- *Advertisement closure (ad closure):* The advertisers make the payment transaction to a publisher after the ad is posted, based on the payment contract.

An Enterprise Process for Ad Publishers

Figure 16.6 shows a general enterprise workflow for advertising publishers, which consists of the following steps.

- *Ad space catalog:* The first step is to create and maintain ad space catalogs. An ad space catalog consists of a number of pages with ad spaces. Each ad

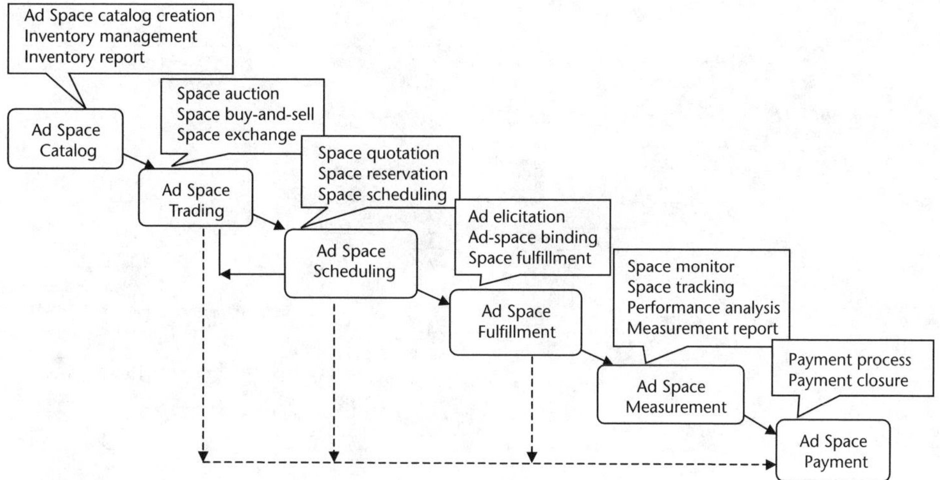

Figure 16.6   An enterprise process for ad publishers.

space has its basic attributes, including its location, posting size, schedule, payment method, and current booking status. They are stored as an ad space catalog.

- *Ad space trading:* In the second step, ad space sales agents interact with the advertisers to sell ad spaces. There are three basic trading models: (1) buy-and-sell, (2) space-auction, and (3) space-exchange. In the buy-and-sell model, ad publishers sell ad space schedules to advertisers using the first-come-first-serve approach. In the space-auction model, ad space trading deals are settled through a predefined auction bidding process. In the space-exchange model, two different publishers reach ad space exchange deals between themselves, which increases their ad space utilization by posting the other party's ads.

- *Ad space scheduling:* WSPs or mobile portal companies create and update the posting schedules for each ad space, based on ad trading deals. It is always a good idea to provide a systematic solution to interact with advertisers by allowing them to search, book, purchase, and confirm various ad space schedules.

- *Ad space fulfillment:* Contracted vendors deliver and post wireless ads for the advertisers, based on ad posting schedules.

- *Ad space measurement:* Advertising vendors monitor and collect ad traffic data to measure the performance of the advertising.

- *Ad space payment:* The advertising payment transaction is generated after the ad posting, according to the predefined ad posting contract.

An Enterprise Process for Advertising Service Vendors

Advertising service vendors need an enterprise process and its operational workflow to support customer services and operations. Figure 16.7 shows one example that demonstrates the interactions between service vendors and other par-

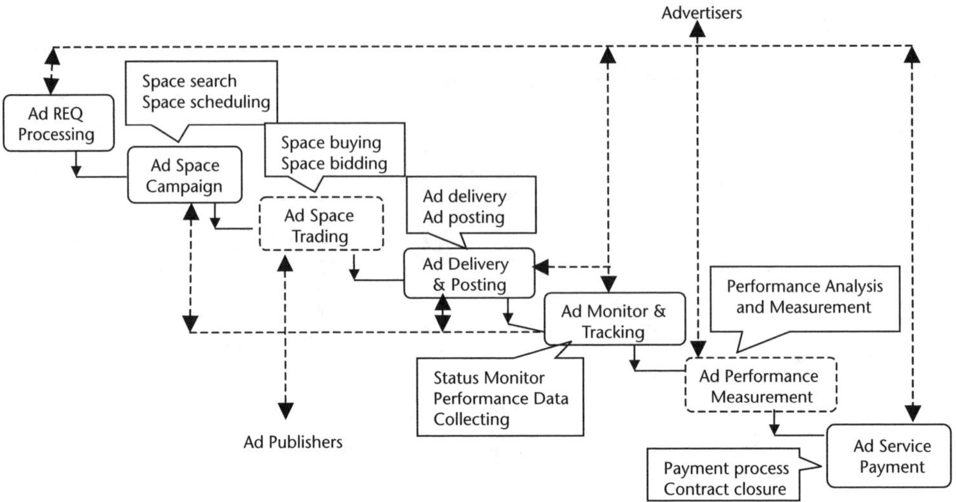

Figure 16.7   An enterprise workflow for advertising service vendors.

ties, including advertisers and publishers. The enterprise process consists of the following steps.

- *Ad request processing:* Advertisers must select advertising service vendors to post wireless ads, and then send them ad posting requests. Each request must provide the detailed requirements for ad posting, including posting schedule, planning budget, targeted consumer groups, advertising content, and potential publisher.

- *Ad space campaign:* Service vendors search for available ad spaces of different ad publishers for each ad request, and launch its advertising campaign to meet the requirements of each request. The ad posting spaces and schedules are then booked and documented as the booking records between a publisher and the service vendor.

- *Ad space trading:* Ad publishers confirm or reject the requested ad spaces and schedules. The corresponding ad trading operation is carried out for each posting request, after publishers confirm its ad space and posting schedule. This results in a contract for ad posting between the service vendor and the publisher.

- *Ad delivery and posting:* Ad service vendors deliver wireless ads to mobile users through the publishers, based on the contracted schedules. Different ad delivery techniques and presentation methods are used to post mobile ads.

- *Ad monitor and tracking:* Service vendors monitor and track the ad posting states, collect the data, and report the results to the advertiser.

- *Ad performance measurement:* Ad service vendors may provide advertisers with ad performance analysis and evaluation services. The vendors use diverse methods to analyze the collected performance data. A detailed discussion of this service is given in a later section.

- *Ad service payment:* The last step consists of a payment process for each ad posting contract whenever its ad posting service is completed.

16.2.2 Wireless Ad Targeting

Wireless ad targeting refers to the use of effective methods to deliver wireless ads to the mobile users who are interested in receiving them. Its major purpose is to increase the effectiveness of wireless advertisements by posting ads to end users at the right time and location. Wireless ad targeting studies and applies cost-effective methods to address the following two challenges:

- Delivery of wireless ads that meet mobile end user interests and needs;
- Posting of wireless ads to potential consumers at the right time and location.

Using an effective targeting approach in advertising has four major benefits for vendors and advertisers.

- The access rate of posted ads will be increased.
- Vendors and advertisers will receive more ad responses from mobile users.
- More purchasing transactions will be made.
- The relationship between the customer and the merchant will be enhanced.

There are four basic targeting strategies in wireless advertising. They are given here.

- *Profile-based targeting:* This strategy selects and delivers ads to end users based on user profiles, channels, and interests. For example, an SMS message about a new movie coupon can be sent to a user who likes to see movies.
- *Location-based targeting:* This strategy selects and delivers ads to end users based on their current locations, using a location-detecting solution.
- *Content-based targeting:* This strategy selects and posts ads on a mobile device based on the current mobile page accessed by the user. For instance, specific automobile sale ads will be posted on a mobile device whenever its user accesses a mobile portal to look for automobile sale information.
- *Hybrid ad targeting:* This strategy combines a number of targeting strategies in an intelligent way to find the most appropriate ads to deliver to the interested mobile users at the right time and location. For example, whenever a Toyota brand lover accesses a mobile portal, Toyota automobile sale ads from local dealers will be sent to the user's mobile device based on the user's current location.

Implementing a targeting solution for wireless advertising requires four types of information: (1) wireless ad information, (2) ad space information, (3) mobile user information, and (4) location-based information. This information should be collected, stored, and maintained in a repository using a systematic solution, in order to target ads in real time.

There are three basic approaches to implementing ad targeting solutions. The first approach is known as the static procedure-based approach, in which a selected targeting strategy is implemented using a predefined algorithm, based on the current data stored in an information repository. This approach is appropriate for the profile-based targeting strategy. The major drawback of this approach is its inflexibility, since it cannot react to real-time user interactions and mobile locations. The second approach is known as the dynamic procedure-based approach, in which a targeting strategy is implemented using a predefined dynamic algorithm. It operates based on the real-time dynamic data (e.g., user's responses, and mobile locations). This approach can be easily used to support a location-based targeting strategy. The third approach, which is known as an intelligent-based procedure method, supports a hybrid targeting strategy and is based on given business rules. These digital rules are stored as a part of a knowledge repository to support reasoning and decision-making. A set of procedure-based algorithms can be implemented, based on the resulting decisions. The major advantages of this approach are its flexibility and intelligence, although it is more complicated and costly to implement.

16.2.3 Wireless Ad Delivery

There are two basic approaches to deliver wireless ads to mobile users [5].

- *Wireless pull strategy:* In the pull approach, wireless ads are delivered to mobile device users only when they select or request the ads. For example, when a consumer requests information about all movie theaters within a specific area, the movie ads will be delivered to the consumer. The pull ads are less intrusive; however, they carry a higher functional cost, since they require more complicated systematic solutions to track the reception of wireless ads. The wireless pull strategy provides the users with the flexibility to interact with their mobile devices. For example, consider that the mobile device users need to receive discount coupons on items in a local shop. They could interact with the wireless ads by providing their current location and submitting a request.
- *Wireless push strategy:* In the push approach, wireless ads usually operate based on user profiles and personal interests. For example, movie advertisements posted by theaters can be sent to the consumers' mobile devices. The drawback is that the consumers might find these push ads intrusive and react negatively.

Some early applications of wireless advertising provided local entertainment information, stock quotes, dining and restaurant reservations, and wireless coupons. Wireless advertising will continue to be introduced to various other wireless applications, such as wireless information services, chatting, gaming, and entertainment.

16.2.4 Wireless Advertisement Tracking

Wireless ad tracking refers to the monitoring and tracking of the posting status of ads. Ad tracking can use a systematic solution, similar to tracking online advertising. The major purpose of wireless ad tracking is to allow advertisers to check and monitor ad delivery and posting statuses. Advertisers can collect, track, and analyze ad posting and delivery data, as well as end user reactions and responses. This collected data is very useful for evaluating advertisement performance through different vendors. An advertisement service agency needs to provide ad tracking as one of the services to wireless advertisers, so the advertisers can monitor the ad postings in their contracted mobile ad spaces.

Wireless ad tracking should monitor the following classes of data.

- Mobile user profiles, including registration information, user interests, and channels;
- Dynamic user access information, such as access contents, access locations, and time;
- Mobile device information, such as device IDs and browser information;
- Wireless ad profiles, including ad ID, ad space ID, and its posting service company;
- Ad delivery information, such as delivery counts and the number of receptions;
- Ad posting information, including ad impression counts and posting status;
- Mobile user reactions, such as ad selected or clicked counts, ad select-through or click-through counts, user direct feedback, and direct transaction counts leading from ads.

Wireless ad tracking has the advantage of collecting more personal and location-oriented information from mobile accesses, as compared to online ad tracking. A wireless tracking solution must provide the answers to the following basic questions.

- Where and when has a wireless ad been delivered and posted? How many mobile users have received the ad? How many of them have viewed the ad?
- Has a wireless ad been successfully delivered and posted based on a contracted schedule? Has the ad reached to the targeted user groups?
- What is the responding traffic for a posting ad? How many trading transactions are generated by a posted ad?

Clearly implementing a wireless ad tracking solution must depend on the underlying business model, wireless technology, and ad delivery approach. A wireless ad tracking solution must include three functional parts.

- *Ad-tracking repository:* This collects, stores, and maintains diverse ad tracking data.
- *Ad-tracking client component on mobile devices:* This interacts with the ad-tracking module at the server side to track ad posting data, collect the reactions of users to events and responses, and send them back to the server.
- *Ad-tracking component of the server:* This communicates with the client-based ad tracking component to store and analyze the collected wireless ad data. A basic reporting function should be implemented to provide different statistical reports.

AvantGo implements an ad response tracking system. It automatically collects responses from ads, and sends automated text or HTML e-mail responses to people who respond to the ad. The response can include additional product information, confirmation, and links to Web sites. Ad response tracking from AvantGo enhances the advertising effectiveness by providing accessible online reports.

16.2.5 Wireless Advertising Payment

There are a number of different payment models for electronic advertising [11, 12]. They are discussed here.

- *Pay-by-impression:* A common pay-by-impression model is the cost-per-thousand impressions (CPM) model. It computes advertising cost based on the number of ad impressions, and assumes that there is value to every ad impression. This model has been widely adopted in online advertising. This model can be used for home page, location-based, and category-based ads in wireless advertising.

- *Pay-by-click:* A common pay-by-click model is known as click-through model, which calculates the advertising cost of a posting ad based on the number of ad-clicks by viewers. Advertisers see this as a definitive measure for the advertising costs. This model is not suitable for a brand-building ad, although it could be very useful for ad icons (or banners) posting on mobile pages, coupons, and location-based ads on mobile devices.

- *Pay-by-message:* This model has been used for e-mail–based Internet marketing. It charges advertisers based on the number of e-mail messages that have been sent, received, and answered. Consumers do not like to receive unsolicited e-mail ads, although some reports have indicated this to be one of the effective ways to reach consumers and get direct responses. When alter messages (both text and audio messages) are wirelessly sent to mobile users, a similar payment model could be adopted.

- *Pay-by-meter:* This model also known as flat fee. It charges a flat advertising fee for an ad space, based on its posting duration time and not on its performance. This was the first payment model used in online advertising. This model could be very reasonable for brand-building and interstitial ads.

- *Pay-by-transaction:* This payment model is entirely based on ad performance. It computes the advertising costs based on the number of the leading sale transactions that result from a posting ad. It could be useful for location-based and category-based wireless ads.

- *Pay-by-display:* Online game-based advertising has used this model to compute advertising payment. It charges an advertiser based on the total number of ad displays for a game ad space. Wireless mobile game vendors can use a similar payment model to compute the advertising costs.

- *Hybrid payment model:* A hybrid model usually is a complicated model that is defined using the other payment models. Vendors use a hybrid payment model to link with their predefined business strategies to deal with different types of advertisements. This model is currently widely used in the United States. Hybrid pricing accounts for an estimated 51% of advertising deals [12].

16.2.6   Performance Measurement in Wireless Advertising

Performance measurement in advertising refers to the performance evaluation of advertisements after posting. Advertisers post their ads through various advertising vendors, and evaluate the ad performance, because they want to check the effectiveness of the posted ads through the different vendors. To perform a quantitative measure, they want the answers to the following questions.

- What kinds of consumer groups have been reached by a posted ad? Has a posted ad reached to the targeted consumer groups?
- How many consumers have seen the ad? How many of them responded to the ads? What are their responses and reactions?
- How many of consumers have made direct purchases through a posted ad? What is the impact on consumers in brand-building?
- Is the selected vendor the right channel to reach to the targeted consumer groups? Which one has a better performance?

Traditional media, such as TV, radio, and newspapers, measure advertising performance in a very limited format. For example, the most common way to conduct ad performance measurement is based on a marketing survey, which is costly and inefficient. A systematic solution can be developed to support performance measurement of wireless ads, as can be done for online advertising. We must define and select a set of performance metrics for ad performance measurement before developing a systematic solution. Table 16.6 lists a typical set of performance evaluation metrics for wireless advertisements.

A performance measurement solution can be developed using these ad performance metrics, and then added into an advertising service system to support performance analysis and evaluation. WSPs or advertisement service agencies can provide advertisers with diversified ad performance evaluation reports. A key measurement is to understand the effect of advertising on the useable broadband wireless of the link.

16.3   Major Players and Their Solutions

There are a number of major business players in wireless advertising platforms, tools, and services. AvantGo and SkyGo provide wireless service platforms. Vindigo, I3Moble, and Infiniq specialize in advertising portals. Others provide wireless advertising and marketing solutions, such as WindWire, Wireless Opinion, DoubleClick, AdForce, and InPhonic/Avesair, and so forth. This section only discusses four of them—SkyGo, AvantGo, Vindigo, and Avesair.

16.3.1   SkyGo

SkyGo (http://www.skygo.com) lets carriers, publishers, and advertisers enhance consumer relationships and generate new revenue through wireless marketing and m-commerce. SkyGo uses the media-dependent, outsourced media sales revenue model. It seeks to subsidize content and subscriber fees to increase and retain its

Table 16.6 Set of Performance Metrics for Measuring Wireless Ads

Measurement Category	Advertising Performance Metrics	Descriptions
Ad posting metrics	Advertisement page impression count	Total count of ad page displays from a vendor
	Advertisement impression count	Total count of ad postings for an ad in an ad space
	Advertisement display time	Total ad display time for an ad by a vendor
Ad response metrics	Ad-click rate (only for PDAs/Pocket PCs)	Ratio of the number ad clicks to ad impression count
	Ad-select rate (only for mobile phones)	Ratio of the user selection count to ad impression count
	Ad-response rate (only for direct responses)	Ratio of user response count to ad impression count
	Ad-reception rate (only for ad alters)	Ratio of user reception count to total ad-delivery count
Ad delivery metrics	Ad-delivery reach rate	Ratio of number of delivered ads to number of received ads
	Ad-delivery frequency	Ad delivery frequency in a given time period
	Ad-delivery impression rate	Ratio of number of delivered ads to number of posting ads
Ad impact metrics	Ad click-through rate (for PDA/Pocket PCs)	Ratio of ad posting count to total click-through counts
	Ad select-through rate (for mobile phones)	Ratio of ad posting count to total select-through counts
	User trading rate for posting ad	Ratio of ad posting count to number of trading transactions
	Brand awareness rate (for brand building ads)	Ratio of users who are aware of ad to number of reached users

subscriber base while generating significant revenue. SkyGo provides technology, media, and consulting services to help companies communicate effectively with mobile consumers. It also offers technology platforms and services for delivering targeted marketing to mobile devices, which can open a new revenue channel that lets advertisers interactively deliver subsidized mobile content and services. SkyGo provides the required functional features to support ad management, delivery, targeting, tracking, and reporting. It is designed to accommodate different mobile devices, including mobile phones, pocket PCs, PDAs, and pagers [13]. SkyGo's Mobile Advertising Platform (SkyGo MAP) is a family of mobile marketing tools that helps marketers and wireless carriers to build relationships with mobile consumers and to deepen brand awareness, generate response, and foster customer loyalty.

16.3.2 AvantGo

AvantGo (http://www.vindigo.com), founded in 1997, is one of the major players in mobile enterprise software. It addresses the growing need to make workforces more mobile. AvantGo uses the media-dependent, in-house media sales revenue model. It seeks to expand the carrier's business function to include media sales, by controlling and drawing revenue from the wireless Internet property. AvantGo develops solutions based on its Dynamic Mobility Model (DMM) [14]. This model leverages existing open standards to support the transitions between online and off-line use, sync, surf, and push, regardless of the challenges presented by multiple-device operating systems, wireless services, and data access models. AvantGo's solutions can automate business processes, improve the exchange of information between companies and their employees and customers, increase business efficiencies, and make customer interactions more effective.

AvantGo's DMM and technology allow developers to write applications that run properly in both wired and wireless environments. AvantGo's M-Business Server Application Edition delivers the power of enterprise applications, XML Web services, and the Internet to mobile devices. Wireless advertising vendors can significantly reduce development time with this technology to provide a low-cost wireless Internet enterprise solution. AvantGo's M-Business Server and client technology allow development engineers to easily create advertising application solutions that deliver wireless ads to a variety of wired and wireless mobile devices.

16.3.3   Vindigo

Vindigo (http://www.vindigo.com) focuses on creating a platform for delivering important location-based information and services. The global consulting firm Ovum estimates that location-based services will deliver $20 billion of revenue by 2006 [15]. Vindigo uses the diversified revenue model, and seeks to create a compelling offering that draws and retains subscribers. Vindigo is one of the leading players in providing mobile content and marketing platforms. It has a track record of successfully deploying robust location-based services with a working advertising model for handheld devices. Vindigo's next generation client-server architecture will attempt to function well in real-time wireless and synchronized environments. Vindigo believes that any mobile application must have intelligence in making use of the contexts such as location, time, schedules, contacts, and personal preferences.

Vindigo's technology consists of two parts: server technology and client technology [16]. Its server architecture supports the user subscription Web site, customer service Web site, content management Web site, ad management Web site, geocoding engine, and directions engine. A common database access layer and location-based logic layer works across all wireless and mobile services. A common wireless query server sends data to wireless browsers, using XML over HTTP, or WML over WAP. Vindigo's client technology supports three types of mobile devices: wireless PDAs, synced PDAs, and WAP phones. For WAP phones, a WML-based thin client is designed to support wireless Web users' accesses to Vindigo's services through a standard Web browser. A thick client software is designed in Vindigo's technology for synced PDAs. It contains all the information for delivering the Vindigo's location-based services. The thick client updates its information from a Vindigo server when the user synchronizes his or her device with a host PC connected to the Internet. This off-line operation enables the client to provide full functionality and fast response times, regardless of the availability or quality of network connectivity. Vindigo's wireless PDA client frequently connects to its server to retrieve information over a wireless link. However, it also runs many user interface features local to the device, such as maps and directions, to avoid the latency of a slow wireless network.

16.3.4 Avesair

Avesair (http://www.avesair.com) is a leading marketing and commerce provider. It makes products that operators, marketers, and content providers can use to deliver a range of targeted services, based on subscriber preferences and contextual information. Avesair uses the media-dependent, in-house media sales revenue model. Avesair has developed relationships with many wireless industry companies, including Nokia, 23snap, Enpocket, Profilium, and SignalSoft. Avesair recently acquired WindWire, an industry supplier of wireless media services. This acquisition adds WindWire's customers to Avesair's North American and international customer list, including AT&T Wireless and Go2Systems.

16.4 Summary

Wireless advertising is becoming a new trend in advertising and marketing. Wireless service providers, advertisers, and advertisement service businesses should understand the challenges, opportunities, and required new cost-effective solutions to deliver diverse ads to the rapidly growing number of wireless mobile devices. This chapter helps readers to understand wireless advertising. It provides them with a comprehensive tutorial to cover the basics of wireless advertising, including basic concepts, business models, applications and benefits, technical challenges, and business issues. By focusing on the engineering perspectives of wireless advertising, this chapter covers the required processes, domain-specific knowledge, and technical methods in wireless advertising solutions. Developing new solutions in wireless advertising requires engineers to consider the following:

- The various business models in wireless advertising, and choice of the proper model;
- The proper enterprise processes and workflows that support wireless advertising operations among different parties;
- The domain-specific knowledge and methods in wireless advertising, including ad targeting, delivery, trading, tracking, and performance measurement;
- The pros and cons of different wireless technologies described in Part II;
- The well-defined software engineering methods and principles discussed in Part IV.

References

[1]   Pastore, M., "Wireless Aims for Widespread Appeal," Cyberatas, 2000, cyberatlas. internet.com/markets/wireless/article/0,,10094_587701,00.html.

[2]   WindWire, "Fast-to-Wireless: Capabilities and Benefits of Wireless Marketing and Advertising," 2000, http://www.imapproject.org/imapproject/downloadroot/public3/ftw_report.prf.

[3]   Mickelson, A., "Wireless Advertising Goes on Trial," http://www.wirelessinternetmag.com/news/0180/0108_ideas_trial.htm.

[4]   SkyGo, Inc., "SkyGo Study Illustrates Wireless Advertising Is an Effective Branding and Direct Response Vehicle," March 5, 2001, www.skygo.com/news/pr01_0305.html.

[5]   Kannan, P. K., A-M. Chang, and A. Winston, "Wireless Commerce: Marketing Issues and Possibilities," *Proc. 34th Annual Hawaii Int. Conference on System Sciences (HICSS-34)*, 2001, p. 9015.

[6]   Yunos, H. M., J. Z. Gao, and S. Shim, "Wireless Advertising's Challenges and Opportunities," *IEEE Computer*, Vol. 36, No. 5, May 2003, pp. 30–37.

[7]   AvantGo, Inc., "My AvantGo Mobile Advertising Services," 2001, http://avantgo.com/products/businesses/marketing_commerce/pdf/myavantgo_advertising.pdf.

[8]   Shim, R., "Is the Wireless Web Ready for Handhelds?" January 11, 2001, http://www.news.com.com/2009-1040-250881.html.

[9]   Saunders, C., "Studies: Mobile Ad Market to Grow, Amid Risks," November 1, 2001, http://www.internetnews.com/IAR/articile.php/915121.

[10] Wireless Advertising Association, "WAA Advertising Standards Initiative Draft Standards," June 26, 2001, http://www.mmaglobal.com/press/.

[11] Gao, J. Z., et al., *Online Advertising—Taxonomy and Engineering Perspectives*, Technical Report, San Jose State University, 2002.

[12] Chaplin, J., "Mobile Marketing," December 10, 2002, http://www.broadcastpapers.com.

[13] SkyGo, Inc., "Ideas & Strategies for Implementing Mobile Marketing," September 2001, http://www.wirelessdevnet.com/library/SkyGo_White_Paper.pdf.

[14] AvantGo, Inc., "Wireless Advertising Solutions," 2001, http://avantgo.com/products/businesses/marketing_ commerce/pdf/wireless_ad.pdf.

[15] "Ovum Predicts Mobile Location Services Driven by M-Commerce Will Generate $20 Billion by 2006," January 23, 2001, http://www.cellular.co.za/news_2001/01232001-ovum_predicts_mobile_location_ services.htm.

[16] Caceres, R., et al., "Mobile Computing Technology at Vindigo," *IEEE Wireless Communications*, No. 1, 2002, pp. 50–55.

CHAPTER 17

Mobile Payment Systems

The rapid advances in wireless networking communication and mobile technology are making a big impact on people's daily lives. With a significant increase in the number of mobile device users, more wireless information services and mobile commerce applications are needed [1]. Wireless payment and mobile transaction services have become an essential part in wireless information services and commerce applications, such as mobile banking, wireless trading, mobile portal, peer-to-peer payments, and e-shopping on mobile devices.

According to the Wireless World Forum, mobile payment on wireless devices will provide very good business opportunities in the coming years.

China, Japan, and Europe are the three fastest growing markets in wireless communication services due to their significant increase in mobile device users. The United States and Canada will be the highest potential growth markets for value-added wireless services and mobile information applications, since wireless communication infrastructures and mobile technologies are mature. The two countries still need to cope with many business and technical issues in cross-network interoperability and standardization.

The creation of wireless payment solutions for mobile devices provides good business opportunities for businesses, and brings new technical challenges and issues to wireless service vendors and merchants. A number of types of electronic payment solutions for Internet-based application and commerce already exist, but there are still new business issues and challenges in wireless payment, due to a lack of study and experience in wireless payment. Quite a few books and papers discuss businesses, management, and technology aspects of wireless commerce [1–3], but very few papers address the importance, issues, and solutions of wireless payment [4–6].

We discuss wireless payments in terms of its concept, business models, opportunities, challenges, solutions, and major players. This chapter discusses wireless payment systems, including functional features, types, issues, and current practices.

17.1 Wireless Payment

Durlacher [7] defines mobile-commerce payment (or m-payment) as any transaction with a monetary value that is conducted via a mobile telecommunications network.

Wireless payment refers to the study and application of wireless-based electronic payment solutions for mobile commerce to support point-of-sale and/or point-of-service payment transactions on mobile users' devices, such as cellular telephones, smart phones, personal digital assistants (PDAs), and mobile terminals. A wireless payment system is a mobile commerce system that processes electronic commerce payment transactions supporting mobile commerce applications in wireless networks and wireless Internet infrastructures.

Wireless-based merchants, mobile content vendors, and wireless information and commerce service providers generally can use wireless payment systems. They can use the payment systems to process and support payment transactions driven from wireless-based commerce applications. These applications can include wireless-based trading systems, mobile portals, wireless information, and commerce service applications.

Wireless-based payment systems can be viewed as one type of electronic payment system. Similar to current online payment systems, wireless-based payment systems also provide automatic payment transaction processing. Unlike online payment systems, where payment transactions are driven by Web user accesses and processed by Internet-based payment systems, wireless-based payment systems process transaction requests from mobile devices and location-oriented terminals.

17.1.1   Basic Requirements of Wireless-Based Payment Systems

The basic requirements of mobile payment transactions (see Figure 17.1) include the following characteristics.

- *Simple and convenient:* The payment method must be easy for mobile users to make payments through mobile devices.
- *Fast and efficient:* Users must get fast response and quick payment processing.
- *Secure:* Buyers, sellers, and all involved parties should have the assurance that their payment accounts will be protected from unauthorized use and that their transactions will be secure.
- *Universal acceptance:* Consumers should be allowed to shop and pay anyone, anywhere, using mobile payment solutions and underlying wireless networking.

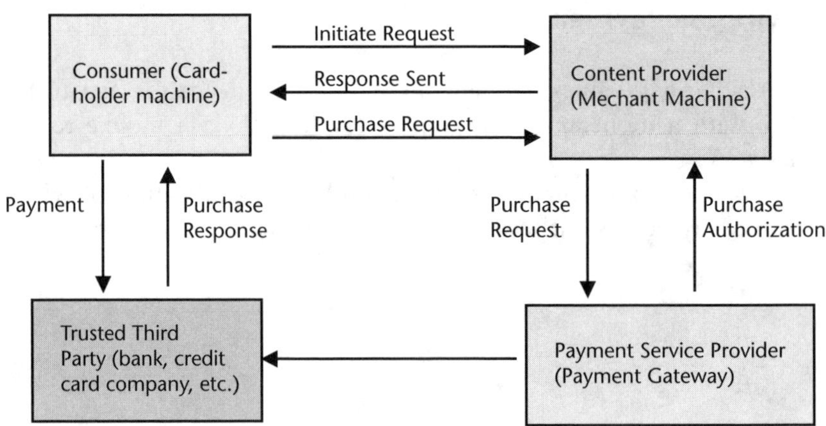

Figure 17.1 Mobile payment transaction.

The key roles to be managed in mobile payments are content provider, authentication provider, payment authorization, settlement provider, and consumer, according to the Telecom Media Networks [5].

- *Consumer:* The consumer is the person who owns a mobile device and buys content or services from the content provider through his or her mobile device.
- *Content provider:* A content provider is an individual (merchant) or some organization that sells either electronic or physical content (products or services) to consumers.
- *Trusted third party:* The trusted third party (TTP) is the company that performs the authentication and authorization of the transaction parties and settlement. It could be a telephone company (telco), bank, or a credit card company.
- *Payment service provider:* The payment service provider (PSP) is the central entity responsible for the payment process. It enables the payment message initiated from the mobile device to be routed to and cleared by the TTP. This service generally includes an "e-wallet" application that enables payers to store their payment details, such as credit card account numbers and shipping addresses, on a provider's secure server, so that they do not need to type in all the information required for each sale on small and difficult-to-use mobile keypad devices. The PSP may also act as a clearinghouse to share the revenues between all the partners involved in the payment process. It could be a telephone company, a bank, a credit card company, or a start-up company. A telephone company could be positioned at the same time as PSP, TTP, and content provider.

17.1.2 Payment Schemes

There are three popular payment schemes: cash payment, prepayment, and postpayment. Three types of online electronic payment solutions have been developed to support these payment schemes. They are:

- Credit card–based electronic payment solutions, in which consumers use credit cards to make automatic payments;
- Electronic check–based solutions, in which consumers use electronic checks to make their payments in an automatic manner;
- Electronic cash–based systems, in which consumers use digital cash to make payments in a systematic way.

The three types of online electronic payment solutions can intuitively be extended to or adopted in wireless Internet payment systems. Potential mobile payment falls into several distinct categories, such as content type (digital goods, voting); content value (micropayments, macropayments); transaction type (pay-per-view, pay-per-unit); and transaction settlement method (credit, debit).

17.1.3 Mobile Payment Process

Telecom Media Networks [5] identified four key roles in mobile payment. They are the content provider, the authentication provider, the payment authorization and settlement provider, and the consumer. The main process of m-payment can be divided into the following phases.

- *Registration:* The consumer needs to open an account with the PSP to enable the payment service through a particular payment method. The PSP will require confirmation during this phase from the TTP that handles the relationship with the customer.
- *Transaction:* The consumer indicates the desire to purchase some content. The content provider forwards the purchase request to the PSP. The PSP then requests authentication and authorization from the TPP. The PSP informs the content provider about the success of the purchase demand. The content provider then delivers the purchased content.
- *Clearing and settlement:* The settlement can take place in real time during the purchase or at a later time. Real-time settlements can be conducted via a pre-paid account if the TTP is a telco, or directly through a bank account if the TTP is a bank. If paid at a later time, then the PSP sends the billing information to the TTP. The TTP sends the bill to the consumers, gets the money back, and forwards it to the TTP. The PSP is then responsible for computing the revenues of each entity and distributing the funds.

17.1.4 Security Solutions for M-Payment Systems

- *PIN or password:* PINs and passwords are among the easiest ways for mobile payment authorization. The PIN uses numbers, and the password uses a combination of letters and numbers. The user can enter a PIN when making payment via mobile phone, or a password when making payment on the payment Web site. A password is more secure because it is harder to be cracked. However, the user enters a PIN and not a password on a mobile phone, because it is easier to input with a numeric keypad.
- *Caller identification (Caller ID):* Caller ID is a service provided by many mobile telephone service providers that allows a receiver of a call to see the number of the caller. If the mobile number of the user does not match the mobile number in the user account ID, then the transaction is cancelled. A caller ID is not easy to fake or change, because it is only generated at the caller party's exchange and not at the mobile phone.
- *Callback:* Callback, in conjunction with a PIN, is another way to authenticate and authorize the user for mobile payment. It authenticates the person who is making a payment using a certain mobile phone number by asking that person to enter his or her PIN when he or she is called back.
- *WAP-WTLS:* The wireless transport layer security (WTLS) is the security layer protocol in the WAP architecture. Most mobile payment systems discussed in this chapter use the WAP for users to access their account information, and use the secure WTLS connection for data integrity and protection.

The WTLS layer operates above the transport protocol layer. It provides the upper-level layer of WAP with a secure transport service interface that preserves the lower-level transport service interface. WTLS also provides an interface for managing secure connections. The primary goal of the WTLS layer is to provide privacy, data integrity, and authentication between two communicating applications.

17.2   Different Types of Wireless-Based Payment Systems

Current wireless-based payment systems could be classified into the following types.

17.2.1   Account-Based Payment Systems

Each customer in an account-based payment system is associated with a specific account maintained by the TTP, such as a bank or telco. In prepaid transactions, this account is directly linked to the consumer's savings account. The consumer maintains a positive balance in this account, which is debited when a prepaid transaction is processed. In post-paid transactions, the charges from a transaction are accrued in the consumer's account, and the consumer is then periodically billed for the balance of the account by the TTP.

Account-based payment systems can be classified into four categories.

Mobile Phone–Based Payment Systems

A mobile phone–based payment system enables customers to purchase and pay for goods or services via mobile phones. Some systems also allow users to send or receive money with their account via the mobile phone. The mobile phone is used as the personal payment tool in connection with the remote sales. A phone card–based payment system has the advantage over the traditional card-based payment, since the mobile phone replaces both the physical card and the card terminal. Payments can take place far away from both the recipient and the bank. Having a mobile phone within the reach of a GSM network is the only requirement.

Smart Card Payment Systems

This type of payment systems uses a smart card, with an embedded microcircuit that contains memory, a microprocessor, and an operating system for memory control. The smart card is a secure storage location for secret information. It can be used for electronic identification, electronic signature, encryption, payment, and data storage. However, a smart card can only perform calculations when connected to an external source of computing energy. Its security is based on public key cryptography, and any attempt to directly access its data is more likely to destroy the card, since it only can be accessed through its reader. The widely used SIM cards are typical examples of smart cards. SIM cards can be used to customize mobile phones, regardless of the communication standards. A SIM card could potentially be used to store payment information, such as card numbers, or used to authenticate a customer making a payment. "Dual-slot" phones are also available, which contain a slot into which the customer can insert their credit card (smart card). This communicates with the SIM card to obtain a card authorization. Other "dual-chip" phones have a second small slot for taking just the chip element of a payment card, which can work independently of the SIM card. SIM cards will continue to provide enhanced functionality beyond the basic communications link. Payments and authentication are going to become increasingly important applications. Mobile operators, with their large customer base and existing SIM/smart card infrastructure, are going to become a key player in the marketplace.

Credit Card Mobile Payment Systems

This type of mobile payment system allows customers to make payments on mobile devices using their credit cards. These payment systems enhance the existing credit card–based financial infrastructure by adding wireless payment capability for consumers on mobile devices. The existing SET secure protocol, developed by Visa and MasterCard for the secure transfer of credit card transactions, has been extended, and is known as 3D SET to support mobile payment for mobile device users [5]. Current SET-based payment systems have not been widely adopted because it was inconvenient to both the cardholder and the merchant. Visa started a new initiative to solve this problem, which is known as "3D mo-del" (three domains) for secure mobile credit card transactions. It looks at the activity between the following domains: (1) the acquirer domain (including merchants and their bank); (2) the issuer domain (including cardholders and their bank); and (3) the interoperability domain (including cardholders' banks and merchants' banks), which is covered by the original SET protocol. The 3D-SET protocol is developed to meet the needs of the three domains. It has two major advantages over the SET protocol. First, the 3D-SET protocol reduces the effort of performing a SET payment on behalf of the cardholder, and it allows the cardholder to use their certificate from any mobile device access. Second, the 3D-SET protocol frees the merchant from the chargeback problem, because it does not require the cardholder to download software. Berlecon Research warns that the future of 3D-SET is the danger, since issuing banks might want to shift the costs to cardholders.

Mobile Point-of-Sale

Payment-mobile point-of-sale (POS) payment systems enable customers to purchase products on vending machines or in retail stores with their mobile phones. Many companies equip mobile devices with POS payment systems. It is designed to complement existing credit and debit card systems, by enabling all mobile phone users to turn their phones into the preferred instrument of payment. Mobile POS systems can be applied to two different payment applications. The first type is known as automated point-of-sale payments. They are frequently used over ATM machines, retail vending machines, parking meters, toll collectors, and ticket machines, to allow mobile device users to purchase goods, such as snacks, parking permits, and movie tickets. The second type is known as attended point-of-sale payments, such as those in retail store counters and taxis, in which mobile users make payments using mobile devices with assistance from a service party, such as a counter clerk or taxi driver.

There are four leading industry groups developing open standards for local mobile transactions in mobile POS payments. They are: (1) the Infrared Data Association's (IrDA) Infrared Financial Messaging Special Interest Group (IrFM SIG), (2) the Mobile Electronic Transaction Forum (MeT Forum), (3) Bluetooth Special Interest Group's (Bluetooth SIG) Short Range Financial Transaction Study Group (SRFT SG), and (4) the National Retail Federation (NRF).

17.2.2   Mobile Wallets

Mobile wallets (m-wallets) are the most popular type of mobile payment options for transactions. Like e-wallets, they allow a user to store billing and shipping information that the user can recall the information with one click while shopping from a mobile device [5].

The main types of mobile wallet schemes in the market are client wallet and hosted wallet [8]. Client wallets are stored on a user's device in the form of a SIM Application Toolkit card that resides in a mobile phone. Since the wallet resides on the hardware, it is difficult to update, and the user's sensitive financial information is potentially compromised if the device is lost or stolen. Hosted wallets are hosted on a server. This gives the service provider much greater control over the functionality it delivers and greater security of the data and transactions. Hosted wallets can be self-hosted wallets or third party hosted wallets.

The MasterCard Global Mobile Commerce Working Group proposed "Remote Wallet Server Architecture." The payment account information in the server-based m-wallet scheme is stored on a financial institution's remote server. The cardholder identifies himself/herself via an ID and password through the mobile device, and approves the transaction. The merchant then charges the purchase to the buyer's payment account and processes the transaction through the acquiring bank and MasterCard network. The payment account information itself is never transmitted from the mobile device, making the transaction more secure than schemes in which the payment account information is transmitted over the air.

17.2.3   SET-Based Mobile Wallet

Server-based mobile e-wallets using SET technology are already being used, providing secure transaction capability for merchants and cardholders. There are three phases in the payment architecture. The first phase is the initiation phase, in which the merchant server sends a payment initiation message to a cardholder device. The second phase is the cardholder device–SET wallet server interaction phase. The cardholder device forwards the merchant's initiation message, so that the wallet server either receives or is able to retrieve the SET wake-up message. The cardholder approves the transaction, and the wallet server authenticates the cardholder. The third phase is the SET transaction phase, in which the SET wallet server and the merchant SET server conduct a SET transaction. SET will eventually support multiple cardholder authentication schemes, including SET certificates, PIN, chip cryptograms, digital IDs, and non-SET certificates (MasterCard, 2002).

17.3 Major Players in Wireless Payment Systems

There are a number of major players in wireless payment systems. In this section, we discuss two groups of major players: (1) wireless payment, including Paybox and Ultra's M-Pay; and (2) mobile wallet, including Encorus PaymentWorks and SNAZ.

17.3.1 Paybox

Paybox (http://www.paybox.net/) was developed by Paybox.net AG, owned by Deutsche Bank, in May 2000. It works like a debit card. Each payment is debited from the user's bank account only after users have authorized the transaction by entering his/her Paybox PIN on the mobile phone. It enables users to purchase goods, services, and make bank transactions with their mobile phones. All transactions are conducted over the existing GSM mobile phone network.

Paybox uses the electronic bill presentment and payment (EBPP) protocol, which helps companies to optimize the payment process costs. The traditional invoice is replaced with a link, which leads the recipient to a prefilled electronic remittance form. The invoice recipient is called after entering his or her mobile phone number or Paybox security number, and can authorize the settlement of the bill by entering his or her four-digit Paybox PIN. The exact sum is then debited from the bill recipient's account, and transferred to the account of the invoice issuer. Electronic bill presentment will be particularly useful to delivery service, telecommunications, utility, publishing, and medical companies, who would like to deploy a more efficient billing system.

The Paybox user is clearly authenticated by entering his PIN, since he or she must register in advance. Paybox informs the customer of the trader's name when the payment is confirmed, which increases security.

17.3.2 Ultra's M-Pay

Ultra's M-Pay (http://www.ultra.si/) is a mobile POS payment system based on Ultra's patented payment terminal using voice to transfer authorization data. It is independent of both the network and the network-operator (i.e., the same terminal works in GSM/GPRS/UMTS terminals). The user's phone transfers data, which simplifies the terminal design and allows the terminal to focus on safe payment authorization.

The M-Pay system enables the mobile operators and financial institutions to engage in m-commerce without having to invest heavily. The existing mobile phone network carries financial data, and the processing centers authorize credit card processing. The customers are required to have a relation with at least one mobile operator, and open an account with at least one financial institution. The customer calls the M-Pay access point, which is typically installed within the mobile operator's premises, to access the service.

The payment process of the M-Pay system involves the following parties.

- M-Pay access points enable telecommunication operators to ensure reliable and secure multiple connections to the M-Pay center, customer identification, and sufficient data throughput.
- Billing operators (e.g., banks, card authentication systems, consisting of issuers, acquirers, and merchants) provide monetary transaction control (e.g., debt collection from customers), and provide goods or services through the M-Pay terminal that is installed at the merchant.
- The M-Pay center is connected with M-Pay access points for transaction authorization.

M-Pay uses the security protocol PIN to authenticate users. The user's identity is defined on a SIM card in the mobile phone, and is further secured by entering a special PIN, either on a phone or payment terminal.

17.3.3   Encorus PaymentWorks

Encorus Technologies (http://www.encorus.com/) describes PaymentWorks Mobile as a software application for enabling payment transactions from cellular phones, the Internet, WAP-enabled mobile devices, and PDAs. PaymentWorks Mobile is specifically designed to address the needs of mobile network operators and other payment providers who process millions of transactions daily. PaymentWorks is designed to be secure and scalable. It consists of several components that can be combined to meet the individual needs of payment and PSP/mobile wallet providers, such as mobile wallet server, merchant component and modules for payment routing, virtual consumer cards, and merchant billing.

The wallet server is the central component of PaymentWorks Mobile. It is installed and hosted by the mobile network operator, processes and administers customer data, and handles payment transactions. All relevant customer data needed to process payments, such as address, billing information, and preferred payment methods (e.g., credit card, telephone bill, direct debit), is stored in the wallet server. The data is stored in a secure customer-specific wallet that the customer can modify from an Internet PC or mobile device.

17.3.4   SNAZ

The SNAZ (http://www.snaz.com/) mobile wallet (m-wallet) makes single-click wireless shopping a reality. The m-wallet retains on the server the customer's user name, password, multiple billing, and shipping details provided by the user during the one-time registration process. SNAZ's mobile wallet can easily interact with any transaction system to execute transactions. Only one user name and password pair is needed to transact with multiple merchants in the SNAZ network. The SNAZ mobile wallet allows users to have total control of their baskets. Users can save multiple items from multiple merchants into lists, and e-mail their lists to family and friends. Items are removed only when the user decides to remove them, not when they log-off. The SNAZ m-wallet also allows users to search for products within the SNAZ merchant network, and use the price-compare option to save on the desired products.

17.4   Payment Models, Challenges, and Issues

Many business, technical, and interoperability challenges exist in mobile payment systems, according to Telecom Media Networks [5]. Current wireless vendors and mobile commerce service businesses must cope with these challenges to meet the expectations of the mobile commerce market.

17.4.1   Business Challenges

There are a number of business challenges in mobile payment systems.

Business Model

The telecommunications industry must face the issue of whether or not telcos should collaborate with banks to address this business opportunity. Studies from Forrester Research show that most retailers would favor a joint venture including a financial company as a payment provider. Previous joint experiences between telcos and banks have not been very successful.

Cost

Cost could slow the m-payment development process. A return on the investment in m-payment systems is not likely to be achieved within the first two years. This does not necessarily mean that one should wait for a more mature market. Successful early adopters will gain significant competitive market advantages that may be impossible to overcome.

Customer Apathy

One of the main reasons for mobile commerce's slow start is customer apathy. According to Forrester's research [9], most consumers around the world are uncomfortable with the idea of mobile payment, even though different wireless security solutions are used. The result indicated the consumers' fear of an unknown medium, due to their lack of understanding about wireless security and current wireless payment solutions. Critical applications are needed to enable people to make a payment more efficiently and quickly in the future.

17.4.2   Technical Challenges

The major technical challenges in wireless payment are discussed here.

Security

Security is a very crucial issue for the m-payment method to gain widespread acceptance. It is the consumer's top concern, because they will have little confidence if the payment method cannot provide guarantees on authenticity, confidentiality, and integrity. Security can be viewed from five aspects: confidentiality, authentication, integrity, authorization, and nonrepudiation. Confidentiality protects the payment details (e.g., a consumer's personal particulars and password) against passive monitoring. Authentication ensures that the consumer and content provider are the entities that they claim to be. Integrity protects payment details from being modified from the time they are sent to the time they are received. Authorization ensures that only authorized consumers are allowed to participate in the payment transaction. Nonrepudiation guarantees that consumers cannot falsely claim that they did not participate in the transaction. This provides benefits for merchants and payment service providers.

Accessibility

Accessibility is considered as a combination of convenience, speed, and ease of use. Convenience indicates the capability of the payment method to pay for any type of content, from any location in the world, using any device. Some payment methods might require consumers to upgrade their existing handsets, or to be preregistered with a company. Speed is the amount of time spent on payment. Consumers care about speed if they have to pay for network access based on time. Ease of use is especially important for micropayments. Accessibility also strongly depends on the devices' capabilities and the quality of the network.

17.4.3 Interoperability Challenges

Interoperability means that applications will work on different networks. Interoperability challenges underpin any global payment system, ensuring that any participating payment product can be used at any participating merchant location. Mobile operators' principal concerns revolve around standardization and interoperability. Operators want payment to be seamless, allowing them to compete on services and applications.

Standardization

Telecom Media Networks notes that a wide variety of technologies for mobile payments exists, ranging from simple premium-charged SMS solutions for mobile content, to advanced dual-slot phone technology for real-world technology. Payment across networks is currently not standardized, and the issue is complicated because different players have different needs and concerns, and many different solutions exist.

These are a few of the existing standards for the mobile payment. the Universal Mobile Telecommunication System (UMTS); the EMV standard, developed by Europay, MasterCard, and Visa International; Jalda, developed by Ericsson and Hewlett Packard; and the Global Mobile Commerce Interoperability Group (GMCIG) standard for m-wallet.

17.5 Conclusion

The rapid increase of the number of mobile device users creates many new demands and business opportunities in wireless information and mobile commerce services. We need cost-effective mobile payment solutions to provide various mobile contents, diverse information services, and mobile commerce applications. Wireless service vendors and mobile content providers must study different business models in mobile payments, and develop and/or deploy new wireless payment systems to support mobile devices to make payments anywhere and at any time.

The economic viability of business models based on mobile data remains suspect, due to the lack of clear business models available for mobile payments and revenue sharing. Wireless payment vendors need more time to study and compare different payment models and schemes, through actual deployment experience in wireless service and mobile commerce applications, since these business models are still under development. Usage charges based on the airtime of wireless communication is no longer an appropriate payment method in the wireless world. New wireless payment methods and electronic solutions are strongly in demand for mobile commerce. A successful wireless payment system has three needs: (1) easy mobile payment interfaces with good interoperability and strong accessibility to diverse mobile devices; (2) secure and reliable payment transactions, based on a standardized wireless payment protocol; and (3) effective security solutions to provide authentication, authorization, and integrity of involved parties, and encryption of wireless communication messages.

Successful payment methods and solutions will continue to meet these challenges, particularly in mobile accessibility, security, standardization, and interoperability. Mobile consumers need more time to understand the advantages and conveniences of mobile commerce and mobile payment. The success of mobile payment will ultimately be driven by the success of mobile applications and services.

References

[1]   Yunos, H. M., J. Gao, and S. Shim, "Wireless Advertising's Challenges and Opportunities," *IEEE Computer*, Vol. 36, No. 5, May 2003.

[2]   Deitel, H. M., et al., *Wireless Internet & Mobile Businesses: How to Program*, Upper Saddle River, NJ: Prentice Hall, 2002.

[3]   Sadeh, N. M., *M-Commerce: Technologies, Services, and Business Models*, New York: John Wiley & Sons, 2002.

[4]   Lin, Y., M. Chang, and H. Rao, "Mobile Prepaid Phone Services," *IEEE Personal Communications*, Vol. 7, No. 3, June 2000, pp. 6–14.

[5]   Telecom Media Networks, "Mobile Payments in M-Commerce," September 2002, http://www.cgey.com/tmn/pdf/MobilePaymentsinMCommrce.pdf.

[6]   Duncan, J., "M-Payment—A Key Ingredient for Successful M-Commerce," 2002, http://www.wmrc.com/businessbriefing/pdf/wireless2002/technology/eOneGlobal.pdf.

[7]   Durlacher Research, Ltd., *Mobile Commerce Report*, 2004, http://www.durlacher.com/fr-research-reps.html.

[8]   Hennessy, D., White Paper from Valista, "The Value of the Mobile Wallet," 2004, http://www.valista.com/downloads/whitepaper/mobile_wallet.pdf.

[9]   Forrester Report, "TU Internet Tier," 1999.

Selected Bibliography

Atlantic Payment, "Payment Based on 3D SET," March 2002, http://www.atlanticpayment. com/3DSET.htm.

Bellare, M., et al., "VarietyCash: A Multi-Purpose Electronic Payment System," 1998, http://www.cs.ucsd.edu/users/mihir/papers/vc.pdf.

Fourati, A., et al., "A SET Based Approach to Secure the Payment in Mobile Commerce," *Proc. of 27th Annual IEEE Conference on Local Computer Networks (LCN'02)*, Tampa, FL, November 6–8, 2002.

Huang, Z., and K. Chen, "Electronic Payment in Mobile Environment," *Proc. of 13th International Workshop on Database and Expert Systems Applications (DEXA'02)*, Aix-en-Provence, France, September 2–6, 2002.

MerchantSeek, "Wireless Payment Processing," 2004, http://www.merchantseek.com/ mcommerce.htm.

Paybox.net, "Mobile Payment Delivery Made Simple," 2002, http://www.paybox.net/ publicrelations/ public_relations_whitepapers.html.

Peirce, M., "Multi-Party Electronic Payments for Mobile Communications," October 31, 2000, http://www.cs.tcd.ie/publications/tech-reports/reports.01/TCD-CS-2001-26.pdf.

Schuba, M., and K. Wrona. "Payment Made Feasible Using SET," 2004, http://www. medialabeurope.org/ people/k-rona/pub/mskw.pdf.

Thales, "Smart Cards for Payment Systems," 2001, http://www.thalesesecurity.com/Smart_ cards_for_payment_systems.pdf.

Van Thanh, D., "Security Issues in Mobile Ecommerce," *Proc. of 11th International Workshop on Database and Expert Systems Applications (DEXA'00)*, London, United Kingdom, September 6–8, 2000.

Vilmos, A., and S. Karnouskos, "SEMOPS: Design of a New Payment Service," 2003, http:// www.semops.com/uploadfiles/SEMOPS_MCTA2003.pdf.

Zheng, X., and D. Chen, "Study of Mobile Payments System," 2003, http://csdl.computer.org/ dl/proceedings/cec/2003/1969/00/19690024.pdf.

About the Authors

Jerry Zeyu Gao is an associate professor in the Department of Computer Engineering at San Jose State University. His research interests include component-based software engineering and Web service, IT technology, and mobile commerce. He has published two other software engineering books, including *Testing and Quality Assurance for Component-Based Software* (Artech House, 2003) and 55 technical papers in IEEE/ACM journals, magazines, and international conferences. He is the cochair for the First and Second IEEE International Workshops on Mobile Commerce and Services.

Simon Shim is an associate professor in the Department of Computer Engineering at San Jose State University. His research interests include electronic commerce, mobile commerce, security servers, and databases. He has published more than 65 technical papers in IEEE/ACM journals, magazines, and international conferences. He cochaired the IEEE International Workshops on Mobile Commerce and Services. He received an M.S. from Rensselaer Polytechnic Institute and a Ph.D. from University of Minnesota, both in computer science.

Hsing Mei is an associate professor in the Department of Computer Science and information engineering at Fu Jen Catholic University, Taiwan. He is also a visiting professor at the Inter-University Institute of Macau. His research interests include Web computing, mobile wireless software, and distributed systems.

Xiao Su is an assistant professor in the Department of Computer Engineering at San Jose State University. She received a B.E. in computer science and engineering from Zhejiang University in China and an M.S. and a Ph.D. in computer science from the University of Illinois at Urbana-Champaign. Her research interests include network security, multimedia computing and networking, media coding and content processing, and mobile computing. She has published more than 20 articles in IEEE- and ACM-sponsored journals and conferences. She has served as publication chairperson, publicity chairperson, and technical program committee member for several well-respected international conferences.

Index

.NET, 41, 108
.NET Compact Framework (CF), 42–43
2G networks: D-AMPS, GSM, cdmaOne,
       155–57
2.5G networks: GPRS, EDGE, 160
3G networks: CDMA2000, WCDMA,
       147,162
4G wireless WAN networks, 165

A

Accessibility, 11, 204, 397
Account-based payment system, 391
Ad space, 374–76
    campaign, 376
    catalog, 374
    fulfillment, 375
    measurement, 375
    payment, 375
    scheduling, 375
    trading, 375
Ad monitor and tracking, 376, 378
Ad delivery and posting, 376, 378–79, 382
Ad service payment, 376
Ad performance measurement, 376
Active Server Pages (ASP), 102, 108
Add-on peripherals, 32
Add-on software, 34
Architecture. *See* System architecture and
       software architecture
Attack. *See* Wireless security attack
Authentication, 274, 268
Audio compression, 124
ASP.NET 108
Assisted Global Positioning System (A-GPS),
       33
Asynchronous calls, 36
Asynchronous connectionless link (ACL), 179

B

Baseband layer, 178
Battery limitation and energy management, 27
Bearer networks, 59
Binary Runtime Environment for Wireless
       (BREW), 48–50
Block cipher standards, 263–67

Bluetooth, 145, 173,175
Bluetooth control center (BCC), 187
Bluetooth Device Address (BD_ADDR), 176
Bluetooth device discovery and connection
       setup, 182
Bluetooth network architecture, 176
Bluetooth network encapsulation protocol
       (BNEP), 181
Bluetooth radio layer, 177
Bluetooth SIG (Special Interest Group), 145
BREW, 48–51
BREW Delivery System (BDS), 50
BREW Software Development Kit (BREW
       SDK), 49
Business model, 369
    diversified revenue model, 370
    media-dependent, outsource media sales
       revenue model, 370
    safe income, no upside revenue model, 369
    voice-subsidized model, 369

C

Camera phones 32
Candidate Enabler Release, 78
Cascading style sheets (CSS), 86
CDMA, 154
CDMA Development Group (CDG), 167
CDMA2000, 163
cdmaOne, 157, 159
Cell capability enhancement, 150
Cell splitting, 151
Cellular digital packet data (CDPD), 158
Cellular phone, 31. *See* Mobile phone
Cellular mobile network. *See* 3G
CGI (common gateway interface), 101
Channel allocation, 152
Circuit switching, 154
CLDC connection framework, 94
Client capabilities and personal profiles
       (CC/PP), 57
Code division multiple access (CDMA), 154,
       159, 163
Codec, 123
Coexistence Mechanisms Task Group (TG2),
       173

Communication capability enhancement, 31
Compact HTML (cHTML), 90, 241
Composite capabilities/preferences profile
        (CC/PP), 89
Configurable system, 41
Confidentiality, 268, 273
Connected device configuration (CDC), 47
Connected limited device configuration
        (CLDC), 94
Content presentation, 349
Control points, 191
Controlled devices, 191
CSS, 86
CSS Mobile Profile (CSS-MP), 60, 86

D

Data object modeling, 273
Data integrity, 269, 275
Device constraint, 346
Device coordination, 188–94
Device emulator, 38
Device management (DM), 77
Device/service discovery, 182
DSSS (direct sequence spread spectrum), 154
DTD (document type definition), 66
Dynamic behavior analysis and modeling, 220

E

ECMAScript, 70
Embedded Linux, 44
External application interface (EAIF), 362
Electronic commerce (e-commerce), 344
Electronic commerce system (e-commerce
        system), 13
Embedded Linux Consortium (ELC), 45
Engineering wireless-based application
        systems, 14–20, 197–337
Enhanced Data Rates for Global/GSM
        Evolution (EDGE), 160
Enhanced messaging service (EMS), 35, 73
Equipment identity register (EIR), 150
Event and Notification Interface, 190
eXecute-In-Place (XIP), 22
Extensible Markup Language, 109
Extended service areas, 143–44
eXtensible Stylesheet Language (XSL), 86
eXtensible Stylesheet Language
        Transformation (XSLT), 87
External functionality interface (EFI), 32

F

FDMA, 153
Flash memory, 27

Flat-file database, 38
Form factors, 348
Forward error correction code (FEC), 179
Freedom of Multimedia Access (FOMA), 165
Frequency-division multiple access (FDMA),
        62
Frequency-division duplex (FDD), 164
Function analysis and modeling. *See* System
        function analysis and modeling

G

GCF (generic connection framework), 94
General Packet Radio Service (GPRS), 74, 161
Generic Access Profile (GAP), 246
Generic object exchange profile (GOEP), 185
Global Positioning System (GPS), 33
Geography Markup Language (GML), 89
Global System for Mobile Communications
        (GSM), 157, 158, 300–3
GPS phone, 33
GSM Association, 167

H

Handheld Device Markup Language. *See*
        HDML
Handoff (handover, or automatic link
        transfer), 151
Handoff detection, 151
Hands-Free Profile (HFP), 186
HDML, 90, 313
Header translation, 63
High Rate WPAN Task Group (TG3), 173
Host control interface (HCI), 180
HTTP proxy, 62
Human interface device profile, 181
Human interface device protocol, 181

I

IEEE 802.15, 172
IEEE 802.11, 132
IEEE 8.02.16, 146
i-mode, 98
i-mode infrastructure, 98
i-mode application system architecture, 239
Infrared Data Association (IrDA), 133
Instant messaging (IM), 334
Interactive multimedia presentation framework
        (IMPF), 117
Interoperability, 55, 397
Interactive mobile service (IMS), 109
Internet Engineering Task Force (IETF), 123
International Mobile Telecommunication-2000
        (IMT-2000), 162

International Telecommunication Union (ITU),
   147
IS-95A, 155,159
ISMA (Internet Streaming Media Alliance),
   122

J

J2ME, 46, 94, 201, 245
J2EE, 254
Java Platform, 46
Java Card, 46
Java Community Process (JCP), 187
Java EE (Java Platform, Enterprise Edition),
   46, 47. *See* J2EE
Java ME (Java Platform, Micro Edition) 48,
   94. *See* J2ME
Java SE (Java Platform, Standard Edition),
   46–47, 94
Java Servlets, 101
Java Virtual Machine (JVM), 46,188
JavaScript, 70–71
JavaServer Pages (JSP), 102
Jini, 188
JSR-82, 187

L

Limited screen space, 28
LMP, 179
Location-based services (LBS), 5, 349
Location Interoperability Forum (LIF), 77
Logical link control and adaptation protocol
   (L2CAP), 177
Low Rate, Long Battery Life Task Group
   (TG4), 173

M

Managed code, 43
Media access control (MAC), 131–37
Microbrowser, 9, 241
Microdrive, 27
MIDlet, 95–97
MIDP (mobile information device profile), 95
Mobile-aware application architecture, 242
Mobile-transparent application architecture,
   244
Mobile interactive application architecture,
   245
Mobile application platform, 45
Mobile client, 81
   smart mobile client, 314–15
   thick mobile client, 314, 316
   thin mobile client, 314–15

Mobile commerce (m-commerce)
Mobile commerce systems, 341, 344. *See*
   Electronic commerce systems,
   Wireless commerce systems
Mobile device, 26
   mobile phone, 31, 34
   PDA, 26
   smart phone, 26
Mobile game device, 32
Mobile information management (MIM), 34
Mobile Internet, 23, 244
Mobile payment process, 390
Mobile payment system. *See* Wireless payment
   systems
Mobile technology, 252
MMSE, 352, 354
MMS client, 352, 356
MMS content delivery, 357
Multimedia messaging service (MMS), 351–63
MMS Protocol, 352
MMS proxy, 352
MMS server, 352
Mobile client design, 311–37
Mobile client development process, 321
Mobile Linux, 44–45
Mobile operating system architecture, 36
Mobile operating systems, 23, 27, 35
Mobile station (MS), 300
Mobile streaming architecture, 110
Mobile switching centers (MSCs)
Mobile-assisted handoff (MAHO), 152
Mobile-controlled handoff (MCHO), 152
Mobile wallet, 393, 395
Mobile Web Client, 90
Mobility, 10,144
Mobility management, 151
MP3 phone, 34
MPEG, MPEG-4, 123–24
Multicast, 122
Multimedia messaging service (MMS), 74, 351
Multimedia messaging service center (MMSC),
   74
Multimedia presentation, 354
MSISDN addressing, 354

N

Nano-X Window, 44
Navigation Discovery, 62
Near-field communication (NFC), 174
NFC Forum, 173
Nintendo DS, 33
NTT DoCoMo, 98

O

Object Analysis and Modeling, 212, 216
Object Exchange Protocol (OBEX), 181
Object Push Profile, 185
Object store, 42
Open Mobile Alliance (OMA), 75, 76
Open mobile operating systems, 35
OMA, 76
Orthogonal frequency division multiplex access (OFDMA), 154

P

Packet switching, 154
Palm OS, 25–29, 313, 372
Payment scheme, 389
Parked Member Address (PM_ADDR), 176
Persistent data storage (flat-file versus relational database), 37
Personal Area Networking Profile, 187
Personalization, 344
Piconet, 145, 176–77
Pictogram, 61
PKI portal, 57
Point-of-sale (POS), 392
Portability, 11, 102
Portable Media Player (PMP), 172
Provisioning, 61
Public key cryptography, 270
PUSH marketing 344
Push Over-The-Air (Push OTA) Protocol, 60
Push Proxy Gateway (PPG), 67, 75

Q

Quality of service (QoS), 121
Quality factors, 233

R

Radio frequency (RF) 180
Radio frequency communications (RFcomm), 180
RC4, 267
Reliability, 11, 18, 55
Requirements engineering process, 200
RIM Blackberry device, 159
RSA 271

S

SAX (Simple API for XML), 86, 88, 89
Scalability, 121, 233
Scatternet, 176
Schematron, 85
Secret key cryptography, 260

Security, 259–307
Security attack, 275–81
Security threat, 285, 293, 303
Self-configuration, 191
Serial port profile (SPP), 184
Service aggregator, 194
Service discovery, 192
Service Discovery Application Profile (SDAP), 184
Service discovery protocol (SDP), 180
Service lookup, 62
Service provider, 194
SGML (standard generalized markup language), 81
Short message service (SMS), 72
Simple service discovery protocol (SSDP), 192
Smart Card Payment System, 391
Smart mobile application architectures, 249
SMIL (synchronized multimedia integration language), 110
Standardization, 233, 373, 397
Software architecture, 227, 230
    architecture deployment, 232
    architecture design process, 231
    architecture evaluation, 233
    architecture verification and validation, 232
    conceptual architecture, 231
    logical architecture, 231
    meta architecture, 231
Solution providers, 118–19
Standardization organization, 167
Stream cipher, 267
Symbian OS, 39
Synchronization profile, 185
SyncML, 77, 82, 89
SyncML Initiative, 77
System analysis and modeling 207
    function analysis and modeling, 211
    object analysis and modeling, 212
    system data analysis and modeling, 212
    system dynamic behavior analysis and modeling, 220
    user analysis and modeling, 209
System architecture, 227, 230, 234

T

Telephony Control Protocol (TCP), 180
Telephony Control Protocol Specification (TCS), 180
Third Generation Partnership Project (3GPP), 192

Third Generation Partnership Project 2
(3GPP2), 168
Time division multiple access (TDMA), 153
Transaction interface, 190
Transport layer security (TLS), 58
Trusted third party (TTP), 389

U

UAProf (User Agent Profile), 71
uiOne—BREW User Interface, 50
Ultrawideband, 145
UML Sequence Diagram, 223
UML Class Diagram, 220
UML State Diagram, 221
Unicast, 122
Unimodal mobile interface, 318
Unimodal mobile client. *See* Unimodal mobile
interface
multimodal mobile client, 318
multimodal mobile interface. *See*
Multimodal mobile client
Universal Plug-and-Play (UPnP), 191
Unreliable wireless environment, 20
Usability, 345, 347
User analysis and modeling, 200
User identity, 344

V

Video Distribution Profile (VDP), 187
Virtual calendar (vCal), 185
Visitor location register (VLR), 150

W

WAE User-Agent, 60
WAP 1, 2, 53–78
WAP 1, 53
WAP Application Environment (WAE), 68
WAP Forum, 53
WAP gateway, 56, 63–66
WAP Messaging Service, 72–74
WAP Programming Model, 56
WAP protocol stack, 57
WAP proxy gateway, 62
WAP service architecture, 59
WAP push mechanism, 66
WCDMA (wideband code-division multiple
access), 163–64
WCSS (WAP/Wireless CSS), 86
Web proxy, 62
Well-formed XML document, 82
Windows CE/Mobile, 41
Wireless ads, 366
brand-building ads, 367

category ads, 367
game-based ads, 369
homepage ads, 367
interstitial ads, 367
location-based ads, 366
mobile coupons, 369
short message ads, 368
Wireless advertising payment, 380
pay-by-click, 380
pay-by-display, 380
pay-by-impression, 380
pay-by-message, 380
pay-by-meter, 380
pay-by-transaction, 380
Wireless advertising, 365–84
Wireless ad targeting, 377
content-based targeting, 377
hybrid ad targeting, 377
location-based targeting, 377
profile-based targeting, 377
Wireless ad delivery, 378
performance measurement in wireless
advertising, 381
wireless advertisement tracking, 378
Wireless Advertising Association (WAA), 366
Wireless application architectures, 227–57
smart mobile application architectures,
249–50
wireless Internet-based application system
architecture, 239–47
wireless enterprise application architectures,
251–57
Wireless Application Protocol. *See* WAP
Wireless Session Protocol (WSP), 57
Wireless Transaction Protocol (WTP), 57
Wireless Transport Layer Secruity (WTLS), 58
Wireless Datagram Protocol (WDP), 58
Wireless market, 3
Wireless messaging, 5
Windows Media Audio (WMA), 34
Windows Mobile, 5
Wireless-based application systems, 13
Wireless-based e-commerce systems, 13
Wireless-based entertainment systems, 14
Wireless-based enterprise application systems,
14
Wireless-based information messaging systems,
13
Wireless location-based application systems,
14
Wireless-based personal groupware, 14
Wireless-based portal systems, 14

Wireless-based real-time application systems, 14

Wireless Internet application systems, 13

Wireless network, 129–95
  1G networks, 156
  2G networks, 156
  2.5G networks, 157
  3G networks, 163
  4G networks, 165
  GSM, 157, 160
  GPRS, 72, 74

Wireless local area network (WLAN), 131–47

Wireless personal area network (WPAN) 171–95

Wireless wide area network (WWAN), 149–68

Wireless markup language (WML), 68, 355

Wireless Multimedia Forum (WMF), 122

Wireless Multimedia Streaming Standard, 122

Wireless payment, 387

Wireless portal, 11

Wireless Profiled HTTP (WP-HTTP), 58

Wireless Profiled TCP (WP-TCP), 58

Wireless push strategy, 378

Wireless pull strategy, 378

Wireless security, 259–81

Wireless security attacks, 275–81

Wireless security solutions, 285–307

Wireless Session Protocol (WSP), 57

Wireless Telephony Application (WTA), 61

Wireless Transaction Protocol (WTP), 57

Wireless Transport Layer Security (WTLS), 58, 244

Wireless World Research Forum (WWRF), 168

WLAN (wireless local area network), 131–47

WML, 68

WMLScript, 70–71

WPAN/Bluetooth Task Group (TG1), 175

X

Extensible Markup Language. *See* XML

XHTML, 68, 90–93, 356

XHTML Mobile Profile (XHTML-MP), 68, 92

XML (eXtensible Markup Language), 81–89

XML-based markup language, 89

XML DTD, 84

XML schema, 84

XML styling, 86

XML syntax, 82

XML validation, 83

Xpath, 87

XSL, 87

XSLT (XSL Transformations), 60, 87

Recent Titles in the Artech House Computing Library

Achieving Software Quality through Teamwork, Isabel Evans

Action Focused Assessment for Software Process Improvement, Tim Kasse

Advanced ANSI SQL Data Modeling and Structure Processing, Michael M. David

Advanced Database Technology and Design, Mario Piattini and Oscar Díaz, editors

Agent-Based Software Development, Michael Luck, Ronald Ashri, and Mark d'Inverno

Agile Software Development, Evaluating the Methods for Your Organization, Alan S. Koch

Agile Systems with Reusable Patterns of Business Knowledge, A Component-Based Approach, Amit Mitra and Amar Gupta

Building Reliable Component-Based Software Systems, Ivica Crnkovic and Magnus Larsson, editors

Business Process Implementation for IT Professionals and Managers, Robert B. Walford

Data Modeling and Design for Today's Architectures, Angelo Bobak

Developing Secure Distributed Systems with CORBA, Ulrich Lang and Rudolf Schreiner

Discovering Real Business Requirements for Software Project Success, Robin F. Goldsmith

Engineering Wireless-Based Software Systems and Applications, Jerry Zeyu Gao, Simon Shim, Hsing Mei, and Xiao Su

Future Codes: Essays in Advanced Computer Technology and the Law, Curtis E. A. Karnow

Global Distributed Applications with Windows® DNA, Enrique Madrona

Guide to Standards and Specifications for Designing Web Software, Stan Magee and Leonard L. Tripp

Implementing and Integrating Product Data Management and Software Configuration, Ivica Crnkovic, Ulf Asklund, and Annita Persson Dahlqvist

Internet Commerce Development, Craig Standing

Knowledge Management Strategy and Technology, Richard F. Bellaver and John M. Lusa, editors

Managing Computer Networks: A Case-Based Reasoning Approach, Lundy Lewis

Metadata Management for Information Control and Business Success, Guy Tozer

Multimedia Database Management Systems, Guojun Lu

Open Systems and Standards for Software Product Development, P. A. Dargan

Practical Guide to Software Quality Management, Second Edition, John W. Horch

Practical Insight into CMMI®, Tim Kasse

Practical Software Process Improvement, Robert Fantina

A Practitioner's Guide to Software Test Design, Lee Copeland

The Requirements Engineering Handbook, Ralph R. Young

Risk-Based E-Business Testing, Paul Gerrard and Neil Thompson

Secure Messaging with PGP and S/MIME, Rolf Oppliger

Software Configuration Management, Second Edition, Alexis Leon

Software Fault Tolerance Techniques and Implementation, Laura L. Pullum

Strategic Software Production with Domain-Oriented Reuse, Paolo Predonzani, Giancarlo Succi, and Tullio Vernazza

Successful Evolution of Software Systems, Hongji Yang and Martin Ward

Systematic Process Improvement Using ISO 9001:2000 and CMMI®, Boris Mutafelija and Harvey Stromberg

Systematic Software Testing, Rick D. Craig and Stefan P. Jaskiel

Testing and Quality Assurance for Component-Based Software, Jerry Zeyu Gao, H. -S. Jacob Tsao, and Ye Wu

Workflow Modeling: Tools for Process Improvement and Application Development, Alec Sharp and Patrick McDermott

For further information on these and other Artech House titles, including previously considered out-of-print books now available through our In-Print-Forever® (IPF®) program, contact:

Artech House
685 Canton Street
Norwood, MA 02062
Phone: 781-769-9750
Fax: 781-769-6334
e-mail: artech@artechhouse.com

Artech House
46 Gillingham Street
London SW1V 1AH UK
Phone: +44 (0)20 7596-8750
Fax: +44 (0)20 7630-0166
e-mail: artech-uk@artechhouse.com

Find us on the World Wide Web at: www.artechhouse.com